Horst Aichinger **Radiation Exposure and Image Quality**
Joachim Dierker **in X-Ray Diagnostic Radiology**
Sigrid Joite-Barfuß
Manfred Säbel Physical Principles and Clinical Applications

Springer

Berlin
Heidelberg
New York
Hong Kong
London
Milan
Paris
Tokyo

Horst Aichinger
Joachim Dierker
Sigrid Joite-Barfuß
Manfred Säbel

Radiation Exposure and Image Quality in X-Ray Diagnostic Radiology

Physical Principles and Clinical Applications

With 199 Figures and 78 Tables

Springer

Horst Aichinger, Dr. rer. nat.
Unterfarrnbacher Str. 32
90766 Fürth
Germany

Joachim Dierker, Dr. rer. nat.
Hegenigstr. 56
91056 Erlangen
Germany

Sigrid Joite-Barfuß, Dipl.-Phys.
Friedhofstr. 4 a
91058 Erlangen
Germany

Manfred Säbel, Prof. Dr. rer. nat.
Universität Erlangen-Nürnberg
Institut für Diagnostische Radiologie
Universitätsstr. 21–23
91054 Erlangen
Germany

ISBN 3-540-44287-1
Springer-Verlag Berlin Heidelberg New York

Library of Congress Cataloging-in-Publication Data
Radiation exposure and image quality in x-ray diagnostic radiology: physical principles and clinical applications / H. Aichinger … [et al.]. p. cm. Includes bibliographical references and index. ISBN 3540442871 (hbk.: alk paper) 1. Radiography, Medical – Image quality. 2. Radiation dosimetry. 3. Radiography, Medical – Safety measures. 4. Medical physics. I. Aichinger, H. (Horst), 1935– RC78.R227 2003 616.07'572–dc21

Springer-Verlag Berlin Heidelberg New York a member of BertelsmannSpringer Science + Business Media GmbH

http://www.springer.de

© Springer-Verlag Berlin Heidelberg 2004
Printed in Germany

Cover design: E. Kirchner, Heidelberg
Typesetting: AM-production, Wiesloch
Printing and bookbinding: Strauss, Mörlenbach

21/3150/Wd – 5 4 3 2 1 0
Printed on acid-free paper

Preface

X-ray diagnostic radiology is a major world-wide activity. In Europe approximately 250 million X-ray examinations are performed annually, and in the United States of America a similar level of radiological activity is undertaken. This results in the fact that the largest contribution to radiation exposure to the population as a whole is known to be from man-made radiation sources arising in the form of diagnostic X-rays (UNSCEAR 2000). It is also known that exposures vary widely, due to differences in X-ray techniques and the level of skill of the operator of the equipment. Consequently radiation protection of the patient is a major aim in modern health policy. The two basic principles of radiation protection of the patient as recommended by the International Commission on Radiological Protection (ICRP 1991) are justification of practice and optimisation of protection.

Justification is the first step in radiation protection. It is generally accepted that no diagnostic exposure is justifiable without a valid clinical indication. In comparison with the associated radiation risk, every examination must result in a net benefit for the patient. Justification also implies that the necessary result cannot be achieved with other methods which would be combined with lower risks for the patient.

Once a diagnostic X-ray examination has been clinically justified, the subsequent imaging process must be optimised. The optimal use of X-rays involves the interplay of three important aspects of the imaging process (CEC 1996):

- The diagnostic quality of the radiological image
- The radiation dose to the patient
- The choice of imaging technique

In respect of diagnostic radiology, the ICRP does not recommend the application of dose limits to patient irradiation but draws attention to the use of dose reference levels as an aid to optimisation of protection in medical exposure.

The *principles of justification and optimisation* are largely translated into a legal framework by the Medical Exposure Directive (MED) 97/43/Euratom (CEC 1997). Concerning optimisation, the MED states explicitly in Article 4, 1(a) that:

> All doses due to medical exposure for radiological purposes except radiotherapeutic procedures … shall be kept as low as reasonably achievable consistent with obtaining the required diagnostic information, taking into account economic and social factors

This concept is known as the ALARA principle (as low as reasonably achievable).

In Article 4, 2(a), the MED states that:

> Member States shall promote the establishment and the use of diagnostic reference levels for radiodiagnostic examinations

Diagnostic reference levels are defined in Article 2 of the MED as:

> Dose levels in medical radiodiagnostic practices…for typical examinations for groups of standard-sized patients or standard phantoms for broadly defined types of equipment. These levels are expected not to be exceeded for standard procedures when good and normal practice regarding diagnostic and technical performance is applied.

The authors of this book have been engaged for a long time in the solution of problems concerning optimisation in X-ray diagnostic radiology, three of us (Aichinger, Dierker, Joite-Barfuß) as physicists in development working groups of Siemens Medical Solutions and one (Säbel) as a medical physicist in a university hospital. During this time a lot of data has been accumulated which could be useful for physicists and engineers in the medical radiodiagnostic industry, for medical physicists and for diagnostic radiologists working on problems of image quality, patient-dose estimation and especially the establishment of diagnostic reference levels. In this connexion the publication of the MED was the trigger point for our decision to collect and publish this material.

In particular, the figures and tables included should enable a medical physicist to:

- Calculate the scatter-free characteristic quantities of the primary radiation beam
- Estimate patient doses (organ dose, effective dose)
- Optimise radiographic and fluoroscopic exposure parameters in relation to the medical indication
- Take into account the influence of scattered radiation on image quality and dose

The data provided for the solution of these tasks are partly based on X-ray spectra, measured on diagnostic X-ray tube assemblies, and supplemented by the results of measurements on phantoms and simulation calculations.

Taking into consideration that mammography screening is increasingly established, X-ray mammography is treated in some detail. On the other hand, concerning computed tomography, the presentation is confined to dose estimation. An essential reason for this is that one of our colleagues (Kalender 2000) has written an excellent book on computed tomography, which covers the aspect of image quality in detail.

Our book consists of three parts:
1. "Physical Principles" reviews some information on radiation physics, dosimetry and X-ray diagnostic technique, which will be useful for the understanding of the figures and tables, but it is not intended to replace standard books on the physics of diagnostic radiology
2. In "Clinical Applications" the material presented is applied to some clinical problems concerning radiation exposure of the patient, image quality and optimisation of imaging equipment
3. The "Supplement" includes all the figures and tables which are necessary for the calculations. It is complimented by a CD-ROM that includes the most important data of the "Supplement" and a lot of additional useful files

A detailed description of the contents of the different parts of the book is given in the "Introduction".

In the preparation of this book, we asked for and received the help of a number of well-qualified people: We are grateful to John H. Hubbel, Ph.D., from the National Institute of Standards and Technology, Gaithersburg, USA, for his support in the selection of the most up-to-date photon interaction coefficients; and to G.T. Barnes, Prof. Ph.D., from the University of Alabama in Birmingham, USA, for information in respect of his work about scattered radiation in mammography. Moreover we thank T. v. Volkmann, Ph.D., from Kodak, for the provision of the data of screen-film combinations and M.A.O. Thijssen, Ph.D., from the University Hospital in Nijmegen, Netherlands, for information about his CDRAD- phantom. Last, but not least, we thank I. Aichinger for her preparation of some of the figures in this book.

Horst Aichinger
Joachim Dierker
Sigrid Joite-Barfuß
Manfred Säbel

References

CEC (Commission of the European Communities) (1996) European guidelines on quality criteria for diagnostic radiographic images. EUR 16260. CEC, Luxemburg

CEC (Commission of the European Communities) (1997) Council directive 97/43/EURATOM of 30 June 1997 on health protection of individuals against the dangers of ionising radiation in relation to medical exposure. Official Journal of the European Communities, L 180/22

ICRP (International Commission on Radiological Protection) (1991) 1990 recommendations of the ICRP. Publication 60. Ann ICRP 21(1–3)

Kalender WA (2000) Computed tomography. Publicis, Munich

UNSCEAR (United Nations Scientific Committee on the Effects of Atomic Radiation) (2000) Sources and effects of ionizing radiation. United Nations, New York

Contents

Part I Introduction 1

I.1 Introduction 3

Part II Physical Principles 5

II.1 Production and Measurement of X-rays 7
II.1.1 Production 7
II.1.2 Measuring Photon Spectra
of Diagnostic X-rays 11
    II.1.2.1 X-ray Spectrometer 11
    II.1.2.2 Measuring 12
    II.1.2.3 Corrections for Detector
    Influences. 12
    II.1.2.4 Preparing for Calculations
    with X-ray Spectra 13

II.2 Interaction of Photons with Matter 15
II.2.1 Exponential Law of Attenuation. 15
II.2.2 Interaction Processes. 15
    II.2.2.1 Compton Effect. 16
    II.2.2.2 Photoelectric Effect 18
    II.2.2.3 Rayleigh Scattering 20
II.2.3 Total Coefficients for Attenuation,
Energy Transfer and Energy Absorption 20

II.3 Radiation Field and Dosimetric Quantities 27
II.3.1 General Radiation Field Characteristics 27
II.3.2 Characteristics of the Radiation Field
in X-ray Diagnostic Radiology 29
II.3.3 Dosimetric Quantities for the Description
of Radiation Exposure 30
II.3.4 Effect of Filtration 31

II.4 Penetration of X-rays. 35
II.4.1 Attenuation by the Patient. 36
    II.4.1.1 Simulation of Attenuation
    by the Patient. 36
    II.4.1.2 Basis of Exposure Tables. 38

II.4.2 Attenuation by the Patient Support and the
Components of the Spot-Film Device. 39
    II.4.2.1 Patient Support. 40
    II.4.2.2 Antiscatter Grid 40
    II.4.2.3 Ionisation Chamber 40
    II.4.2.4 Geometry. 41
II.4.3 Important Differences Between Over-Table
and Under-Table X-ray Units 43

II.5 Scattered Radiation 45
II.5.1 Measurement of Scattered Radiation 45
II.5.2 Properties of Scattered Radiation. 47
    II.5.2.1 General X-ray Diagnostics 47
    II.5.2.2 Mammography. 48
II.5.3 Techniques to Reduce Scattered Radiation
at the Image Receptor 49
    II.5.3.1 Air-Gap Technique. 49
    II.5.3.2 Slot Technique 49
    II.5.3.3 Antiscatter Grid 50
II.5.4 Quantities for the Description
of the Characteristics of Grids 51
II.5.5 Derived Quantities for the Description
of the Efficiency of Antiscatter Grids 52
II.5.6 Application Limits for Focused Grids. 54

II.6 Image Receptors 57
II.6.1 Characteristics of Phosphor Screens 57
II.6.2 Digital Image Receptors 60
II.6.3 Automatic Exposure Control 61
    II.6.3.1 Direct Technique. 61
    II.6.3.2 Fluoroscopy and Indirect Technique . 62
II.6.4 Image Receptor Dose/Dose Rate
(System Dose/Dose Rate) 63

II.7 Image Quality and Dose 69
II.7.1 Contrast. 70
II.7.2 Sharpness 72
II.7.3 Noise. 74
II.7.4 Signal-to-Noise Ratio 75
II.7.5 Detective Quantum Efficiency. 76
II.7.6 Other Image-Quality Figures 77
II.7.7 Dependence of Image Quality
on Exposure Parameters. 79

Part III Clinical Applications 83

III.1   Evaluation of Dose to the Patient 85
III.1.1 Specific Dosimetric Quantities
    Used for Patient-Dose Evaluation 85
III.1.2 Quantities Influencing Patient Dose 85
III.1.3 Initial Dosimetric Quantities
    for the Estimation of Organ Doses 88
III.1.4 Determination of the
    Initial Dosimetric Quantities 88
    III.1.4.1 Radiography 88
        III.1.4.1.1 Determination of the
        Entrance Surface Air Kerma K_E
        from the Value K_m, Measured
        at Focus Distance r_m 88
        III.1.4.1.2 Determination of the
        Entrance Surface Air Kerma K_E
        from the Air Kerma-Area Product P_F 88
        III.1.4.1.3 Determination of the
        Entrance Surface Air Kerma K_E
        from the Entrance Surface Dose K_O . 89
        III.1.4.1.4 Determination of the
        Entrance Surface Air Kerma K_E
        from the Radiation Output of an
        X-ray Tube Assembly Y_{100} and the
        Radiographic Parameters 89
        III.1.4.1.5 Determination of the
        Entrance Surface Air Kerma K_E from
        the Image Receptor Dose K_B or the
        Sensitivity of the Image Receptor S . 89
    III.1.4.2 Fluoroscopy 90
    III.1.4.3 Mammography 90
    III.1.4.4 Computed Tomography 90
III.1.5 Estimation of Organ Doses from the
    Initial Dosimetric Quantities 90
    III.1.5.1 Estimation of Organ Doses H_T
        by the Use of Conversion Factors . . 91
        III.1.5.1.1 Radiography
        and Fluoroscopy 91
        III.1.5.1.2 Mammography 91
        III.1.5.1.3 Computed Tomography . 94
    III.1.5.2 Estimation of the Absorbed Dose
        to an Organ D_{org} by the Use of
        Tissue-Air Ratios or Relative
        Depth Doses 94
III.1.6 Estimation of Effective Dose 96
III.1.7 Uncertainties in Patient-Dose Estimation . . . 97

III.2   Scattered Radiation 99
III.2.1 Influence on Image Quality and Dose 99
III.2.2 The Various Types of Antiscatter Grids 100
    III.2.2.1 Employment of Grids
        in Paediatrics 101
    III.2.2.2 Employment of Grids
        in Mammography 104
    III.2.2.3 Employment of Grids
        in Digital Radiography 105

III.3   Optimisation of Image Quality and Dose 109
III.3.1 Image-Quality Figure
    for General X-ray Diagnostics 109
III.3.2 Image-Quality Figure in Angiography 111
III.3.3 Image-Quality Figure in Mammography . . . 112

Part IV Supplement 117

IV.1   X-ray Spectra 123
IV.1.1 General X-ray Diagnostics 123
IV.1.2 Mammography 130

IV.2   Interaction Coefficients 135
IV.2.1 Elements 135
IV.2.2 Compounds and Mixtures 139

IV.3   Characteristics of the Primary Radiation Beam . . 143
IV.3.1 General X-ray Diagnostics 143
IV.3.2 Mammography 166

IV.4   Characteristics of the Imaging Radiation Field . . 177
IV.4.1 General X-ray Diagnostics 177
IV.4.2 Mammography 190

IV.5   Miscellaneous 193
IV.5.1 Penetration and Absorption of X-rays 193
IV.5.2 X-ray Detectors 193
IV.5.3 Image-Quality Figures 197

IV.6   Patient-Dose Estimation 199
IV.6.1 General X-ray Diagnostics 199
IV.6.2 Mammography 203

Subject Index 207

Part I   Introduction

I.1   Introduction

An introduction to the structure and contents of the three main parts of the book (Part II "Physical Principles", Part III "Clinical Applications", Part IV "Supplement") is given in the following:

1. "Physical Principles". In Chap. II.1, a short description of the function of an X-ray tube and the production of X-rays is given. Then some problems concerning the measurement of photon spectra of diagnostic X-rays are discussed.

The exponential law of attenuation is introduced in Chap. II.2. Then the different interaction processes of photons with matter that are relevant for diagnostic radiology are discussed together with the corresponding interaction coefficients. Emphasis is laid on the different mechanisms responsible for the transfer of radiation energy to matter and the production of scattered radiation.

Chapter II.3 reviews the physical quantities, which describe the X-radiation field and the radiation exposure of the patient in X-ray diagnostic radiology. These quantities can also be used as source material within the process of optimising image quality. In this respect the employment of additional filters with the objective to reduce patient dose will also be discussed.

Intermediate layers in the radiation beam between the patient and the image receptor (e.g. table top, antiscatter grid, ionisation chamber of the automatic exposure control system, film cassette cover) attenuate the imaging radiation and therefore give rise to increased radiation exposure of the patient and possibly to a reduction in image quality. In Chap. II.4 the characteristics of the intermediate layers behind the patient and the beam attenuation by the patient itself are described.

In Chap. II.5 the quantitative evaluation of scattered radiation will be discussed and then the characteristics of scattered radiation, in relation to the imaging parameters shown. Secondly the techniques to reduce its deleterious effect on image quality will be demonstrated.

In Chap. II.6 a short review of the most important detector materials (CsJ, Gd_2O_2S, Se, $CaWO_4$) is given and the characteristics which are responsible for their different energy response to X-radiation are discussed. The knowledge of the energy dependence of the image receptor's sensitivity is needed for the production of an opti-

mum image and an accurate determination of the patient dose.

The image quality obtained with an imaging system can be described and quantified by the characteristics contrast, sharpness and noise. The corresponding physical quantities used for description are at first presented in Chap. II.7. Then the derived quantities signal-to-noise ratio and detective quantum efficiency are introduced. Then some other image quality figures are described, followed by a short discussion of the dependence of image quality on exposure parameters.

2. "Clinical Applications". At first Chap. III.1 presents a short discussion of quantities influencing patient dose. For the estimation of organ doses, clearly defined initial dosimetric quantities are required which can be easily measured with readily available instruments of sufficient precision and accuracy. These quantities are fixed and their determination is described. Finally the most important methods for the estimation of organ doses from the initial dosimetric quantities are discussed.

In Chap. III.2 it will be explained that which method of scatter reduction technique is selected is of great importance for image quality and dose . In comparing different grid types for example, one must take into consideration not only the geometrical characteristics of the grid design but also the materials used for their cover and interspaces. Moreover the introduction of digital radiography necessitates the reassessment of the grid design, because in digital imaging it is possibly not always necessary to increase the incident exposure when a grid is employed. Examples for these statements will be given especially for scatter reduction in paediatrics and mammography.

Finally in Chap. III.3 some examples for the optimisation process in X-ray diagnostic radiology are presented. Especially the application of image quality figures to optimisation problems in angiography and mammography is demonstrated.

3. "Supplement". Part IV completes the topics discussed in the preceding parts of the book by providing an extensive collection of relevant X-ray spectra and data which are of importance in each radiographic exami-

nation unit. This information opens the possibility to evaluate and improve image quality and take simultaneously into consideration the radiation exposure of the patient.

4. CD-ROM. The book includes a CD-ROM with a comprehensive database, mostly as Excel files, which can be used in the readers' programs: X-ray spectra for conventional diagnostics and mammography; interaction coefficients of the most important elements, compounds and mixtures; characteristics of the primary and the imaging radiation field; data from various antiscatter grids; dosimetric data, which are needed for the evaluation of patient dose; and a lot of further helpful information.

Part II   Physical Principles

II.1   Production and Measurement of X-rays

X-rays are produced when a beam of fast electrons strikes a target. The electrons lose, on this occasion, most of their energy in collisions with atomic electrons in the target, causing ionization and excitation of atoms. In addition they can be sharply deflected by the electric field of the atomic nuclei, thereby losing energy by emitting X-ray photons.

First a short description of the function of an X-ray tube and the production of X-rays is given. Then some problems concerning the measurement of photon spectra of diagnostic X-rays are discussed.

II.1.1   Production

The X-ray tube assembly used in diagnostic radiology is made up of an oil-filled housing containing an insert – the X-ray tube – which is an evacuated envelope of glass (or a metal-ceramics construction) within which are mounted a cathode with a filament and an anode (see Fig. II.1.1). The housing is constructed in such a form that it protects the environment from electric shocks, X-ray leakage and mechanical failure of the insert.

If an exposure is made, the filament is heated by passage of an electric current. It produces hereby a narrow beam of electrons, which are accelerated by the simultaneously applied potential difference (i.e. the tube voltage) U of about 20–150 kV between the cathode and the anode and strike the anode. Electrons which run through a potential difference of e.g. 150 kV gain a speed which is equal to more than 60% of the speed of light (see Fig. II.1.2).

The electrons interact with the material of the anode, slow down and stop. Most of the energy absorbed from the electrons appears in the form of heat and only a small amount emerges as X-rays. For a better heat dissipation, the anode disc rotates therefore during exposure and the ring-shaped area which is hit by the electrons defines the so called focal spot track. If the efficiency of conversion of electron energy into X-rays is defined by:

$$\eta = \frac{P_{rad}}{U \cdot I} \tag{II.1.1}$$

where P_{rad} is the X-ray power, I is the tube current (and $U \cdot I$ is the electric power), then the relation:

$$\eta = k \cdot U \cdot Z \tag{II.1.2}$$

approximately holds (Morneburg 1995, p. 89). Z is the atomic number and k is a constant, which was experimentally determined to be about $1.1 \cdot 10^{-9}$ V^{-1}. For example, assuming $U = 60$ kV and tungsten ($Z = 74$) as anode material, η is about 0.5%.

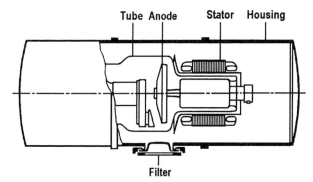

Fig. II.1.1. Principal design of a rotating-anode X-ray tube

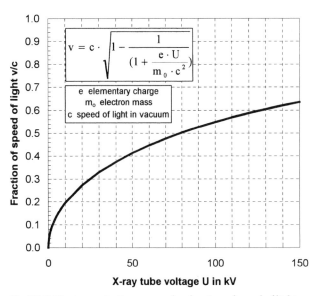

Fig. II.1.2. Electron velocity expressed as fraction of speed of light vs X-ray tube voltage

Some of the X-rays produced pass through the exit windows of the insert and the tube housing, forming the primary radiation beam. X-rays that are emitted from the focal spot in other directions are absorbed by the housing. The X-ray tube assembly is provided with a multi-leaf collimator so that the size of the primary radiation field may be varied as necessary.

The design of the filament and the electron optics that guide the electrons from the cathode to the anode is very important, because the sharpness in the image may be limited by the size of the X-ray source (focal spot), and the radiation output from the tube is determined by the tube current striking the anode (see Eq. II.1.1). Mostly the filament is constructed from a spiral of tungsten wire, which is set in a nickel block (Webb 1988, p. 33). This block supports the filament and is shaped to create an electric field that focuses the electrons into a slit beam.

As a rule, X-ray tubes for diagnostic radiology have a rotating anode with a bevelled edge, which is at a steep angle to the direction of the electron beam (see Fig. II.1.3a). The exit window accepts X-rays that are approximately at right angles to the electron beam so that the X-ray source as viewed from the image receptor in the direction of the central ray (line focus principle) appears to be approximately square. But, due to the angulation of the anode surface, throughout the imaging plane a considerable variation of the size, shape and orientation of the focal spot projections results (see Fig. II.1.3b). The length l of the focal spot is mainly determined by the filament length and the anode angle Θ, the width b of the focal spot by the diameter of the filament coil and the width of the focusing slot. This behaviour demonstrates the difficulty of defining the focal spot size (see IEC 1993).

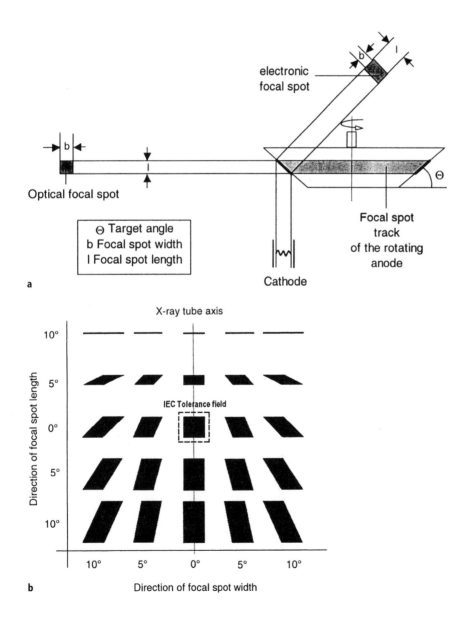
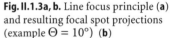

Fig. II.1.3a, b. Line focus principle (**a**) and resulting focal spot projections (example $\Theta = 10°$) (**b**)

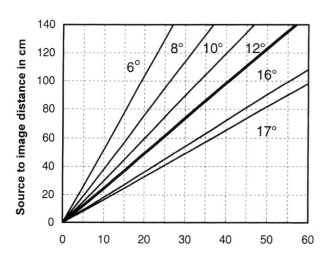

Fig. II.1.4. Interconnection between image field size, source-to-image distance and anode angle

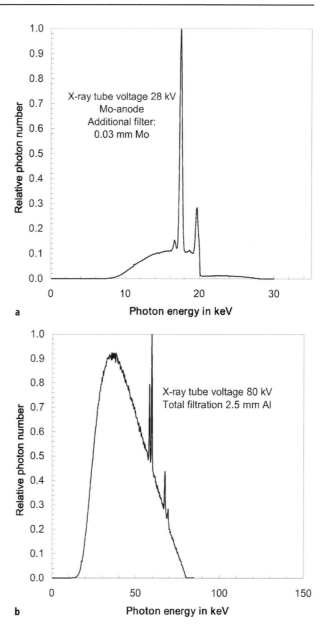

Fig. II.1.5a, b. Two typical X-ray spectra: **a** Mo anode, tube voltage 28 kV, target angle 22°, angle selected for measurement 16° and 30 μm Mo filter; **b** W anode, tube voltage 80 kV, target angle 10°, measurement in the direction of the central axis and total filtration 2.5 mm Al

The choice of the anode angle (or target angle) Θ (see Fig. II.1.3a) will depend upon the application, with the angle being varied according to the requirements of the dimensions of the radiation field, the focal spot sizes and the tube output. For general-purpose units, an angle of about 12° is appropriate (see Fig. II.1.4).

As a rule, the X-ray tube will have two filaments and in some cases the anode disc two bevels at different angles, so that either a small or a large focal spot may be selected.

It has already been mentioned above that most of the energy in the electron beam is deposited in the target in the form of heat. Consequently a high atomic number Z and melting point T_m are essential for the anode material. Therefore the anode is usually constructed from tungsten (Z = 74, T_m = 3410°C). For special applications, where a low-energy X-ray beam is required (predominantly X-ray mammography), a molybdenum (Z = 42, T_m = 2617°C) or a rhodium target (Z = 45, T_m = 1960°C) is also used.

An X-ray spectrum represents the distribution of the number of photons created as a function of their energy E. The shape of the emitted X-ray spectrum above all will depend upon the anode material, the tube voltage applied and the effects of any filters placed in the primary X-ray beam (see Chap. II.3.4 and figures and tables in Part IV.1.1 and IV.1.2). In investigations regarding the patient exposure and image quality or special X-ray exposure techniques (e.g. dual energy), the influence of the wave form of the x-ray tube voltage and of the target angle on the X-ray spectrum are of great importance as well. The waveform depends especially on the type of X-ray generator used (Dendy and Heaton 1999, p. 26; Porubszky 1986).

Figure II.1.5 shows two typical X-ray spectra. The first one is for a tube with a molybdenum target (Fig. II.1.5a)

and the second one for a tube with a tungsten target (Fig. II.1.5b). Both spectra consist of two components, the continuous spectrum and the line or characteristic spectrum.

The effect of the waveform of the tube voltage on the continuous spectrum is shown in Fig. II.1.6.

One can see from Fig. II.1.6 that the waveform of the X-ray tube voltage has influence not only on the intensity of the emitted X-radiation, but also on the position of the maximum in the continuous spectrum. The position of the characteristic lines with respect to the energy, however, is constant, only their intensity is influenced by the waveform.

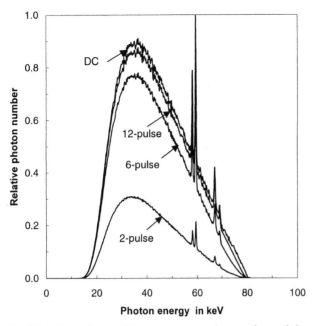

Fig. II.1.6. Dependence of X-ray spectra on the waveform of the X-ray tube voltage (2-pulse, 6-pulse, 12-pulse and DC generator)

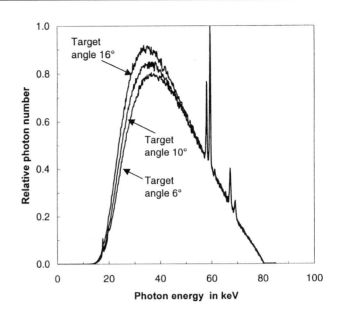

Fig. II.1.7. Dependence of X-ray spectra on the target angle (6°, 10° and 16°)

The effect of the target angle on the X-ray spectrum is pointed out in Fig. II.1.7. A smaller target angle leads to a reduced number of photons emitted in the direction of the central line, and furthermore the maximum of the X-ray spectrum is shifted to somewhat higher energies. These effects increase when the X-ray tube gets older, resulting in a more roughened focal spot track. The line spectrum again is always fixed with respect to the energy, only the intensity of the lines depends on the target angle.

The continuous spectrum results from the deceleration of the electrons in the electric field of the nucleus of the target atoms. These X-rays are known as "bremsstrahlung" or braking radiation. The amount of energy lost by the electron in such an interaction is very variable and hence the energy of the X-ray photon:

$$E = h\nu \qquad\qquad (II.1.3)$$

where $h = 6.626 \cdot 10^{-34}$ Js is Planck's constant and ν the frequency of the corresponding electromagnetic wave, can take a wide range of values. The upper limit of the photon energy results from the fact that an electron, very occasionally, loses all its kinetic energy to the production of a photon; it is given by:

$$E_{max} = e \cdot U \qquad\qquad (II.1.4)$$

where e is the elementary charge ($1.6 \cdot 10^{-19}$ C). The low-energy cut-off is caused by the fact that low-energy photons are easily attenuated, e.g. by the anode material, the

insulating oil, the window (made of glass or beryllium) of the X-ray tube and any added filtration, so that the X-ray intensity emerging is negligible.

The line spectrum is caused by incoming electrons interacting with bound orbital electrons in the target. If an incoming electron has sufficient kinetic energy to overcome the binding energy, it can remove the bound electron, creating a vacancy in the shell. The probability of this happening is greatest for the innermost shells. The vacancy is then filled by an electron from an outer shell falling into it and the excess energy is emitted as a photon (or a cascade of Auger electrons; see Chap. II.2.2.2). Thus, if for example the vacancy is created in the K shell, it may be filled by an electron falling from either the L shell, the M shell or outer shells. Even a free electron may fill the vacancy, but the most likely transition is from the L shell. The orbital electrons occupy well-defined energy levels and these energy levels are different for different elements. When an electron falls from one outer energy level to another inner energy level, a photon is emitted that has an energy equal to the difference between the two energy levels in that atom and hence is characteristic of that element. The K series of lines for tungsten will range from 58.0 (for a transition from the L shell to the K shell) to 69.5 keV (if a free electron fills the K-shell vacancy). The corresponding values for molybdenum are 17.4 keV and 20.0 keV. Transitions to the L shell are of no practical importance in diagnostic radiology, since the maximum energy change for tungsten is 11 keV and photons of this energy are practically completely absorbed before they leave the X-ray tube.

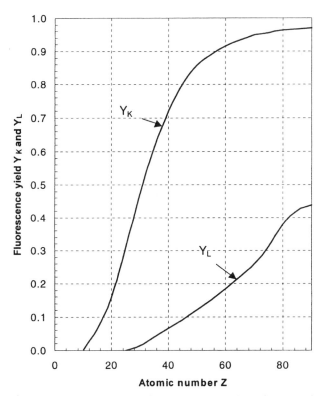

Fig. II.1.8. Fluorescence yields for the K and L shell as a function of atomic number Z (adapted from Attix 1986, p. 146)

The number of characteristic X-ray photons emitted, expressed as a fraction of the number of primary vacancies created in the corresponding atomic electron shell, is known as the fluorescence yield Y. The fluorescence yields for K and L shell (Y_K and Y_L, respectively) are plotted as a function of atomic number in Fig. II.1.8. It can be seen that Y_K is more than 90% for high atomic number materials. Y_L is negligible for low atomic number materials and reaches about 40% in the high Z range.

II.1.2   Measuring Photon Spectra of Diagnostic X-rays

The X-ray spectra used in this book have been measured on X-ray source assemblies for general diagnostic radiology containing different X-ray tubes with target angles between 6° and 16° (tungsten anode) and a glass envelope, as well as on X-ray tubes for mammography with anode materials molybdenum and tungsten and a target angle of 22°. In case of the diagnostic X-ray tubes the measurements were made in the direction of the central axis, in case of the mammography tubes in the direction of 16°, i.e. in a direction which corresponds to a point of the breast support which lies about 6 cm from the chest wall edge. Mammographic spectra with the anode material rhodium were measured on a fine structure tube. We

have also measured spectra of a diagnostic X-ray source assembly with a Be window and 12° anode angle, used for reference purposes, for example the measurement or calculation of the inherent filtration. All spectra were measured with minimum possible total filtration, those for mammography and at the reference X-ray source assembly, with 1 mm Be filtration.

II.1.2.1   X-ray Spectrometer

The X-ray spectrometer used was equipped with a highly pure Ge-detector. The main components of the spectrometer used were:

- Detector
- Preamplifier with pulsed optical feedback
- Main amplifier (with pile-up rejecter and life time corrector)
- Multi-channel analyser

The detector, a highly pure planar Ge crystal, had a sensitive area of 100 mm^2 and a thickness of 7.5 mm. With a 2 kV bias, the depletion depth reached the whole detector volume. While detecting an X-ray photon via photoelectric effect (see Chap. II.2.2.2) the resulting photoelectron loses its energy, by collision with the Ge crystal lattice, step by step with a mean ionization energy of 2.9 eV per collision, creating electron-hole pairs, which are collected at the electrodes. The number of the created electron-hole pairs is proportional to the energy of the absorbed photon.

A charge-sensitive preamplifier with pulsed optical feedback was used to amplify the collected charge with minimum noise contribution. With capacity feedback, the output DC-voltage level increased with the collected charge and had to be discharged. To avoid a resistive feedback, which is an additional noise source, the discharge was maintained by a flash of a luminescence diode to the entrance field effect transistor (FET) of the preamplifier at a certain voltage level, which is called pulsed optical feedback. The Ge-detector and parts of the preamplifier including the entrance FET were mounted in a cryostat, which was cooled with liquid nitrogen. The entrance window of the detector consisted of 0.033 mm Be.

The main amplifier performed pulse-forming by differentiating and integrating the output signal of the preamplifier to get a usable signal required from the multi-channel analyser. It also performed detection and rejection of pulse pile up at higher count rates with lifetime correction.

The pulse height at the output of the main amplifier was proportional to the collected charge and therefore proportional to the energy of the detected photon. The energy resolution of the spectrometer depends on the noise of detector and electronics and the quantum noise. The latter may be reduced by effective charge collection with an adequate integrating time constant (about 1–4 μs). The energy resolution of the X-ray spectrometer (full width at half-maximum, FWHM) was lower than 0.5 keV at 122 keV (^{57}Co).

The multi-channel analyser accumulates the single detector pulses, sorting them into channels according to their pulse height. With a sufficient collecting time, one obtains a pulse-height spectrum, which must be converted to an energy spectrum by an energy calibration of the pulse-height scale with γ radiation and X-ray fluorescence radiation. Suitable radionuclides for energy calibration in the energy range up to 150 keV are:

^{133}Ba	31.0, 35.0, 80.9	keV
^{57}Co	122.1, 136.5	keV
^{241}Am	59.5	keV

With an adequate adjustment of the pulse amplification, a definite energy grid may be selected. We used 0.2 keV/channel in the range up to 150 keV and 0.1 keV/channel for mammography spectra.

II.1.2.2   Measuring

For quantitative measurements of the photon spectra of diagnostic X-ray source assemblies, accurate controlling of the exposure parameters is required. These are X-ray tube voltage and current, measuring time and solid angle, defined by collimator diaphragm and focus distance. Especially in the case of soft X-radiation (low filtration), temperature and air pressure have to be measured in order to correct the X-ray attenuation in air.

An oil-filled tank containing a voltage divider with 1% accuracy and a connected electrometer, as well as a multi-range microammeter at the anode side, arranged between the high-voltage transformer and X-ray source assembly, was used to control tube voltage and tube current. Thereby the primary voltage for the tube-voltage and tube-current (filament heating) control transformers, which supplied the high-voltage transformer, had been stabilised with a line power stabiliser. With this configuration the tube voltage and tube current could be controlled within 1% accuracy.

Because of the finite focal-spot size, the measuring diaphragm (0.1 mm diameter; for tube voltages above 70 kV we used a crossed-slit collimator), which, together with the focus distance, defines the solid angle, had to be well adjusted to the focal spot. With a narrow collimated beam and low tube current (down to 20 μA) we achieved a sufficiently low count rate of about 1000 counts/s, so the accumulation of a single spectrum took up to 1 h. The raw spectra were transferred together with the exposure parameters via serial interface to a desktop computer.

II.1.2.3   Corrections for Detector Influences

The measured spectra had to be corrected against distortions caused by the Ge-detector. Because of the reasonably good energy resolution of the detector, a correction of the line broadening, caused by statistical fluctuations in pulse height, was not carried out. The main interactions of photons in the detector that cause distortions were:

- Photoelectric absorption with escape of a fluorescent K-shell photon
- Compton scattering with escape of the scattered photon
- No interaction (photon passes through the detector)

In all other cases, the incident photon deposited its whole energy in the detector and was therefore counted with its true energy. All effects depend on photon energy (see Chap. II.2.2) and the detector dimensions and had to be measured or calculated individually. For the corrections used (see Fig. II.1.9), Monte Carlo-calculated Ge-detector response is available from Seltzer (1981). Further hints to the correction procedure are given by Israel et al. (1971) and Seelentag and Panzer (1978).

The photo peak efficiency ε(E) of the detector is the ratio of the number of incident photons, which deposited their whole energy in the detector, to the total number of incident photons with energy E. The photo peak efficiency may be determined by comparing the measured photon numbers with well-known γ-line photon numbers or approximately by calculation of the relative energy absorption in the detector, using energy absorption coefficients (Israel et al. 1971). For the entire spectrum, the number of counts (without Compton and K-escape counts) has to be multiplied by 1/ε.

In the case of Compton scattering, an incident photon with energy E produces a Compton electron and a scattered photon (see Chap. II.2.2.1). If the scattered photon escapes the detector only the Compton electron is absorbed and counted in the Compton continuum in the energy range below the Compton edge E_C, where:

$$E_C = \frac{E}{(1+\frac{m_0 c^2}{2 \cdot E})} \qquad (II.1.5)$$

If the scattered photon is absorbed within the detector the event is detected as photo absorption at energy E (photo peak). To correct the spectrum, the energy-dependent Compton/peak fraction was determined using single-line photon sources and heavily filtered X-rays. The correction starts at the high-energy end of the spectrum, removing the expected Compton fraction from the ener-

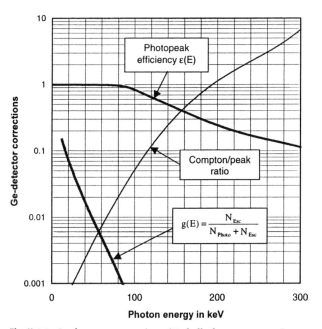

Fig. II.1.9. Ge-detector corrections (K-shell photon escape, Compton-escape and no interaction)

gy range below the Compton edge E_C, assuming a rectangular Compton electron distribution.

Fluorescent K-radiation may occur after photoabsorption of incident photons with energy E above the Ge-K-edge at 11.1 keV (see Chap. II.1.1). Depending on detector dimensions, parts of the K-radiation may escape the detector and the incident photon is counted at energy $E-E_K$, where E_K is the energy of the characteristic X-ray photon. Especially for the correction of Ge-K-escape in mammography spectra, the separation in K_a (9.9 keV) and K_b (11 keV) escape lines, with a relative frequency of 6/7 and 1/7, respectively, was performed. The Ge-K-escape fraction:

$$g(E) = \frac{N_{Esc}}{(N_{Photo} + N_{Esc})} \qquad (II.1.6)$$

where N_{Photo} and N_{Esc} is the corresponding number of photons, has been measured with X-ray or γ-line sources. The correction starts at 60 keV, with minimal escape fraction going to lower energies according to:

$$N_c(E) = \frac{N_m(E) - g(E+E_K) \cdot N_C(E+E_K)}{(1 - g(E))} \qquad (II.1.7)$$

where $N_c(E)$ and $N_m(E)$ are the corrected and measured counts at energy E.

II.1.2.4 Preparing for Calculations with X-ray Spectra

In the corrected spectra, measured with an energy grid of 0.2 keV/channel (0.1 keV/channel for mammography spectra), the photon fluence Φ_E was normalised to a tube charge of 1 mA s and a solid angle of 1 sr (steradian). To correct uncertainties in the measuring parameters, especially the solid angle, for all spectra of a series, the calculated air kerma (see Chap. II.3.1) was compared with measured air kerma values under the same conditions (X-ray tube voltage, tube charge, total filtration). Deviations between calculated and measured air kerma have been corrected. In spite of correction of the Compton continuum, the low-energy tail of the spectrum may contain counts which influence the calculated radiation quality. To determine the low-energy limit for spectrum integration, the calculated half-value layer (HVL) was compared with measured values. The start energy for calculations with diagnostic spectra (2.5 mm Al total filtration) is about 12 keV and, for mammography spectra, about 5 keV.

For further calculations we used the X-ray spectra with an energy grid ΔE of 1 keV/channel, summing up five channels in the range E–ΔE and placing the result at energy channel E. Therefore energy-dependent tables with energy grid 1 keV/channel such as mass-attenuation coefficients, have been calculated for energies E–0.5 keV.

The measured spectra exist in series with constant anode angles from 6° to 16° and X-ray tube voltages (constant potential with max. ripple of 3%) from 40 kV up to 150 kV in steps of about 10 kV. The spectra series with anode angles 8° and 12° have been created by linear interpolation of the spectral content from neighbouring spectra. For mammography the spectra have been measured for Mo-, Rh- and W-anodes at an anode emission angle of 16° (see Chap. II.1.2) from 20 to 50 kV usually in 5 kV steps. Spectra for tube voltages between have been linear interpolated from the spectral content of neighbouring spectra with extrapolation to the maximum energy (see spectra shown in Chap. IV.1.1, IV.1.2 and in the corresponding data files on the CD-ROM).

References

Attix FH (1986) Introduction to radiological physics and radiation dosimetry. Wiley, New York

Dendy PP, Heaton B (1999) Physics for diagnostic radiology. Institute of Physics Publishing, Bristol

IEC (International Electrotechnical Commission) (1993) X-ray tube assemblies for medical diagnosis – characteristics of focal spots. IEC publication 60636, Geneva

Israel HI, Lier DW, Storm E (1971) Comparison of detectors used in measurement of 10 to 300 keV X-ray spectra. Nucl Instrum Methods 91:141–157

Morneburg H (ed) (1995) Bildgebende Systeme für die medizinische Diagnostik. Publicis MCD, Erlangen

Porubszky T (1986) Calculation concerning voltage ripple of X-ray generators. Phys Med Biol 31:371–381

Seelentag WW, Panzer W (1979) Stripping of X-ray bremsstrahlung spectra up to 300 kV$_p$ on a desk type computer. Phys Med Biol 24:767–780

Seltzer SM (1981) Calculated response of intrinsic germanium detectors to narrow beams of photons with energies up to ~300 keV. Nucl Instrum Methods 188:133–151

Webb S (ed) (1988) The physics of medical imaging. Institute of Physics Publishing, Bristol

As an X-ray beam passes through material, there exist three alternatives for each photon:

1. It can penetrate the material without interacting
2. It can interact with material and be completely absorbed by depositing its energy
3. It can interact and be scattered from its original direction, possibly depositing a part of its energy

This behaviour leads to the purely exponential attenuation of a mono-energetic photon beam. First this exponential law of attenuation is briefly introduced. Then the different interaction processes of photons with matter that are relevant for diagnostic radiology are discussed together with the corresponding interaction coefficients. Emphasis is laid on the different mechanisms responsible for the transfer of radiation energy to matter and the production of scattered radiation.

II.2.1   Exponential Law of Attenuation

When a parallel beam of a number N_0 of mono-energetic photons passes a flat plate of material of thickness d, the number N of photons transmitted is given by:

$$N = N_0 \cdot e^{-\mu \cdot d} \qquad \text{(II.2.1)}$$

where μ is the linear attenuation coefficient. Equation II.2.1 is generally called the exponential law of attenuation. It holds also for example for the intensity of the photon beam. Since μ is dependent on the material's density, the quantity usually tabulated is the mass attenuation coefficient μ/ρ, in which the dependence on the density has been removed.

The linear attenuation coefficient μ includes the contributions of the different photon interaction processes. Since these contributions are generally calculated in terms of atomic cross sections, the relation (Hubbell 1999):

$$\frac{\mu}{\rho} = \sigma_{tot} \cdot \frac{N_A}{A} \qquad \text{(II.2.2)}$$

where σ_{tot} is the total atomic cross section, seems to be useful for the following considerations: $N_A = 6.0221367 \cdot 10^{23}$ atoms/mol is Avogadro's number and A is the relative atomic weight (in grams per mole).

II.2.2   Interaction Processes

Generally there are five types of interactions with matter by photons which are considered in radiological physics:

1. Compton effect
2. Photoelectric effect
3. Rayleigh scattering
4. Pair production
5. Nuclear photoeffect

For the production of an electron/positron pair, a photon energy of at least 1.02 MeV is needed and photonuclear interactions are only significant for photon energies above a few million electron volts. Therefore, only the first three processes are relevant for X-ray diagnostic radiology. From these the first two are the most important, as they result in the transfer of energy to electrons, which then impart that energy to material in many (mostly small) coulomb-force interactions along their tracks. Moreover the Compton effect results in the emission of scattered radiation, which gives rise to a lot of problems in medical imaging with X-rays. Rayleigh scattering is elastic; the photon is merely redirected through a small angle with no energy loss.

Consequently, for the diagnostic X-ray energy range:

$$\mu = \mu_{pe} + \mu_{incoh} + \mu_{coh} \qquad \text{(II.2.3)}$$

where μ_{pe} is the photoelectric attenuation coefficient, μ_{incoh} is the Compton (or incoherent) attenuation coefficient and μ_{coh} is the Rayleigh (or coherent) attenuation coefficient. The corresponding relation holds for the total atomic cross section σ_{tot}:

$$\sigma_{tot} = \sigma_{pe} + \sigma_{incoh} + \sigma_{coh} \qquad \text{(II.2.4)}$$

The three interaction processes and their contributions to σ_{tot} and μ, respectively, are now shortly discussed. According to Attix (1986), the description of the different interaction processes can be subdivided into two aspects: kinematics and cross section. The first relates the energies and angles of the participating particles in an interaction; the second predicts the probability that an interaction event will occur.

II.2.2.1 Compton Effect

The Compton effect or inelastic scattering may be thought of most easily in terms of classic mechanics in which the photon makes a billiard ball-type collision with an atomic electron, with both energy and momentum conserved. It is customary to assume that the electron struck by the incoming photon is initially unbound and stationary. These assumptions are certainly not rigorous, in as much as the electrons all occupy various atomic energy levels, thus are in motion and are bound to the nucleus. Under conditions where these influences cannot be ignored (e.g. low photon energies), they are taken into account by corrections, e.g. the incoherent scattering function (Hubbell 1999). In the following the initial motion and binding of the electron will be ignored.

Figure II.2.1 schematically shows a Compton interaction: A photon of energy $h\nu$ and momentum $h\nu/c$ (where c is the speed of light in vacuum) is colliding with an electron. The electron has no initial kinetic energy or momentum. After the collision the electron departs at angle Θ, with kinetic energy T. The photon scatters at angle Φ with a lower energy $h\nu'$. The relativistic treatment of the collision kinetics is based upon conservation of both energy and momentum. The following set of equations provides a complete description of the kinematics of the Compton effect:

$$T = h\nu - h\nu' \tag{II.2.5}$$

$$h\nu' = \frac{h\nu}{[1+(\dfrac{h\nu}{m_0 c^2})(1-\cos\Phi)]} \tag{II.2.6}$$

$$\cot\Theta = (1+\frac{h\nu}{m_0 c^2}) \cdot \tan\frac{\Phi}{2} \tag{II.2.7}$$

where $m_0 c^2 = 511$ keV is the rest energy of the electron.

It can be seen from Eq. II.2.6 that for low photon energies ($h\nu \ll m_0 c^2$) the electron receives practically no kinetic energy in the interaction. This means that Compton scattering is nearly elastic in this case. From Eq. II.2.6 follows also that maximum energy transfer T_{max} to the electron occurs for $\Phi = 180°$ (cos 180° = –1). For example, for $h\nu = 100$ keV it can be calculated from Eqs. II.2.5 and II.2.6 that $T_{max} = 28.1$ keV. Finally it can be seen from Eq. II.2.7 that the electrons are always scattered in the forward direction ($\Theta < 90°$).

In order to calculate the cross section for the Compton effect, Klein and Nishina (abbreviation: K-N) in 1928 applied Dirac's relativistic theory of the electron to that problem. At first they received an expression for the differential cross section for photon scattering at angle Φ per unit solid angle and per electron (Attix 1986, p. 132):

$$\frac{d\sigma_e}{d\Omega_\phi} = \frac{r_0^2}{2} \cdot (\frac{h\nu'}{h\nu})^2 \cdot (\frac{h\nu'}{h\nu} + \frac{h\nu}{h\nu'} - \sin^2\Phi) \tag{II.2.8}$$

where $r_0 = e^2/m_0 c^2 = 2.818 \ 10^{-13}$ cm is called the "classical electron radius". Figure II.2.2 shows the differential K-N

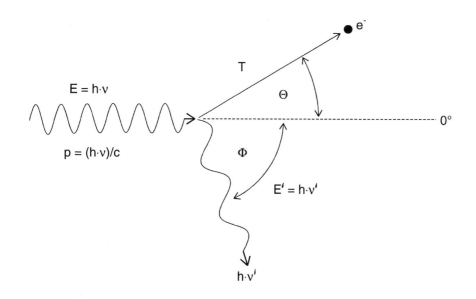

Fig. II.2.1. Kinematics of the Compton effect: Photon-scattering angle Φ; electron-scattering angle Θ

cross section for photon energies of 20, 50, 100 and 150 keV. The forward bias of the scattered photons with increasing energy is apparent.

Integration over all photon scattering angles Φ results in a longish expression (Attix 1986, p. 132) for the total K-N cross section per electron σ_e, which is not given here. Figure II.2.3 shows σ_e as a function of the primary photon energy. It decreases gradually for higher photon energies to approach a $(h\nu)^{-1}$ dependence. Since the electron binding energy has been assumed to be zero, the K-N cross section per atom $\sigma_{incoh, K-N}$ is then given by:

$$\sigma_{incoh,K-N} = \sigma_e \cdot Z \qquad \text{(II.2.9)}$$

From Eq. II.2.2 it follows that the Compton attenuation coefficient is given by:

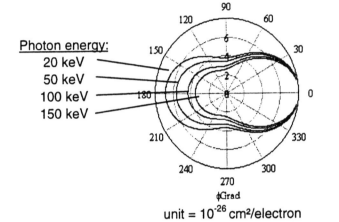

Photon energy:
20 keV
50 keV
100 keV
150 keV

unit = 10^{-26} cm²/electron

Fig. II.2.2. Differential Klein-Nishina (K-N) cross section for photon energies of 20 keV , 50 keV, 100 keV and 150 keV

$$\mu_{incoh} = \rho \cdot N_A \cdot \sigma_e \cdot \frac{Z}{A} \qquad \text{(II.2.10)}$$

Z/A ranges between 0.5 and 0.4 (with the exception of hydrogen, where Z/A = 1), tending to decrease gradually with increasing Z. Therefore, the Compton attenuation coefficient is approximately independent from Z.

In each Compton interaction, the energy of the incident photon is shared between the scattered photon and the recoiling electron. Especially for dosimetric problems (see Chap. II.3), it is of interest to know the overall fraction of the energy, carried by the primary photons, that is given to the electrons, averaged over all scattering angles. This can be done by first modifying the differential K-N cross section $d\sigma_e/d\Omega_\Phi$ in Eq. II.2.8 to obtain a quantity referred to as the differential K-N energy-transfer cross section $d\sigma_{e,tr}/d\Omega_\Phi$:

$$\frac{d\sigma_{e,tr}}{d\Omega_\Phi} = \frac{d\sigma_e}{d\Omega_\Phi} \cdot \frac{T}{h\nu} \qquad \text{(II.2.11)}$$

Integrating over all photon-scattering angles Φ from 0 to 180°, results again in a longish expression (Attix 1986, p. 134), this time for the K-N energy-transfer cross section per electron $\sigma_{e,tr}$, which is also not given here. Figure II.2.3 shows $\sigma_{e,tr}$ as a function of the primary photon energy. It can be seen that the energy transfer to electrons is small for low photon energies. The average energy of the Compton recoil electrons generated by photons of energy $h\nu$ is then given by:

$$T = h\nu \cdot \frac{\sigma_{e,tr}}{\sigma_e} \qquad \text{(II.2.12)}$$

Fig. II.2.3. K-N cross section per electron and corresponding energy-transfer cross section per electron as a function of primary photon energy

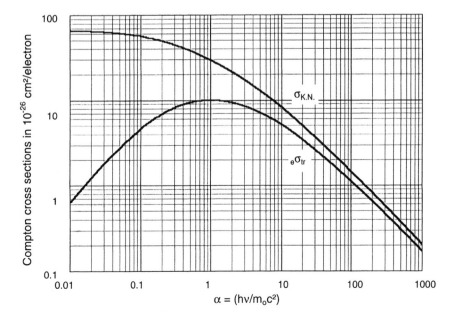

and according to Eq. II.2.10 the Compton energy-transfer coefficient is given by:

$$\mu_{incoh,tr} = \rho \cdot N_A \cdot \sigma_{e,tr} \cdot \frac{Z}{A} \qquad (II.2.13)$$

The contributions of the several kinds of interaction processes to the total linear attenuation coefficient μ and the total energy-transfer coefficient μ_{tr} are summarised in Chap. II.2.3.

As already mentioned, the binding energies of the atomic electrons are ignored in the K-N treatment. Currently they are taken into account by the incoherent scattering function S(x,Z) approach, in which x is a momentum transfer variable related to the incident photon energy $h\nu$ and the deflection angle Φ of the scattered photon. In this approach the differential cross section $d\sigma_e/d\Omega_\Phi$ in Eq. II.2.8 is weighted by S(x,Z) before the integration resulting in σ_e. Hints to existing data for S(x,Z) are also given in Chap. II.2.3.

II.2.2.2 Photoelectric Effect

The photoelectric effect is the dominant process at the lower end of the diagnostic range of X-ray photon energies (see Figs. II.2.4, II.2.5). In this energy range, it predominates over the Compton effect, particularly with respect to the energy transferred to the secondary electrons. In the preceding chapter, it was seen that a photon cannot give up all of its energy in colliding with a free electron, but it can do so in an interaction with a tightly bound electron, such as those in the inner shells of an atom. The kinematics of the photoelectric effect is shown schematically in Fig. II.2.4.

A photon of energy $h\nu$ and momentum $h\nu/c$ is interacting with an atomic shell electron bound by the binding energy E_B. The interaction cannot take place with respect to a given electron unless $h\nu>E_B$ for that electron. The smaller $h\nu$ is, the more likely is the occurrence of the photoelectric effect, provided that $h\nu>E_B$. The photon is totally absorbed in the interaction. The kinetic energy T, given to the electron, is independent of its emission angle Θ and amounts to:

$$T = h\nu - E_B \qquad (II.2.14)$$

since the kinetic energy T_A given to the recoiling atom is almost zero.

The theoretical derivation of the interaction cross section for the photoelectric effect is more difficult than for the Compton effect, because the binding energy of the electron cannot be ignored on principle. There is no simple equation for the differential photoelectric cross section that corresponds to Eq. II.2.8. Published tables of photoelectric interaction coefficients such as the National Institute of Standards and Technology (NIST) of the USA data (see Chap. II.2.3) are based on experimental results, supplemented by theoretically assisted interpolations for other energies and materials than those measured.

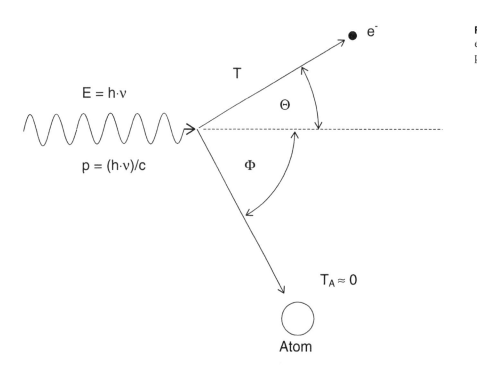

Fig. II.2.4. Kinematics of the photoelectric effect: Emission angle of photo-electron Θ

The interaction cross section per atom for photoelectric effect, integrated over all angles of photoelectron emission, can be written as (Attix 1986, p. 140):

$$\sigma_{pe} \propto k \cdot \frac{Z^n}{(h\nu)^m} \qquad (II.2.15)$$

where k is a constant, $n \approx 4$ at $h\nu = 0.1$ MeV, gradually rising to about 4.6 at 3 MeV, and $m \approx 3$ at $h\nu = 0.1$ MeV, gradually decreasing to about 1 at 5 MeV.

Therefore, in the energy region $h\nu \le 100$ keV, where the photoelectric effect becomes most important, the relations:

$$\sigma_{pe} \propto \frac{Z^4}{(h\nu)^3} \qquad (II.2.16)$$

and, according to Eq. II.2.2:

$$\mu_{pe} \propto \rho \cdot \frac{Z^3}{(h\nu)^3} \qquad (II.2.17)$$

are approximately valid.

Concerning the energy transfer cross section $\mu_{pe,tr}$ for the photoelectric effect, it is evident from Eq. II.2.11 that the fraction of $h\nu$ that is transferred to the photoelectron is:

$$\frac{T}{h\nu} = \frac{(h\nu - E_B)}{h\nu} \qquad (II.2.18)$$

But this is only a first approximation to the total fraction of $h\nu$ that is transferred to electrons at all, because part or all of the binding energy E_B can also be converted into electron kinetic energy through the Auger effect.

The Auger effect is an interaction competing with the emission of characteristic X-ray photons, particularly for elements of low Z (see Chap. II.1.1): An atom in which, e.g. an L electron makes a transition to fill a vacancy in the K shell does not always emit a photon. A different, non-optical transition can occur in which an L electron is ejected from the atom, thereby leaving two vacancies in the L shell. The electron thus ejected from the atom is called an Auger electron. The L shell vacancies then lead to the emission of further Auger electrons more or less simultaneously in a sort of chain reaction.

Consideration of the Auger effect leads, e.g., for photons with energy $h\nu$ larger than the binding energy $E_{B,K}$ of the K shell, to the following expression of the photoelectric energy transfer coefficient (Attix 1986, p. 146):

$$\mu_{pe,tr} = \mu_{pe} \cdot \frac{(h\nu - P_K Y_K \cdot h\nu_K)}{h\nu} \qquad (II.2.19)$$

where P_K is the fraction $\mu_{pe,K}/\mu_{pe}$ of all photoelectric interactions that occur in the K shell, Y_K is the corresponding fluorescence yield (see Chap. II.1.1) and $h\nu_K$ is the mean energy for L to K shell transitions.

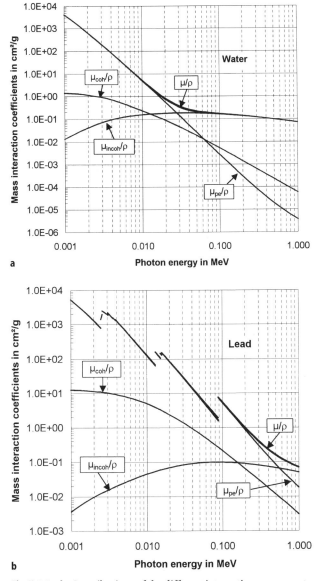

Fig. II.2.5a, b. Contributions of the different interaction processes to the mass attenuation coefficient μ/ρ of **a** water and **b** lead

II.2.2.3 Rayleigh Scattering

At X-ray photon energies other than for mammography, the probability for Rayleigh scattering is much less than for the Compton effect. For this reason Rayleigh scattering has received little attention in diagnostic radiology. Rayleigh scattering is called "coherent" because the photon is scattered by the combined action of the bound electrons. The interaction is elastic in the sense that the photon loses only a negligible fraction of its energy, since the recoil is by the entire atom including the nucleus, rather than by an individual electron as in the Compton effect. Therefore, Rayleigh scattering contributes nothing to the transfer of photon energy to matter.

The differential cross section per atom for coherent scattering at angle Φ per unit solid angle is (Hubbell 1999):

$$\frac{d\sigma_A}{d\Omega_\Phi} = \frac{r_0^2}{2} \cdot (1+\cos^2\Phi) \cdot [F(x,Z)]^2 \qquad (II.2.20)$$

where r_0 is again the classical electron radius as in Eq. II.2.8 and $F(x,Z)$ is the atomic form factor. Also, as in the incoherent scattering function $S(x,Z)$, x is again a momentum transfer variable related to the incident photon energy $h\nu$ and the deflection angle Φ of the scattered photon.

The effect of the atomic form factor is that the angular dependence of coherent scatter is highly forward peaked.

Integration over all photon-scattering angles Φ results in the total atomic cross section for coherent scattering σ_{coh}. Its dependence on $h\nu$ and Z has the form (Attix 1986, p. 153):

$$\sigma_{coh} \propto \frac{Z^2}{(h\nu)^2} \qquad (II.2.21)$$

which results, according to according to Eq. II.2.2 in:

$$\mu_{coh} \propto \rho \cdot \frac{Z}{(h\nu)^2} \qquad (II.2.22)$$

II.2.3 Total Coefficients for Attenuation, Energy Transfer and Energy Absorption

Since the Compton attenuation coefficient μ_{incoh} (Eq. II.2.10), the photoelectric attenuation coefficient μ_{pe} (Eq. II.2.17) and the Rayleigh (or coherent) attenuation coefficient μ_{coh} (Eq. II.2.22) are dependent on the density ρ, which can vary considerably for a given element or compound, for compilation purposes this dependency is removed by tabulating the mass attenuation coefficient μ/ρ (see Chap. II.2.1). Current compilations of the mass attenuation coefficient are derived from theoretical or semi-empirical values of the cross sections for the individual processes. From Eqs. II.2.2, II.2.3 and II.2.4, for the diagnostic X-ray energy range:

$$\frac{\mu}{\rho} = (\sigma_{pe} + \sigma_{incoh} + \sigma_{coh}) \cdot \frac{N_A}{A} = \frac{\mu_{pe}}{\rho} + \frac{\mu_{incoh}}{\rho} + \frac{\mu_{coh}}{\rho}$$

$$(II.2.23)$$

At present there exist essentially two updated photon data libraries:

1. EPDL97 (Cullen et al. 1997)
2. XCOM (Berger and Hubbell 1987)

The EPDL97 database provides the atomic cross sections, the incoherent scattering function S and the atomic form factor F. Values of μ/ρ for different elements and compounds must be calculated from these data. The XCOM database, which has been updated to 1999 (Hubbell, personal communication), provides values of μ/ρ for elements and compounds. Comparisons of the two databases carried out by different authors (Boone and Chavez 1996; Hubbell 1999; Schmidt 2001) have demonstrated that there are only minor deviations at low photon energies.

For the calculations carried out within the scope of this book, the XCOM data seemed more suitable. This database is also available on the internet under the address http://physics.nist.gov/PhysRefData/contents.html. Because the data have essentially be collected and prepared by the Ionizing Radiation Division of NIST, the XCOM database is often called the NIST database (Hubbell 1999). The latter term is also used in this book. Tables in Chap. IV.2 include data for: Al, Si, Cu, Se, I, Gd, Pb, air, H_2O, polymethylmethacrylate (PMMA), CsI, Gd_2O_2S, $CaWO_4$. Table II.2.1 contains a list of elements of which the data are available on the CD-ROM, Table II.2.2 a corresponding list of the compounds and mixtures on the CD-ROM. Figure II.2.5 shows as an example the contributions of the different interaction processes to the mass attenuation coefficient of water and lead.

Table II.2.1. Elements, for which data of μ/ρ (from NIST 2001) are available in data files on the CD-ROM

Z	File	Chemical element name (en)	(de)	Density ϱ[a] [g/cm³]	Z	File	Chemical element name (en)	(de)	Density ϱ[a] [g/cm³]
1	H	Hydrogen	Wasserstoff	0.000084	44	Ru	Ruthenium	Ruthenium	12.41
2	He	Helium	Helium	0.000166	45	Rh	Rhodium	Rhodium	12.41
3	Li	Lithium	Lithium	0.534	46	Pd	Palladium	Palladium	12.02
4	Be	Beryllium	Beryllium	1.848	47	Ag	Silver	Silber	10.5
5	B	Boron	Bor	2.37	48	Cd	Cadmium	Cadmium	8.65
6	C	Carbon	Kohlenstoff	1.7	49	In	Indium	Indium	7.31
7	N	Nitrogen	Stickstoff	0.001165	50	Sn	Tin	Zinn	7.31
8	O	Oxigen	Sauerstoff	0.001332	51	Sb	Antimony	Antimon	6.691
9	F	Fluorine	Fluor	0.00158	52	Te	Tellurium	Tellur	6.24
10	Ne	Neon	Neon	0.000839	53	I	Iodine	Jod	4.93
11	Na	Sodium	Natrium	0.971	54	Xe	Xenon	Xenon	0.005485
12	Mg	Magnesium	Magnesium	1.738	55	Cs	Cesium	Caesium	1.873
13	Al	Aluminum	Aluminium	2.699	56	Ba	Barium	Barium	3.5
14	Si	Silicon	Silizium	2.33	57	La	Lanthanum	Lanthan	6.154
15	P	Phosphorus	Phosphor	2.2	58	Ce	Cerium	Cer	6.657
16	S	Sulfur	Schwefel	1.957	59	Pr	Praseodymium	Praseodym	6.71
17	Cl	Chlorine	Chlor	0.002995	60	Nd	Neodymium	Neodym	6.9
18	Ar	Argon	Argon	0.001662	61	Pm	Promethium	Promethium	7.22
19	K	Potassium	Kalium	0.862	62	Sm	Samarium	Samarium	7.46
20	Ca	Calcium	Calzium	1.55	63	Eu	Europium	Europium	5.243
21	Sc	Scandium	Scandium	2.989	64	Gd	Gadolinium	Gadolinium	7.895
22	Ti	Titanium	Titan	4.54	65	Tb	Terbium	Terbium	8.229
23	V	Vanadin	Vanadium	6.11	66	Dy	Dysprosium	Dysprosium	8.536
24	Cr	Chromium	Chrom	7.18	67	Ho	Holmium	Holmium	8.795
25	Mn	Manganese	Mangan	7.44	68	Er	Erbium	Erbium	9.066
26	Fe	Iron	Eisen	7.874	69	Tm	Thulium	Thulium	9.321
27	Co	Cobalt	Kobalt	8.9	70	Yb	Ytterbium	Ytterbium	6.73
28	Ni	Nickel	Nickel	8.902	71	Lu	Lutetium	Lutetium	9.84
29	Cu	Copper	Kupfer	8.96	72	Hf	Hafnium	Hafnium	13.31
30	Zn	Zinc	Zink	7.133	73	Ta	Tantalum	Tantal	16.65
31	Ga	Gallium	Gallium	5.904	74	W	Tungsten	Wolfram	19.3
32	Ge	Germanium	Germanium	5.323	75	Re	Rhenium	Rhenium	21.02
33	As	Arsenic	Arsen	5.73	76	Os	Osmium	Osmium	22.57
34	Se	Selenium	Selen	4.5	77	Ir	Iridium	Iridium	22.42
35	Br	Bromine	Brom	3.14	78	Pt	Platinum	Platin	21.45
36	Kr	Krypton	Krypton	0.003478	79	Au	Gold	Gold	19.32
37	Rb	Rubidium	Rubidium	1.532	80	Hg	Mercury	Quecksilber	13.55
38	Sr	Strontium	Strontium	2.54	81	Tl	Thallium	Thallium	11.72
39	Y	Yttrium	Yttrium	4.469	82	Pb	Lead	Blei	11.35
40	Zr	Zirkonium	Zirkon	6.506	83	Bi	Bismuth	Wismut	9.747
41	Nb	Niob	Niob	8.57	92	U	Uranium	Uran	18.95
42	Mo	Molybdenum	Molybdaen	10.22					
43	Tc	Technetium	Technetium	11.5					

[a] The density of gaseous elements is related to 20°C, 1013 hPa

For the mass energy-transfer coefficient μ_{tr}/ρ, one can write according to Eq. II.2.23:

$$\frac{\mu_{tr}}{\rho} = (\sigma_{pe,tr} + \sigma_{incoh,tr}) \cdot \frac{N}{A} = \frac{\mu_{pe,tr}}{\rho} + \frac{\mu_{incoh,tr}}{\rho} \qquad (II.2.24)$$

Calculations of μ_{tr}/ρ have been carried out by Hubbell and Seltzer (1995). Mainly for dosimetric purposes, one needs the mass energy-absorption coefficient μ_{en}/ρ,

which is related to the mass energy-transfer coefficient μ_{tr}/ρ by:

$$\frac{\mu_{en}}{\rho} = (1-g) \cdot \frac{\mu_{tr}}{\rho} \qquad (II.2.25)$$

where g represents the average fraction of secondary electron energy that is lost in bremsstrahlung production. It can be seen from Table II.2.3 that g is negligible for low-Z materials in the diagnostic X-ray energy range.

Table II.2.2 Compounds and mixtures, for which data of μ/ρ (from NIST 2001) are available in data files on the CD-ROM.

File name	Compound	Density ϱ g/cm³	n	Element	Z	F	File	Relative Mass
AIR	AIR, dry (Density r at 20°C, 1013 hPa)	0.001205	1	C	6	1		0.00012
			2	N	7	1		0.75527
			3	O	8	1		0.23178
			4	Ar	18	1		0.01283
WATER	Water	1.0	1	H	1	2		0.11190
			2	O	8	1		0.88810
PMMA	PMMA $(C_5H_8O_2)_n$ (Acryl glass)	1.18	1	C	6	5		0.59985
			2	H	1	8		0.08054
			3	O	8	2		0.31961
RW_1	RW 1 (Water) Hermann (1985)	0.97	1	H	1	1		0.13190
			2	C	6	1		0.79410
			3	O	8	1		0.03810
			4	Mg	12	1		0.00910
			5	Ca	20	1		0.02680
PC	Polycarbonate $[C_{16}H_{14}O_3]_n + C_{15}H_{16}O_2$	1.2	1	C	6	16		0.75575
			2	H	1	14		0.05549
			3	O	8	3		0.18876
PE	Polythene $(CH_2)_n$	0.92	1	C	6	1		0.85628
			2	H	1	2		0.14372
PS	Polystyrene $(C_8H_8)_n$	1.06	1	C	6	8		0.92258
			2	H	1	8		0.07742
ADIPOSE	Adipose tissue ICRU 44 (1989)	0.92	1	H	1	1		0.1140
			2	C	6	1		0.5980
			3	N	7	1		0.0070
			4	O	8	1		0.2780
			5	Na	11	1		0.0010
			6	S	16	1		0.0010
			7	Cl	17	1		0.0010
BREAST	Breast tissue ICRU 44 (1989)	1.02	1	H	1	1		0.1060
			2	C	6	1		0.3320
			3	N	7	1		0.0300
			4	O	8	1		0.5270
			5	Na	11	1		0.0010
			6	P	15	1		0.0010
			7	S	16	1		0.0010
			8	Cl	17	1		0.0010
LUNG	Lung tissue ICRU 44 (1989)	1.05	1	H	1	1		0.1030
			2	C	6	1		0.1050
			3	N	7	1		0.0310
			4	O	8	1		0.7490
			5	Na	11	1		0.0020
			6	P	15	1		0.0020
			7	S	16	1		0.0030
			8	Cl	17	1		0.0030
			9	K	19	1		0.0020
MUSCLE	Muscle, skeletal ICRU 44 (1989)	1.05	1	H	1	1		0.1020
			2	C	6	1		0.1430
			3	N	7	1		0.0340
			4	O	8	1		0.7100
			5	Na	11	1		0.0010
			6	P	15	1		0.0020
			7	S	16	1		0.0030
			8	Cl	17	1		0.0010
			9	K	19	1		0.0040
BONE	Bone, cortical ICRU 44 (1989)	1.92	1	H	1	1		0.0340
			2	C	6	1		0.1550
			3	N	7	1		0.0420
			4	O	8	1		0.4350
			5	Na	11	1		0.0010
			6	Mg	12	1		0.0020

File name	Compound	Density ϱ g/cm³	n	Components Element	Z	F	File	Relative Mass
			7	P	15	1		0.1030
			8	S	16	1		0.0030
			9	Ca	20	1		0.2250
TISSUE	Soft tissue ICRU 44 (1989)	1.06	1	H	1	1		0.1020
			2	C	6	1		0.1430
			3	N	7	1		0.0340
			4	O	8	1		0.7080
			5	Na	11	1		0.0020
			6	P	15	1		0.0030
			7	S	16	1		0.0030
			8	Cl	17	1		0.0020
			9	K	19	1		0.0030
M_Adipo	Adipose (Hammerstein, 1979)	0.93	1	H	1	1		0.1120
			2	C	6	1		0.6190
			3	N	7	1		0.0170
			4	O	8	1		0.2510
			5	K	19	1		0.0010
M_Gland	Glandular tissue (Hammerstein, 1979)	1.04	1	H	1	1		0.1020
			2	C	6	1		0.1840
			3	N	7	1		0.0320
			4	O	8	1		0.6770
			5	K	19	1		0.0050
Mamm_AG	50% adipose + 50% glandular tissue (Hammerstein, 1979)	0.98	1				M_Adipo	0.5000
			2				M_Gland	0.5000
Micro_Ca	Microcalcification $Ca_5(PO_4)_3OH$ (70%) (Klein, 1979)	2.6	1	Ca	20	5		0.3989
			2	P	15	3		0.1850
			3	O	8	13		0.4141
			4	H	1	1		0.0020
PI	Polyimide $C_{22}H_{10}O_2N_4$ (Kapton)	1.42	1	C	6	22		0.72925
			2	H	1	10		0.02782
			3	O	8	2		0.08831
			4	N	7	4		0.15462
TG_8245	Tube glass 8245	2.31	1				SiO2	0.6856
			2				B2O3	0.1710
			3				Al2O3	0.0520
			4				ZrO2	0.0020
			5				Li2O	0.0050
			6				Na2O	0.0720
			7				K2O	0.0010
			8				As2O3	0.0014
			9				ZnO	0.0100
TG_8330	Tube glass 8330 (Duran)	2.23	1				SiO2	0.8056
			2				B2O3	0.1273
			3				Al2O3	0.0241
			4				Fe2O3	0.0003
			5				ZrO2	0.0004
			6				TiO2	0.0003
			7				Na2O	0.0352
			8				K2O	0.0060
			9				CaO	0.0002
			10				Cl	0.0006
TG_8486	Tube glass 8486 (Suprax)	2.34	1				SiO2	0.7565
			2				B2O3	0.1107
			3				Al2O3	0.0414
			4				Na2O	0.0404
			5				K2O	0.0090
			6				As2O3	0.0020
			7				MnO	0.0010
			8				BaO	0.0280
			9				CaO	0.0100
			10				F	0.0010

File name	Compound	Density ϱ g/cm^3	n	Components Element	Z	F	File	Relative Mass
P200	P200 Polyether sulfone	1.37	1	C	6	12		0.6206
			2	H	1	8		0.0347
			3	O	8	3		0.2067
			4	S	16	1		0.1381
OIL	Insulating and cooling oil for X-ray tube assemblies	0.815	1				C	0.8485
			2				H	0.1515
Al2O3	Al$_2$O$_3$ ceramics	3.7	1	Al	13	2		0.5293
			2	O	8	3		0.4708
AlMg3	Aluminum alloy (DIN 1725-1)	2.66	1				Al	0.9485
			2				Si	0.0040
			3				Fe	0.0040
			4				Cu	0.0010
			5				Mn	0.0050
			6				Mg	0.0310
			7				Cr	0.0030
			8				Zn	0.0020
			9				Ti	0.0015
Al_F1	Al 99.5 DIN 1712–3 (components normalized)	2.7	1				Al	0.9950
			2				Si	0.0013
			3				Fe	0.0020
			4				Cu	0.0003
			5				Mn	0.0003
			6				Mg	0.0003
			7				Zn	0.0003
			8				Ti	0.0005
Al_F2	Al 99.5 DIN 1712-3 (components with maximal Z)	2.7	1				Al	0.9950
			2				Fe	0.0040
			3				Cu	0.0005
			4				Zn	0.0005
Mylar	Polyethylen terephthalate	1.38	1	H	1	1		0.0420
			2	C	6	1		0.6250
			3	O	8	1		0.3330
Lexan	Carbon fiber (BRH-Reference)	1.275	1				H	0.0555
			2				C	0.7557
			3				O	0.1888
Teflon	Teflon (C$_2$F$_4$)$_n$	2.16	1	C	6	2		0.2402
			2	F	9	4		0.7598
Pb_Acryl	Kyowaglass XA Typ H (30% Pb in Acryl)	1.6	1				PMMA	0.7000
			2				Pb	0.3000
PVC	Polyvinyl chloride (C$_2$H$_3$C$_1$)$_n$	1.42	1	C	6	2		0.3844
			2	H	1	3		0.0484
			3	Cl	17	1		0.5673
CaWO4	Calcium tungstate	6.062	1	Ca	20	1		0.1392
			2	W	74	1		0.6385
			3	O	8	4		0.2223
CsJ	Cesium iodide CsI	4.51	1	Cs	55	1		0.5115
			2	J	53	1		0.4885
Gd2O2S	Gadolinium oxisulfide Gd$_2$O$_2$S:Tb	7.34	1	Gd	64	2		0.8308
			2	O	8	2		0.0845
			3	S	16	1		0.0847
BaFBrJ	Storage phosphor BaF(Br.$_{85}$J.$_{15}$)	4.8	1	Ba	56	1		0.5645
			2	F	9	1		0.0781
			3	Br	35	1		0.2792
			4	J	53	0		0.0782
YTaO4	Yttrium tantalate YTaO$_4$	7.57	1	Y	39	1		0.2663
			2	Ta	73	1		0.5420
			3	O	8	4		0.1917

Table II.2.3 Average fraction g in % of secondary electron energy that is lost in bremsstrahlung (adapted from Krieger 1998, p.161)

Electron energy in MeV	PMMA	Water	Air	Bone	Tungsten	Lead
0.01	0.01	0.01	0.01	0.01	0.11	0.12
0.05	0.03	0.03	0.04	0.04	0.54	0.61
0.10	0.05	0.06	0.07	0.08	1.03	1.16
0.15	0.07	0.08	0.09	0.10	1.47	1.66
0.50	0.18	0.20	0.22	0.26	3.71	4.24
1.00	0.32	0.36	0.40	0.46	6.03	6.84

References

Attix FH (1986) Introduction to radiological physics and radiation dosimetry. Wiley, New York

Berger MJ and Hubbell JH (1987) XCOM: photon cross sections on a personal computer. NBSIR 87–3597. NBS, Washington, DC

Boone JM, Chavez AE (1996) Comparison of X-ray cross sections for diagnostic and therapeutic medical physics. Med Phys 23:1997–2005

Cullen DE, Hubbell JH, Kissel L (1997) EPDL 1997: the evaluated photon data library. Lawrence Livermore National Laboratory Report UCRL-50400. vol. 6, rev. 5

DIN (Deutsches Institut für Normung) (1976) Aluminium – Halbzeug. DIN 1712-3. Beuth, Berlin

DIN (Deutsches Institut für Normung) (1983) Aluminiumlegierungen ñ Knetlegierungen. DIN 1725-1. Beuth, Berlin

Hammerstein GR, Miller DW, White DR, Masterson ME, Woodard HQ, Laughlin JS (1979) Absorbed radiation dose in mammography. Radiology 130:485–491

Hermann KP, Geworski L, Muth M, Harder D (1985) Polyethylene-based water-equivalent phantom material for X-ray dosimetry at tube voltages from 10 to 100 kV. Phys Med Biol 30:1195–2000

Hubbell JH (1999) Review of photon interaction cross section data in the medical and biological context. Phys Med Biol 44:1–22

Hubbell JH, Seltzer SM (1995) Tables of X-ray mass attenuation coefficients and mass energy-absorption coefficients 1 keV to 20 MeV for elements Z = 1 to 92 and 48 additional substances of dosimetric interest. NISTIR 5632. Natl Inst Standards Technol, USA

ICRU (1989) Tissue substitutes in radiation dosimetry and measurement. Report 44. ICRU, Bethesda, MD

Klein J (1979) Zur filmmammographischen Nachweisbarkeitsgrenze von Mikroverkalkungen. Fortschr Röntgenstr 131:205–210

Krieger H (1998) Grundlagen. Strahlenphysik, Dosimetrie und Strahlenschutz, vol. 1. Teubner, Stuttgart

NIST (2001) Database at http://physics.nist.gov/PhysRefData/contents.html

Schmidt B (2001) Dosisberechnungen für die Computertomographie. Thesis. University of Erlangen-Nürnberg, Germany

II.3   Radiation Field and Dosimetric Quantities

The physical characteristics of the radiation source and the exposure parameters, which together determine the radiation quality, are the anode material of the X-ray tube and the filtration of the primary radiation beam, the X-ray peak tube voltage selected, its temporal course (e.g. especially at short exposure times or, in pulsed exposure techniques, its rise and drop) and the inherent waveform of the tube voltage (2-, 6-, 12-, multi-pulse or DC). The radiation quality (photon energy spectrum) influences both patient dose and image quality. An increase in the X-ray tube voltage for a certain anode-filter combination at a definite image receptor dose (see Chap. II.3.2) will result in an increased penetration of the X-ray beam and consequently in a reduction of the absorbed dose and the contrast observed in the image.

Mono-energetic radiation would be of great advantage for imaging (Carroll 1994), but the X-ray tube (see Chap. II.1) is the only radiation source, which delivers the radiation intensity (fluence rate) needed in radiography for short exposure times, to avoid motion unsharpness. When realising simulation calculations, particularly with regard to image quality, it is, however, usual to start with considerations of the mono-energetic case (see Chap. III.3) and then to extend the results obtained to real bremsstrahlung.

In this chapter the physical quantities which describe the X-radiation field and the radiation exposure of the patient (see also Chap. III.1) in X-ray diagnostic radiology will be introduced (see Fig. II.3.1).

These quantities can also be used as source material within the process of optimising image quality (see Chap. III.3). In this respect the employment of additional filters with the objective to reduce patient dose will be discussed in Chap. II.3.4.

II.3.1   General Radiation Field Characteristics

The full description of an X-radiation field requires information on the number N and the energy E of the photons as well as on their spatial, directional and temporal distribution (ICRU 1998). In this book only the quantities needed in X-ray diagnostics will be introduced.

The *photon fluence* Φ is the quotient of dN by da, where dN is the number of photons irrespective of their energy incident on a sphere of cross-sectional area da, thus:

$$\Phi = \frac{dN}{da} \tag{II.3.1}$$

with the unit per square metre. The distribution of the *fluence* Φ_E with respect to the number of *photons of energy E* is given by:

$$\Phi_E = \frac{d\Phi}{dE} = \frac{dN}{da \cdot dE} \tag{II.3.2}$$

where Φ_E is the fluence $d\Phi$ of photons of energy between E and E + dE. From Eq. II.3.2 follows for the corresponding *energy fluence* Ψ_E of the photons as:

$$\Psi_E = E \cdot \Phi_E = E \cdot \frac{dN}{da \cdot dE} \tag{II.3.2}$$

where Ψ_E is the radiant energy with respect to photons of the energy between E and E + dE incident on a sphere of cross-sectional area da with unit joules per square metre.

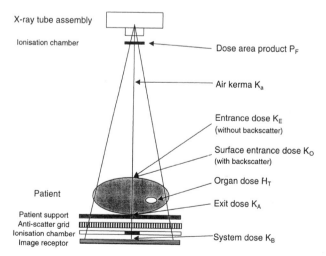

X-ray tube assembly

Ionisation chamber

Dose area product P_F

Air kerma K_a

Entrance dose K_E (without backscatter)

Surface entrance dose K_O (with backscatter)

Organ dose H_T

Patient

Exit dose K_A

Patient support
Anti-scatter grid
Ionisation chamber
Image receptor

System dose K_B

Fig. II.3.1. The dose quantities, which describe the X-radiation field and the radiation exposure of the patient

As a rule for the calculation of the *patient dose*, whichever dose quantity is meant (see Chap. III.1), one starts with the *entrance dose* (without back scatter) or the *dose area product* (DIN 2002). The entrance dose is measured as *air kerma* and the dose area product is essentially the product of air kerma and the corresponding area of the radiation field. The kerma in general is defined as *kinetic energy released in material* by ionising radiation. It is determined as the quotient (ICRU 1998):

$$K = \frac{dE_{tr}}{dm} \tag{II.3.4}$$

where dE_{tr} is the sum of the initial kinetic energies of all the charged ionising particles liberated by uncharged ionising particles (e.g. photons) in material of mass dm. The unit of kerma is the joule per kilogram. The special name of the unit of kerma is the gray (Gy).

The air kerma $K_a(E)$ released by photons of energy E is related to the energy fluence Ψ_E or the fluence Φ_E of photons of energy E by the equation:

$$K_a(E) = \Psi_E \frac{\mu_{tr}(E)}{\rho} = \Phi_E \cdot E \cdot \frac{\mu_{tr}(E)}{\rho} \tag{II.3.5}$$

where $\dfrac{\mu_{tr}(E)}{\rho}$

is the mass energy transfer coefficient of air for X-ray photons with energy E (see Chap. II.2). Corresponding to the *energy-spectrum of photons* Φ_E, the quantity $K_a(E)$ is called the *kerma-spectrum* (see Fig. II.3.2).

Fig. II.3.2 shows as an example the photon- and the kerma-spectrum created with a tungsten anode filtered with 2.5 mm Al at a tube voltage of 80 kV. The kerma in air measured at a given peak tube voltage U_p with an ionisation chamber can be calculated by:

$$K_{a,eU_p} = \int_0^{E=eU_p} \Phi_E \cdot E \cdot \frac{\mu_{tr}(E)}{\rho} \cdot dE \tag{II.3.6}$$

Dosimetric measurements are carried out mostly with ionisation chambers; therefore the charge liberated per unit mass by the X-ray photons within the chamber air volume – the so-called exposure – is recorded. Exposure is the ionisation equivalent of air kerma. Consequently the *exposure X* (ICRU 1998) is the quotient of dQ by dm, where dQ is the absolute value of the total charge of the ions of one sign produced in air when all the electrons deliberated or created by the X-rays in air of mass dm are completely stopped in air, thus:

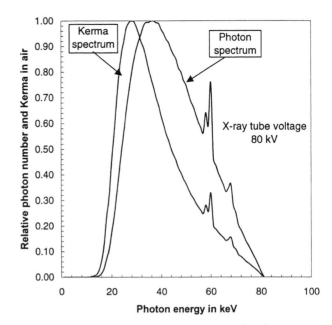

Fig. II.3.2. Photon-spectrum Φ_E and corresponding kerma-spectrum $K_a(E)$ at an X-ray tube voltage of 80 kV (DC-voltage) (The energy distribution of the photon fluence Φ_E in the diagram is shown with an energy gradation of 1 keV)

$$X = \frac{dQ}{dm} \tag{II.3.7}$$

The unit of the exposure is coulombs per kilogram. The ionisation produced by Auger electrons is included in dQ, the ionisation due to photons emitted by radiative processes (i.e. bremsstrahlung and fluorescence photons) is not. The exposure can be expressed in terms of the energy distribution Φ_E of the incident photons:

$$X = \frac{e}{W} \int \Phi_E \cdot E \cdot \frac{\mu_{tr}(E)}{\rho} (1-g) \cdot dE \tag{II.3.8}$$

where W is the mean energy expended in air per ion-pair formed and g represents the average fraction of secondary electron energy that is lost in bremsstrahlung production (see Chap. II.2.3). In the energy range used in diagnostic radiology, g is small, therefore Eq. II.3.8 can be approximated by:

$$X = \frac{e}{W} K_a (1 - \bar{g}) \tag{II.3.9}$$

where K_a is the air kerma for primary photons and $\bar{g}$ the mean value of g averaged over the distribution of the air kerma with respect to the electron energy.

The corresponding time-dependent quantities: the photon flux $\dot{N}$ the fluence rate $\dot{\Phi}$ the energy fluence rate $\dot{\Psi}$ (sometimes also called intensity), the kerma rate $\dot{K}$ are defined in a straight-forward manner(see ICRU 1998), therefore the derivation shall not be discussed here in detail.

II.3.2 Characteristics of the Radiation Field in X-ray Diagnostic Radiology

In diagnostic radiology the dose quantities describing the radiation field are essentially *derived* from the *air kerma* K_a.

The *nominal air kerma rate* $\dot{K}_{a,100}$ is the air kerma rate free in air on the central ray for a focus distance of 100 cm. As a rule it is given for a tube current of 1 mA.

The *radiation output of an X-ray tube assembly* Y_{100} is the quotient of the air kerma $K_{a,100}$ free in air on the central ray for a focus distance of 100 cm and the tube current-exposure time product $I_R \cdot t$:

$$Y_{100} = \frac{K_{a,100}}{I_R \cdot t} \tag{II.3.10}$$

As a rule it is given for a tube current-exposure time product of 1 mAs.

The *entrance surface air kerma* K_E and the *entrance surface air kerma rate* $\dot{K}_E$ is the air kerma (rate) free in air (i.e. without back scatter) at a point in the central beam in a plane corresponding to the entrance surface of a patient (or a phantom). The quantity K_E is also called entrance dose (see Chap. II.3.1 and III.1.2).

The *entrance surface dose* K_O and the *entrance surface dose rate* $\dot{K}_O$ is the air kerma (rate) with back scatter assessed at a point in the central beam on the entrance surface of a patient (or a phantom).

For radiography the *image receptor dose* K_B is the air kerma on the central ray at a point representing the position of the image receptor. Sometimes this dosimetric quantity is also called "system dose". Its value should correspond to an optimised exposure of the image receptor. For conventional film-screen systems, this means an image with an optimum optical density.

For fluoroscopy the *image receptor dose rate* $\dot{K}_B$ is the air kerma rate on the central ray at a point representing the position of the image receptor. Sometimes this dosimetric quantity is also called "system dose rate".

The system dose or the system dose rate often can be measured in practice only approximately, because the specific entrance medium (e.g. the cover of the film cassette or the entrance window of the image intensifier) of the image receptor creates back-scattered radiation (sometimes including characteristic radiation of the phosphor material used). Back-scatter factors of the relevant image receptors are seldom known; it would be therefore be of advantage if they are given, e.g. by the manufacturer.

The *air kerma-area product* P_F is the integral of the air kerma K_a over the area A of the X-ray beam perpendicular to the central ray:

$$P_F = \int_A K_a dA \tag{II.3.11}$$

Sometimes this quantity is also called dose area product (see IEC 2000). In the situation where the air kerma K_a is constant over the area A, which is approximately true for beam areas that are not too large, Eq. II.3.11 reduces to:

$$P_F = K_a \cdot A \tag{II.3.12}$$

hence the name "air kerma-area product". It is measured free-in-air at some distance from the patient so that photons back-scattered from the patient or phantom practically do not contribute to the resulting value. The air kerma-area product has the useful property of being invariant with distance from the X-ray tube focus. Therefore, it can be measured at any plane between the diaphragm housing of the X-ray tube and the patient, as long as the plane of measurement is not too close to the patient. The position of the plane of measurement does not need to be specified. For the measurement of the air kerma-area product, usually a specially designed, large-area parallel-plate ionisation chamber is attached to the diaphragm housing of the X-ray tube assembly. The dimensions of the ionisation chamber are designed to encompass the entire X-ray beam no matter how wide the diaphragms are set.

For computed tomography (CT), the *dose-length product* P_L is the line integral of the air kerma K_a along the z-axis of rotation of the CT scanner for a complete rotation of 360°:

$$P_L = \int_L K_a(z) dz \tag{II.3.13}$$

The determination of this quantity is carried out with special pencil ionisation chambers which are calibrated for the measurement of the dose-length product. The integration length L is then given by the active length of the ionisation chamber (10–15 cm as a rule).

The *slice-averaged axial air kerma* K_{CT} is the mean value of the air kerma free-in-air at the axis of rotation, given by:

$$K_{CT} = \frac{P_L}{s} \tag{II.3.14}$$

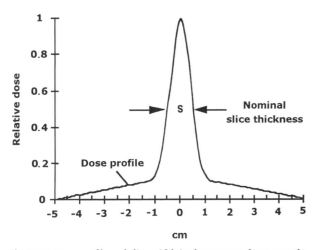

Fig. II.3.3. Dose profile and slice width in the computed tomography -system axis, (Example s = 10 mm)

where P_L is the dose-length product of a single slice and s is the nominal slice thickness. This definition holds for a CT examination with consecutive slices. If the slice interval v is different from the nominal slice thickness s (e.g. for examinations with overlapping slices) Eq. II.3.14 changes to:

$$K_{CT} = \frac{P_L}{v} \qquad (II.3.15)$$

In the literature (Kalender 2000, p. 125; Nagel 2000), the slice-averaged axial air kerma is also called the CT dose index in air ($CTDI_{air}$).

For sequential single-slice scanners (1 slice per rotation of the X-ray tube assembly), the slice collimation corresponds approximately to the nominal slice thickness s. For multiple-slice scanners, the primary radiation beam includes N slices with nominal thickness s. The width of the fan beam (dose profile in the direction of the axis of rotation, see Fig. II.3.3) is then approximately equal to Ns and Eq. II.3.14 changes to:

$$K_{CT} = \frac{P_L}{N \cdot s} \qquad (II.3.16)$$

II.3.3   Dosimetric Quantities for the Description of Radiation Exposure

The fundamental dosimetric quantity is the *absorbed dose D*. The definition of this quantity is based on the quantity *energy imparted* ε which is given by the relation (ICRU 1993):

$$\varepsilon = R_{in} - R_{out} + \Sigma Q \qquad (II.3.17)$$

where R_{in} is the radiant energy incident on the volume of matter in consideration, i.e. – with respect to X-rays – the sum of the energies of all photons and electrons (excluding rest energies) which enter the volume; R_{out} is the radiant energy emerging from the volume, i.e. the sum of energies of all photons and electrons (excluding rest energies) which leave the volume; ΣQ is – in the universally applicable definition of ε – the sum of all changes of the resting mass energy of nuclei and elementary particles in any interactions which occur in the volume and can be therefore neglected in the energy range considered in X-ray diagnostics.

The *absorbed dose D* is defined as the quotient of the mean energy $\overline{d\varepsilon}$ imparted by the ionising radiation to matter of mass dm, thus:

$$D = \frac{\overline{d\varepsilon}}{dm} \qquad (II.3.18)$$

The unit of the absorbed dose is joules per kilogram (gray, Gy). This definition of the absorbed dose as a point function allows the determination of the spatial variation of D. As an example the energy deposited in a mass dm of air (in the case of electronic equilibrium) is given by (see also Eq. II.3.7):

$$D_{air} = X \cdot \frac{W}{e} \qquad (II.3.19)$$

For radiation protection practice, it is necessary to take into account the relative biological effectiveness of the different types of ionising radiation (ICRP 1991). Before the corresponding dose quantities *equivalent dose* and *effective dose* are defined (ICRU 1993; McCollough and Schueler 2000), some information on the biological effects of radiation will be reviewed briefly.

The process of ionisation necessarily changes molecules in the cell. As a consequence, cellular damage may occur. If this damage is not adequately repaired, it may prevent cell survival or reproduction or may result in a viable but modified cell. The two outcomes have different implications for the organism: If the number of cells, lost by an organ or tissue, is large enough, there will be observable harm reflecting a loss of tissue function. The probability of causing such harm will be zero at small doses, but above some threshold level of dose will increase steeply to unity. Above the threshold level, the severity of harm will also increase with dose. This type of effect is called "deterministic" by the ICRP (ICRP 1991). The outcome is very different if the irradiated cell is modified rather than killed. For somatic cells this may result, after a prolonged and variable delay called the latency period, in the manifestation of a cancer. The probability of a radiation induced cancer increases with increments of dose, probably with no threshold. The severity of cancer is not affected by the dose. This kind of radiation effect is called "stochastic". If the radiation-induced damage occurs in a germ cell any resulting effects are expressed in the progeny of the exposed person. This type of stochastic effect is called "hereditary".

Although there have been isolated reports of radiation-induced skin injuries to patients resulting from prolonged, fluoroscopically guided invasive procedures (Wagner et al. 1994), threshold doses for the most sensitive deterministic effects are generally well above the doses received by patients in conventional X-ray diagnostic radiology (Dendy and Heaton 1999, p. 289). Consequently the radiation risk resulting from stochastic effects is predominant in this case.

Now the two dose quantities are introduced: The *equivalent dose* H_T in a tissue or organ is given by:

$$H_T = \sum_R w_R \cdot D_{T,R} \tag{II.3.20}$$

where:

$$D_{T,R} = \frac{1}{m_T} \int_{m_T} D_R \cdot dm \tag{II.3.21}$$

is the mean absorbed dose in the tissue or organ (T) with mass m_T, due to radiation R and w_R is the corresponding radiation weighting factor. The factor w_R is dimensionless. Therefore H_T still has the units joules per kilogram, but is now given the special name sievert (Sv).

w_R is representative of the relative biological effectiveness of a given type and energy of ionising radiation in inducing stochastic effects at low doses. Generally w_R is assumed to be unity for the radiation qualities used in X-ray diagnostic radiology. For mammography it has been proposed recently (Jung 2001) that $w_R = 2$ should be applied due to the soft X-rays used. The *effective dose E* is given by:

$$E = \sum_T w_T \cdot H_T \tag{II.3.22}$$

Table II.3.1 Tissue Weighting Factors (ICRU 1993)

Tissue or organ	Tissue weighting factor, w_T
Gonads	0.20
Bone marrow (red)	0.12
Colon	0.12
Lung	0.12
Stomach	0.12
Bladder	0.05
Breast	0.05
Liver	0.05
Oesophagus	0.05
Thyroid	0.05
Skin	0.01
Bone surface	0.01
Remainder	0.05

where H_T is the equivalent dose in the tissue or organ (T) and w_T the corresponding tissue weighting factor. The factor w_T is again dimensionless and:

$$\sum_T w_T = 1 \tag{II.3.23}$$

The International Commission on Radiological Protection (1991) has specified numerical values for tissue weighting factors, which are shown in Table II.3.1. Corresponding to Eq. II.3.23, these factors describe the fraction of the stochastic radiation risk for the different organs and tissues. In the case of the gonads, it is the risk for hereditary effects; for the other organs and tissues, it is the risk of cancer induction.

II.3.4 Effect of Filtration

It is well known that additional filters (in addition to the minimum filtration required by law) increase the X-ray beam half-value layer (HVL) and lead to decreases in patients' surface entrance dose (skin dose), when both the X-ray tube peak voltage U_p and the system dose at the image receptor input are held constant. The interconnection between the Al HVL and the total Al filtration (or total equivalent Al filtration with respect to beam hardening) of the primary beam for various X-ray tube voltages is shown in Fig. II.3.4.

The HVL is that thickness of a specified material (as a rule given in millimetres Al or Cu) which attenuates, under narrow beam conditions, X-radiation with a particular spectrum to such an extent that the air-kerma rate, exposure rate or absorbed dose rate is reduced to one-half of the value that is measured without the material.

Although the surface entrance dose decreases fairly rapidly with increasing filtration, this generally is not the case for the organs, e.g. the mid-line dose within the primary beam decreases much less rapidly than the surface entrance dose (Behrman and Yasuda 1998).

Mid-line doses outside the radiation field were observed even to increase at higher filtration due to increases in lateral scatter (Ardran and Crooks 1962; Jones and Wall 1985). If instead of the image receptor dose the optical density (when using film-screen systems) or the brightness (or noise) of the image intensifier output screen (when employing indirect technique) is held constant, the energy dependence of the image receptor (see Chap. II.6) must also be considered. Possibly the image receptor dose must be increased (see Fig. II.6.5) because of a lower screen efficiency at higher HVLs (Ardran and Crooks 1962) to obtain constant optical density or brightness (or noise) and therefore also the mid-line dose will be increased. It is obvious that the absorbed dose to organs proximal to the beam entry point track with the surface entrance dose, whereas the dose to organs close to the image receptor is almost independent of beam filtration for a constant exit dose. Doses for organs at intermediate depth show a filtration dependence in between these two extremes.

Additional filtration of the incident X-radiation (without changing the X-ray tube voltage) generally – with the exception of the special situation in mammography – leads to a reduction in image contrast. Fig. II.3.5 shows that in mammography the image contrast, however, is improved with increasing thickness of the K-edge filter located at the X-ray tube output, although the HVL of the incident primary radiation is increased. Fig. II.3.5 shows the situation for a Mo/Mo anode filter system, when the thickness of the Mo-filter is changed from 30 μm to 40 μm and to 50 μm.

Reason for this effect is that especially the radiation above and below the K-edge is attenuated and therefore the HVL behind an object thickness of about 20 mm PMMA and more is reduced with increasing primary beam filtration (see Fig. II.3.6).

Methods to find a compromise between dose and image quality when using additional filtration will be further discussed in Chap. II.7 and illustrated by examples in Chap. III.3. In every case the effect of additional filters on image quality and dose can be derived from the energy distribution of the impinging X-ray photons (see Fig. II.3.7).

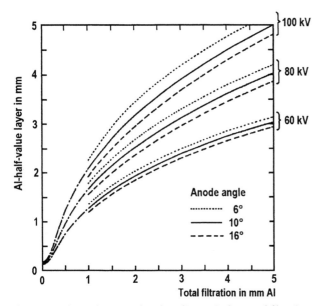

Fig. II.3.4. Relation between the Al quality equivalent total filtration and the Al half-value layer at various X-ray tube voltages and anode angles (DIN 1990)

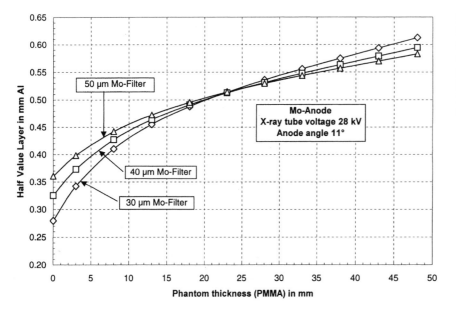

Fig. II.3.5. Reduction of the half-value layer of the imaging radiation results in an improvement of the image contrast with increasing thickness of the K-edge filter (Mo-anode with 30 μm, 40 μm, 50 μm Mo-filter) above a phantom thickness of 23 mm PMMA. (calculated with Ipem 1997)

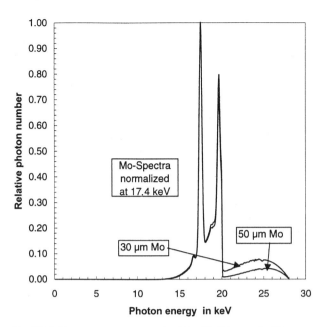

Fig. II.3.6. X-ray spectra generated with a Mo/Mo-anode filter system at 28 kV with increasing thickness of the K-edge filter in mammography (see Fig. II.3.5)

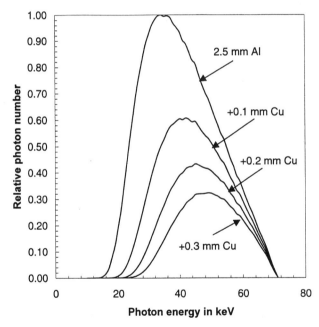

Fig. II.3.7. Effect of additional filtration on X-ray spectrum, applied X-ray tube voltage 70 kV

In general radiography the maximum of the energy distribution with increasing filtration is shifted against higher energies. If the image contrast is to be maintained – in contrast to mammography – the X-ray tube voltage must therefore be lowered.

To obtain the minimum filtration prescribed by law (equivalent to 2.5 mm Al) in general Al is used as filter

material, for additional filters, e.g. in angiography most often copper is used. Zamenhof (1982) has shown that copper offers minimal benefit over Al for digital fluoroscopy applications. In angiography most often a 0.1 mm Cu filter is generally applied to meet the regulations with respect to the maximum allowed entrance dose rate value (87 mGy/min in a distance of 30 cm above the table top in the case of over-table units) defined by the Food and Drug Administration (Food and Drug Administration 1997). In Chap. III.3 will be shown that additional Cu-filters with a thickness of more than 0.3 mm are not practical, because the skin dose will not be reduced much further, whereas the employment of such filters has a great impact on tube load and image quality.

The dose-reduction performance of K-edge filters for fluoroscopic examinations with contrast agents containing barium or iodine during fluoroscopy has been investigated by Gagne et al. (1994) by simulation calculations of quantities such as entrance exposure rate, integral dose rate, contrast, signal-to-noise ratio (SNR), image-quality figure (SNR^2 per dose) and the resulting tube load. In general, either beam-hardening filters (i.e. Al and Cu) and K-edge filters (e.g. rare earth material with K-edges in the range of 45–65 keV) provide a significant reduction in skin exposure attended with a corresponding increased tube load, but the practical implementation of adding filtration to fluoroscopic systems is most simply accomplished with beam-hardening filters rather than K-edge filters. K-edge filters can provide only a slightly better performance when used over a limited range of patient thickness (e.g. in paediatrics) and in special medical indications (Koedooder and Venema 1986; Nagel 1989).

References

Ardran GM, Crooks HE (1962) Dose in diagnostic radiology: the effects of changes in kilovoltage and filtration. Br J Radiol 35:172

Behrman Rh, Yasuda G (1998) Effective dose in diagnostic radiology as a function of X-ray beam filtration for a constant film density. Med Phys 25 (5):780–790

Carroll F (1994) Use of monochromatic X-rays in medical diagnosis and therapy. J X-Ray Sci Technol 4:323

Dendy PP, Heaton B (1999) Physics for diagnostic radiology. Institute of Physics Publishing, Bristol

DIN (Deutsches Institut für Normung) (1990) Klinische Dosimetrie: Röntgendiagnostik. DIN 6809, Part 3. Beuth, Berlin

DIN (2002) Klinische Dosimetrie: Verfahren zur Ermittlung der Patientendosis in der Röntgendiagnostik. DIN 6809 (draft), Part 7. Beuth, Berlin

Food and Drug Administration (1997) Performance standard for diagnostic X-ray systems and their major components 21. CFR (4-1-97) 1020.30–1020.33

Gagne RM, Quinn PW, Jennings RJ (1994) Comparison of beam-hardening and K-edge filters for imaging barium and iodine during fluoroscopy. Med Phys 21:107–121

ICRP (International Commission on Radiological Protection) (1991) 1990 recommendations of the ICRP, Publication 60, Ann ICRP 21(1–3)

ICRU (International Commission on Radiation Units and Measurements) (1993) Quantities and units in radiation protection dosimetry. Report 51. ICRU, Bethesda, MD

ICRU (1998) Fundamental quantities and units for ionising radiation. Report 60. ICRU, Bethesda, MD

IEC (International Electrotechnical Commission) (2000) Kerma-area product meter. Publication 60580. IEC, Geneva

Ipem (The Institute of Physics and Engineering in Medicine) (1997) Catalogue of Diagnostic X-Ray Spectra and other Data. Report No. 78 York Y01 2WR United Kingdom

Jones DG, Wall BF (1985) Organ doses from medical X-ray examinations calculated using Monte Carlo techniques. National Radiation Protection Board Report NRPB-R186 NRPB, Chilton, Didcot, UK

Jung H (2001) Abschätzung von Nutzen und Risiko eines Mammographiescreenings unter ausschließlichem Bezug auf das Strahlenrisiko. Radiologe 41:385–395

Kalender W A (2000) Computed tomography. Publicis MCD, Munich

Koedooder K, Venema H (1986) Filter materials for dose reduction in screen-film radiography. Phys Med Biol 31:585–600

McCollough CH, Schueler BA (2000) Educational treatise: calculation of effective dose. Med Phys 27(5):828–837

Nagel HD (1989) Comparison of performance characteristics of conventional and K-edge filters in general diagnostic radiology. Phys Med Biol 34:1269–1287

Nagel HD (ed) (2000) Radiation exposure in computed tomography: fundamentals, influencing parameters, dose assessment, optimisation, scanner data, terminology. COCIR European Co-ordination Committee of the Radiological and Electromedical Industries, Offizin Hartung Druck, Hamburg

Wagner LK, Eifel PJ, Geise RA (1994) Potential biological effects following high X-ray dose interventional procedures. J Vasc Interven Radiol 5:71–84

Zamenhof RG (1982) The optimisation of signal detectability in digital fluoroscopy. Med Phys 9(5):688–694

II.4   Penetration of X-rays

The attenuation properties of the various kinds of tissue in the patient's body with respect to X-ray photons in the energy range of about 10 keV to 150 keV is determined principally by the photoelectric effect and Compton scattering (see Chap. II.2). Therefore, in X-ray imaging, photons emitted from the focal spot of the X-ray tube enter the patient, where they may be absorbed, transmitted without interaction (primary photons) or scattered (secondary photons). The *radiation image* is formed from the emergent primary photons while impaired by the secondary photons (see Chaps. II.5 and III.2). By the interaction of all these photons with a suitable image receptor, the radiographic image is built up. So the X-ray image consists of a two-dimensional projection of the attenuating properties of the tissues in the three-dimensional volume of the patient's body along the path of the X-ray photons superimposed by scattered radiation.

Intermediate layers in the radiation beam between the focal spot and the patient (e.g. additional filters, compression devices) give rise to *beam-hardening*. Additional filters are of great importance for the resulting patient dose and image quality in the sense of optimisation (see Chap. II.3.4). Often they are characterised by the so-called *Al* or

Cu quality equivalent filtration. The quality of the achieved *primary radiation beam* – at a given anode material, tube voltage and filtration – is normally described by its half-value layer (HVL). In Chap. II.3.4 details of the effect of additional filters on dose and image quality have already been reported. Various possibilities of filter optimisation will be discussed in Chap. III.3.

Intermediate layers in the radiation beam *between the patient and the image receptor* (e.g. table top, antiscatter grid, ionisation chamber of the automatic exposure control system, film cassette cover; see Fig. II.4.1) attenuate the imaging radiation and therefore give rise to increased radiation exposure of the patient and possibly to a reduction in image quality.

The X-ray beam attenuation effect of these layers is mostly characterised by their *attenuation factor* at a specific radiation quality and measuring arrangement (IEC 1999) or their *Al attenuation equivalent*. In the IEC 61223–3–1 document mentioned above, typical values are defined for the attenuation factor (see Table II.4.1). In the following, the characteristics of the intermediate layers behind the patient are discussed in detail, but at first the beam attenuation by the patient itself will be studied.

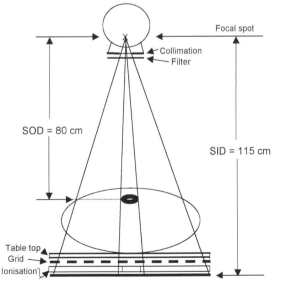

Geometrical magnification:
M = SID/SOD
(example: M = 1.44)

Overtable unit:
Range M = 1.1 ....1.5

Resolution limit
(constant radiant intensity
within the focal spot area):

$$v_g = \frac{M}{F \cdot (M-1)}$$

F = effective focal spot size in mm
SID = Source image distance
SOD = Source object distance

Fig. II.4.1. Principal design of a spot-film device: Intermediate layers between patient and image receptor; the resolution limit v_g depends on the effective focal spot size and the geometrical magnification *M*

Focal spot
Collimation
Filter
SOD = 80 cm
SID = 115 cm
Table top
Grid
Ionisation

Table II.4.1 Typical values for the attenuation ratio of material between the patient and the X-ray image receptor according to IEC 61223-3-1 (1999)

Material (components of x-ray equipment)	Attenuation ratio[a]
Patient support	1.25
Front panel of film changer	1.25
Anti-scatter grid	1.43
Automatic exposure control (AEC)	1.11

[a] Measuring parameters are 80 kV and an annuating layer of 25 mm Al

II.4.1   Attenuation by the Patient

In X-ray imaging, the attenuation of the radiation beam by the patient depends on the exposure parameters applied, which shall therefore be selected with respect to the medical indication. The following explanations and examples illustrate this statement.

The possibility to display tissue structures in the image is caused by their different attenuation characteristics for X-radiation. The differences in the optical density in the radiograph or in the brightness on the monitor arise due to differences in the transmission of the X-ray beam, i.e. the radiation becomes modulated by the structures within the patient's body. The depth of the modulation depends on the difference of the linear attenuation coefficients (see also Chap. II.7.1), the thickness of the tissue structures and the exposure parameters selected. In the ICRU 44 document (ICRU 1989), the mass attenuation characteristics of various tissue substitutes (cf. also Table II.2.2) are given. The depth of the modulation is at the same time reduced by the counteracting secondary radiation, i.e. the scattered radiation.

In radiographs it is often difficult to visualise important image details – even in the absence of scattered radiation – especially in mammography within the very similar tissue compositions. The linear attenuation coefficients of fibroglandular tissue, infiltrating ductal carcinoma and adipose breast tissue (see Fig. II.4.2), decrease rapidly with increasing energy in the energy range of X-radiation used in mammography, and furthermore the difference between them is very small (Hammerstein et al. 1979; Johns and Yaffe 1987; ICRU 1989; Yaffe 1992).

For this reason *low-energy X-radiation* and *steep film-screen systems* (see Chap. II.6) are needed to obtain sufficient contrast for diagnosis. The demand for "soft" X-radiation results simultaneously in higher radiation exposure and therefore radiation protection is of great importance in mammography.

An opposite example is chest radiography: The ribs overlap the lung tissue, therefore *a high X-ray tube voltage* is recommended, e.g. by the *European Guidelines on Quality Criteria for Diagnostic Radiographic Images*

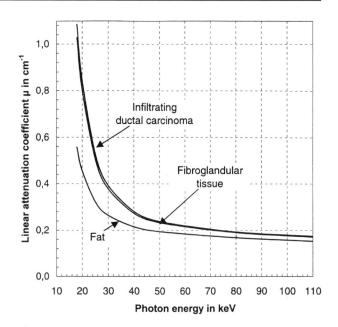

Fig. II.4.2. Linear attenuation coefficients of various breast tissues dependent on the energy of the impinging X-ray photons in mammography

(CEC 1996a), when taking a radiograph to display the lung tissue also behind the highly absorbing ribs. In general the selection of the tube voltage for each X-ray examination is task dependent and therefore so is the attenuation of the X-ray beam by the patient.

II.4.1.1   Simulation of Attenuation by the Patient

When studying imaging quality, sophisticated phantoms are used. Radiologists prefer systems which imitate anatomy or pathology (anthropomorphic phantoms, e.g. phantoms of skull, thorax), physicists prefer phantoms (e.g. CDRAD, Leeds-phantom) with which the physical properties of imaging systems can be tested and some of the characteristics describing image quality (see Chaps. II.7.6 and III.3) can be determined. However, when studying exclusively the physics of the attenuation or scattering process of X-ray photons, as a rule water, PMMA phantoms or phantoms made from tissue-equivalent mixtures are used (see data in Part IV). Figure II.4.3 shows for instance the *primary radiation attenuation* of water as a function of the water-layer thickness for tube voltages in the range of 40–150 kV. The attenuation of the primary radiation can be measured under narrow beam conditions as described, e.g. in Chap. II.5.

The total radiation intensity drops off more slowly with increasing phantom thickness. Figure II.4.4 shows the *attenuation of the total radiation* and *of the primary radiation* at tube voltages of 75 kV and 100 kV and a field

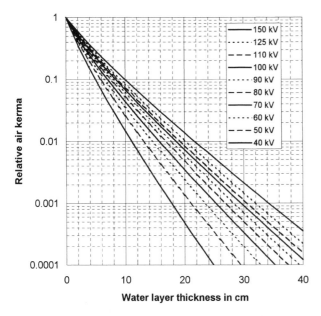

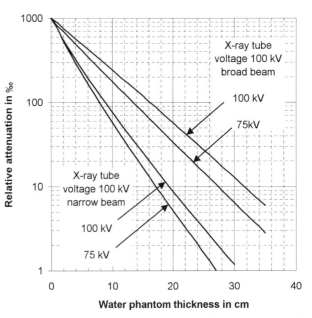

Fig. II.4.3. Primary radiation attenuation of water dependent on water thickness at X-ray tube voltages from 40–150 kV (from left to right, respectively); total filtration 2.5 mm Al

Fig. II.4.4. Comparison of primary and total radiation attenuation of water dependent on water thickness at X-ray tube voltages of 75 kV and 100 kV

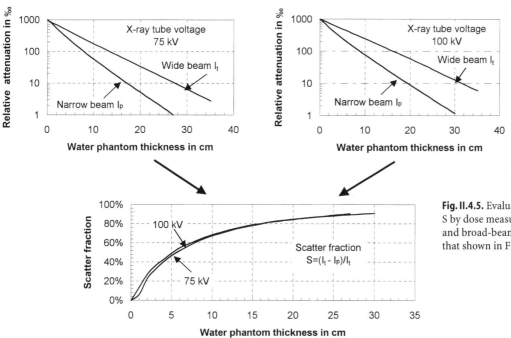

Fig. II.4.5. Evaluation of scatter fraction S by dose measurements under narrow- and broad-beam conditions similar to that shown in Fig. II.5.12 (without grid)

size of 30 cm × 30 cm, determined by measuring the dose (air kerma) at the bottom of the water phantom used.

The decrease in both quantities with the phantom thickness d is very different. Subtracting the readings of the curves measured at the same X-ray tube voltage U and dividing after that their difference by the corresponding total radiation measurements, the percentage of scattered radiation (also called "scatter fraction"; see Fig. II.4.5) behind the various phantom thicknesses d can be evaluated (see also Chap. II.5.1 and Eq. II.5.2).

Already from Fig. II.4.5 can be seen that the scatter fraction behind the phantom is a function of the phantom thickness d but nearly independent of the X-ray tube voltage U applied.

Behind thick phantoms (e.g. 30 cm water), much less than 1% of the impinging primary radiation is available for imaging. But not only the primary photon fluence of the radiation is reduced behind the patient or phantom, the energy distribution of the X-ray photons is also shifted against higher energies (see Fig. II.4.6).

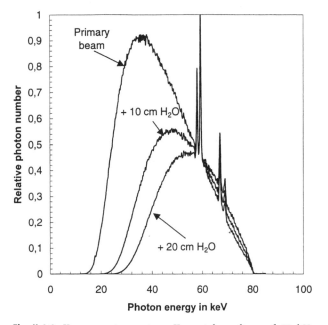

Fig. II.4.6. X-ray spectrum at an X-ray tube voltage of 80 kV (DC-voltage, total filtration 2.5 mm Al) behind 0 cm, 10 cm and 20 cm water

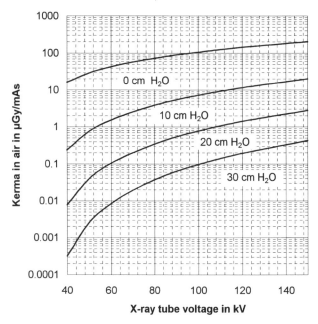

Fig. II.4.7. Air kerma per milliampere-second dependent on X-ray tube voltage (filtration 2.5 mm Al, focus distance 100 cm) behind 0 cm, 10 cm, 20 cm and 30 cm water

Therefore the contrast in radiographs of large patients is not only reduced by the existence of a *high percentage of scatter* (see Chap. II.5.3), but also by the resulting beam-hardening.

II.4.1.2   Basis of Exposure Tables

When considering the air kerma behind a phantom or the patient in relation to the X-ray tube voltage (see Fig. II.4.7), the well-known relation between dose (air kerma K) and the X-ray tube voltage U used can be derived:

$$K \propto U^n \cdot I \cdot t \qquad (II.4.1)$$

where I is the X-ray tube current and t the exposure time necessary for taking the radiograph of the patient. The exponent n in Eq. II.4.1 depends on the X-ray tube voltage U, the filtration of the primary beam and the patient thickness itself.

For a better understanding, Fig. II.4.7 shows the dependence of the air kerma on the X-ray tube voltage U (DC voltage, ripple <1%, filtration 2.5 mm Al, only primary radiation) without scattered radiation for 1 mAs at a focus distance of 1 m filtered by 0 cm, 10 cm, 20 cm and 30 cm water. These filtration values are comparable with the situations given during the exposure of the thorax, the abdomen and the lateral lumbar spine, respectively. The thicker the water phantom is, the steeper is the increase in

the air kerma with increasing tube voltage U. But, in the real case of a patient exposure, also scattered radiation must be taken in account and the increase in the curves takes its course more slowly.

From the gradient of the kerma curves, the exponent n in Eq. II.4.1 can be derived for the relevant exposure situation. When selecting manually the exposure parameters (U, I, and t) for the X-ray examination, as a rule exposure tables are used by the technician (Claasen 1962). In these tables n is assumed to be 4–5. This is an acceptable compromise, whereas n can reach values up to 8 for very high patient thicknesses and low tube voltages (Meiler 1949).

Because of the logarithmic relationship (see Chap. II.6) between the exposure and the resulting optical density OD, when using a film-screen system as image receptor:

$$\Delta OD = \gamma \cdot \log \frac{K_2}{K_1} \qquad (II.4.2)$$

one can write:

$$\log \frac{K_2}{K_1} = n \cdot \log U_2 + \log I_2 + \log t_2 - n \cdot \log U_1 - \log I_1 - \log t_1$$

$$(II.4.3)$$

or

$$\log \frac{K_2}{K_1} = n \cdot (\log U_2 - \log U_1) + (\log I \cdot t_2 - \log I \cdot t_1) \qquad (II.4.4)$$

Table II.4.2 Gradation of the exposure parameters tube voltage (in kV) and tube current time product (in mAs) as exposure points (in EP)

EP	kV	mAs	EP	kV	mAs	EP	kV	mAs	EP	kV	mAs
-10		0.1	0	40	1	10	60	10	20	102	100
-9		0.13	1	41	1.25	11	63	12.5	21	109	125
-8		0.16	2	42	1.6	12	66	16	22	117	160
-7		0.2	3	44	2	13	70	20	23	125	200
-6		0.25	4	46	2.5	14	73	25	24	133	250
-5		0.32	5	48	3.2	15	77	32	25	141	320
-4		0.4	6	50	4	16	81	40	26	150	400
-3		0.5	7	52	5	17	85	50	27		500
-2		0.63	8	55	6.3	18	90	63	28		630
-1		0.8	9	57	8	19	96	80	29		800
									30		1000

Exposure points EP can be calculated by the relations: factor $= 10^{\frac{EP}{10}}$ and EP$=10 \cdot$log (factor) which is equivalent to a gradation according to $(^{10}\sqrt{10})^n$. "Factor" means the ratio between two image receptor dose levels.

where ΔOD is the change in optical density, when increasing the image receptor dose from K_1 to K_2; γ is the gradient (see Chap. II.6) of the film-screen system applied. Equation II.4.4 is the basis of the exposure tables (see Table II.4.2); it means that logarithmic graded exposure parameters U and It (exposure point system) can be summed up or subtracted for the determination and adaptation of the exposure parameters to a specific radiographic examination situation.

One exposure point is thereby defined as an alteration of the tube voltage, the tube current, the exposure time or the tube current time product which results in a change of the image receptor dose by 25%. The gradation is given as $^{10}\sqrt{10}$. By changing the exposure by *three exposure points*, i.e. by a factor of $(^{10}\sqrt{10})^3$, the dose at the image receptor is therefore doubled. Also, in modern automatic exposure control systems, the relationship of Eq. II.4.4 in principle is used for the calculation of exposure parameters.

II.4.2 Attenuation by the Patient Support and the Components of the Spot-Film Device

The attenuation of the imaging radiation by the intermediate layers of the spot-film device (see IEC 1984) between the patient and the X-ray image receptor in international standards is considered very often as a whole (see IEC 1999). For the determination of the attenuation, one measures (see Fig. II.4.8), in the narrow-beam condition, the air kerma K_T above the patient support unit and the air kerma K_B in the image-receptor plane (when using a film-screen combination), with an additional attenuation layer (e.g. 25 mm Al) in the primary beam, and then calculates the attenuation ratio T_R (see also Chap. III.1.2):

$$T_R = \frac{K_T}{K_B} \cdot \frac{r_T^2}{r_B^2} \qquad (II.4.5)$$

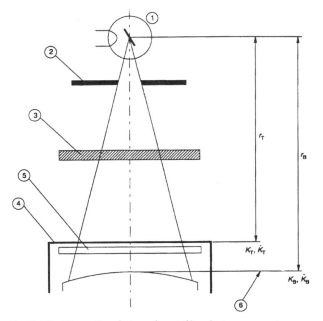

Fig. II.4.8. Attenuation factor of spot-film device; measuring arrangement according to IEC 61223-3-1 (IEC1999). 1, X-ray tube; 2, diaphragm (beam limiting device); 3, additional attenuating layer/phantom 25 mm Al; 4, patient support; 5, antiscatter grid; 6, image receptor plane (radiographic film cassette or X-ray image-intensifier housing); r_T focal spot to patient support distance; r_B focal spot to image receptor plane distance; $K_T(\dot{K}_T)$ transmission kerma (transmission kerma rate); $K_B(\dot{K}_B)$ X-ray image receptor air kerma (air kerma rate), also called system dose (system dose rate)

where r_T and r_B are the corresponding distances from the focal spot.

For *optimising* the properties of an imaging system, the single components of the so-called spot-film device (IEC 1984), however, must be examined in detail.

II.4.2.1   Patient Support

The table top in an X-ray examination unit has to be very stable, made of low-attenuation material and free of structures which possibly give rise to artefacts in the X-ray images. Carbon fibre has been accepted as material which meets these requirements. An Al-attenuation equivalent lower than 1 mm Al can be reached when the table top is designed in layers of carbon fibre and foam (e.g. polystyrene). At an X-ray tube voltage of 80 kV and a filtration layer of 20 cm water, a layer of 1 mm Al increases the entrance dose, i.e. the patient dose, needed for a radiograph by about 10%. Astonishingly these are hard-to-find requirements in national and international documents for the limitation of the attenuation of the radiation beam by the table top. Only in the IEC standard (see IEC 1999) is a typical attenuation factor of 1.25 at a tube voltage of 80 kV and a filtration of the primary beam by 25 mm Al mentioned for patient supports.

II.4.2.2   Antiscatter Grid

The antiscatter grid cannot be considered as a common attenuation layer, e.g. like a patient support, rather it has exceptional physical characteristics. In diagnostic X-ray imaging, it is the most important and widely applied tool to reduce the amount of scattered radiation at the recording system and to enhance image contrast (see Chaps. II.5 and III.2). As a rule the employment of a grid is, however, connected with an increased radiation exposure of the patient. Therefore it is also important to remove the grid in situations when the image quality is not improved by the grid, e.g. in paediatrics (see Chaps. III.2.2.1 and III.2.2.3).

For a patient thickness which is equivalent to a 20-cm water phantom, a moving 12/40 grid (that means a grid with a ratio of r = 12 and a line density of 40 lines per centimetre; see Chap. II.5.4), the patient dose is increased by a *grid exposure factor* (very often called Bucky-factor, B) of about B = 6.

To obtain a Bucky-factor as low as possible, especially the primary radiation transmission T_p of a grid should be as high as possible (see also Chap. II.5.5, Eq. II.5.18). The grid ratio r is of minor importance (see definition in Chaps. II.5.4 and III.2). For example, contrary to the assumption made in international recommendations, grids with a high grid ratio (e.g. r = 15) create, if properly designed, a lower patient dose than grids with a ratio of r = 8 recommended by the CEC document for paediatrics (CEC 1996b). Furthermore several studies have found that grids made with a carbon fibre cover and cotton fibre interspace material result in greater improvements in contrast and lower Bucky-factors than grids made with Al (Sandborg et al. 1993). These statements will be discussed in more detail in Chap. III.2.

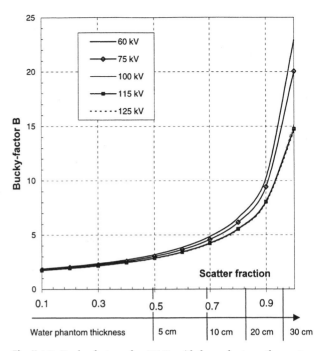

Fig. II.4.9. Bucky-factor of a 12/40-grid dependent on the scatter fraction S at various X-ray tube voltages (geometrical arrangement similar to IEC 2001a)

In national and international standards, the attenuation characteristics of antiscatter grids are determined with respect to a well-defined *measuring arrangement without scatter* (IEC 1999) or together with a water phantom of fixed dimensions and therefore a well-defined scatter fraction (IEC 1978, 2001a). These arrangements hardly meet the situation given in clinical practice, because the grid exposure factor or Bucky-factor is not a constant physical quantity, but depends on the imaging geometry, the patient thickness and the X-ray tube voltage used (see Fig. II.4.9 and Eq. II.5.18). The typical grid attenuation factors given for the scatter-free condition in national or international standards (e.g. in DIN 6809, part 7, 2002, as B = 1.4 to 1.6) and consequently also in installation manuals of X-ray equipment serve their purpose only for the adjustment of X-ray equipment during installation. Together with X-ray examination units, where it is possible to remove the grid, the Bucky-factor relevant in patient examinations can be determined for the exposure condition under consideration, by taking the same exposure with and without the grid and by registration of the corresponding tube current-exposure time products.

II.4.2.3   Ionisation Chamber

In combination with direct radiography (e.g. film-screen technique) mostly an ionisation chamber is used as radiation detector for the automatic exposure control system.

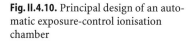 **Fig. II.4.10.** Principal design of an automatic exposure-control ionisation chamber

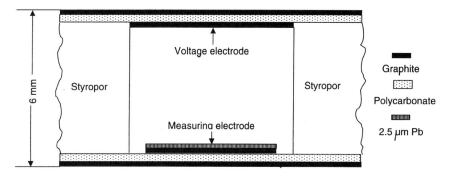

The chamber in most cases is arranged directly behind the antiscatter grid in front of the image receptor (e.g. the film-cassette). Because of this location, the attenuation of the imaging radiation by the ionisation chamber should be as low as possible. Furthermore the chamber must be "shadow free" to avoid artefacts in the radiograph. With the principal design shown in Fig. II.4.10, it is possible to fabricate such a chamber with a very low Al attenuation equivalent filtration (e.g. 0.2 mm Al outside, 0.4 mm Al inside the measuring fields).

A very thin metal layer (e.g. about 2 μm Pb) is deposited by thermal evaporation on the measuring electrode of the ionisation chamber. During exposure, photoelectrons are released from the metal layer which ionise the air volume in the chamber (see Chap. II.6.3). On the other hand the contribution of the direct ionisation of the air in the chamber by the impinging radiation is very low. From this follows the important advantage of the design shown in Fig. II.4.10 that the measuring signal (i.e. the ionisation current) is practically independent from the surrounding conditions (temperature, atmospheric pressure).

The attenuation of the imaging radiation by an ionisation chamber according to the design represented in Fig. II.4.10 at an X-ray tube voltage of 80 kV and measured in combination with a 20-cm water phantom is about 4–5%. According to international standards (IEC 1999), a typical value for the attenuation ratio of 1.11 (i.e. an attenuation of 11%) is mentioned, a value which is only obtained when detectors with an Al cover or detectors which are composed of semiconductor elements are employed.

II.4.2.4 Geometry

The imaging geometry of radiographic installations has great influence on patient exposure and image quality. Both aspects cannot be considered independent of each other. When calculating, e.g. the surface entrance dose of the patient starting from the image receptor dose (system dose), a geometric attenuation factor must be applied (see Chap. III.1, Eq. III.1.13). At a given distance between the focal spot and the patient's entrance surface r_E, the patient

dose increases in X-ray imaging in proportion to the square of the distance r_B between the focal spot and the image receptor, a fact which is important in magnification technique.

The design of each examination unit implies always a certain distance between the patient support and the imaging plane, e.g. caused by the installation of the moving grid and the ionisation chamber. This distance must be considered when one calculates the total equipment attenuation factor $m_{geometry}$ (see Eqs. III.1.4 and III.1.13) or if one compares the performance of different imaging systems. As already mentioned above, the intensity of the primary radiation must be increased according to the inverse square law (see Chap. II.5.3.1) to obtain a constant dose at the image receptor (see Fig. II.5.8), if the image receptor is moved away from the patient. This factor in analogy with the definitions given in Chap. II.5.4. can be interpreted as the "reciprocal primary radiation transmission" $1/T_p$ of the *air-gap technique* (see Chap. II.5.3.1):

$$m_{geometry} = (\frac{f}{f-a})^2 = \frac{1}{T_p} \qquad (II.4.6)$$

where f is thereby the source-to-image distance (SID) and a the distance between the patient output plane and the image plane.

Apart from these dose considerations, with respect to image quality, a distance between patient support and image plane can improve or make worse resolution of small image details dependent on their location within the patient's body. The spatial resolution limit of the image receptor and the focal spot size (see Fig. II.4.1) used are decisive.

Figure II.4.11 shows the visual resolution limit obtained with three types of film-screen combinations of different speed and for an image receptor which consists of an image intensifier, with 40-cm diameter input field size, connected to a digital imaging system, with a 1k image matrix size (1024×1024 pixel), dependent on the geometrical magnification factor M. The focal spot size of 0.6 is assumed to be in accordance with IEC 60636 (IEC 1993). Represented in the figure are the curves for the res-

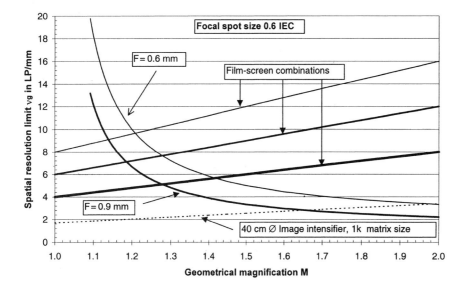

Fig. II.4.11. Spatial resolution dependent on on the geometrical magnification for various film-screen systems and a digital image receptor (1024×1024 pixel matrix size) and a focal spot size of 0.6 mm (*LP/mm* line pairs per millimetre)

olution limit of the focal spot width, which is allowed to lie between 0.6 mm and 0.9 mm. The resolution limit – assuming a constant radiation intensity distribution within the focal spot – is calculated according to (see Fig. II.4.1):

$$\nu_g = \frac{M}{F \cdot (M-1)} \qquad (II.4.7)$$

Magnification factors M in the range of 1.1–1.5 are found in radiographs made with over-table examination units. From Fig. II.4.11 it follows therefore that image details which are located distant from the image receptor will be reproduced with improved sharpness only in the digital images. In the film-screen images, the sharpness of these details is already limited by the focal spot size.

Furthermore the inverse square law also must be considered, if one thinks about an optimum geometry for the performance of the examination unit as a whole. When using an automatic exposure control (AEC) system at a given X-ray tube voltage, the cut-off dose (see Chap. II.6.3) in the AEC detector plane is held constant. Therefore the greater the distance between the detector and the image plane the greater is the dependence of the dose at the image receptor on the patient thickness. The cut-off dose is thereby the dose value at which the AEC system gives the signal for disconnecting of the X-ray tube voltage. Images of patients with thick soft tissue and bones will be underexposed (see Fig. II.4.12). This phenomenon occurs because the scatter fraction in front of the image receptor decreases faster than the corresponding primary fraction with increasing distance of the image receptor from the patient. In the air-gap technique (Sorenson and Floch 1985), advantage is taken of this effect (see Chap. II.5.3.1). In every case the ionisation chamber of the AEC

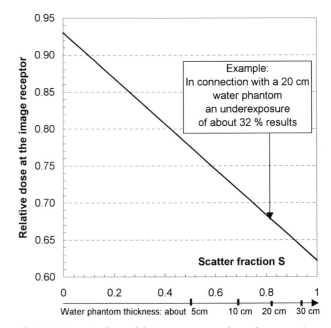

Fig. II.4.12. Dose in front of the image receptor dependent on patient thickness; geometrical arrangement according to Fig. II.5.8

should be located as close as possible to the image receptor.

Finally one must not forget that the cover of the image receptor (e.g. the cassette of the film-screen combination or entrance window of the image intensifier) also has influence on the patient dose. Limits for the maximum permissible attenuation of the imaging radiation by the cassette cover are given in ISO publication 4090 (ISO 2001): "The front cover shall have an absorption not greater than the equivalent of 1.8 mm pure Al (99%) when traversed by radiation of quality RQA 4 (approximately 60 kV; see IEC 61267: 2001)".

Fig. II.4.13. Geometrical configuration of over-table and under-table radioscopic systems

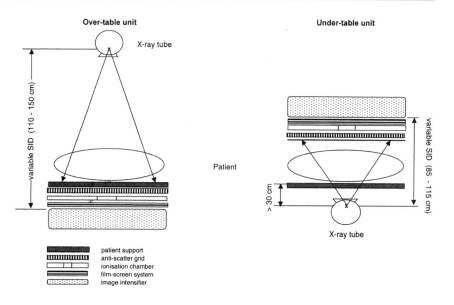

Over-table unit

Under-table unit

X-ray tube

variable SID (110 - 150 cm)

Patient

variable SID (85 - 115 cm)

> 30 cm

X-ray tube

patient support
anti-scatter grid
ionisation chamber
film-screen system
image intensifier

II.4.3 Important Differences Between Over-Table and Under-Table X-ray Units

When considering radiation exposure of patient and staff, the following differences must be taken into consideration (see Fig. II.4.13):

- Over-table units allow larger SIDs in comparison to under-table units and special protective devices are necessary for the staff. The patient support is located between the patient and the image receptor. The radiologists standing beside the patient during examination must therefore be specially protected against the scattered radiation emerging from the patient.
- Under-table units as a rule are used at smaller SIDs in comparison to over-table units. X-ray protection is more favourable to arrange. The spot-film device during the examination is arranged between the patient and radiologist and protects the radiologist against scattered radiation emerging from the patient. The patient support contributes to the filtration of the primary beam.
- Large-area ionisation chambers are used for the measurement of the dose-area product, which can be mounted on the X-ray tube diaphragm housing to intercept the entire X-ray beam and integrate the entrance dose over the whole beam area. Their response is independent of the distance from the focal spot, therefore they can be positioned anywhere between the patient and the X-ray tube (see Chap. II.3). But if high accuracy of the measurement is demanded, the different location of the patient support in the X-ray beam at over- and under-table units must be considered during calibration.

References

CEC (Commission of the European Communities) (1996a) European guidelines on quality criteria for diagnostic radiographic images. EUR 16260. CEC, Luxemburg

CEC (1996b) European guidelines on quality criteria for diagnostic radiographic images in pediatrics. EUR 16261EN. CEC, Luxemburg

Claasen (1962) Wozu Belichtungstabelle nach Punkten? SRW-Nachrichten 16:29

DIN (Deutsches Institut für Normung) (2002) Klinische Dosimetrie: Verfahren zur Ermittlung der Patientendosis in der Röntgendiagnostik. DIN 6809, Part 7 (Draft). Beuth, Berlin

Hammerstein GR, Miller DW, White DR, Masterson ME, Woodard HQ, Laughlin JS (1979) Absorbed radiation dose in mammography. Radiology 130:485–491

ICRU (International Commission on Radiation Units and Measurements) (1989) Tissue substitutes in radiation dosimetry and measurement. Report 44. ICRU, Bethesda, MD

IEC (International Electrotechnical Commission) (1978) Characteristics of anti-scatter grids used in X-ray equipment. Publication 60627. IEC, Geneva

IEC (International Electrotechnical Commission) (1984) Medical radiology – terminology. Publication 60788. IEC, Geneva

IEC (International Electrotechnical Commission) (1988) Medical radiology – terminology. Publication 60788. IEC, Geneva

IEC (1993) X-ray tube assemblies for medical diagnosis – characteristics of focal spots. Publication 60636. IEC, Geneva

IEC (1999) Evaluation and routine testing in medical imaging departments. Part 3-1. Acceptance tests – imaging performance of X-ray equipment for radiographic and radioscopic systems. Publication 61223-3-1. IEC, Geneva

IEC (2001a) Diagnostic X-ray imaging equipment – characteristics of general purpose and mammographic antiscatter grids. Publication 60627. IEC, Geneva

IEC (2001b) Medical diagnostic X-ray equipment – radiation conditions for use in the determination of characteristics. Publication 61267. IEC, Geneva

ISO (International Standards Organisation) (2001) Photography – medical radiographic cassettes/screens/films and hard-copy imaging films – dimensions and specifications. Publication 4090

Johns PC, Yaffe MJ (1987) X-ray characterization of normal and neoplastic breast tissues. Phys Med Biol 32:675–695

Meiler J (1949) Die in der Röntgendiagnostik verwendeten Spannungskurvenformen und ihr Einfluss auf Bildqualität und Röhrenbelastung. Fortschritte auf dem Gebiet der Röntgenstrahlen 72:222–241

Sandborg M, Dance DR, Alm Carlsson G, Persliden J (1993) Selection of antiscatter grids for different imaging tasks: the advantage of low atomic number cover and interspace materials. Br J Radiol 66:1151–1163

Sorenson JA, Floch J (1985) Scatter rejection by air gaps: an empirical model. Med Phys 12(3):308–316

Yaffe M J (1992) Digital mammography. In: Haus AG, Yaffe MJ (eds) Syllabus: a categorical course in physics technical aspects of breast imaging. Radiological Society of North America, Oak Brook, pp 69–84

II.5   Scattered Radiation

In X-ray imaging the X-ray photons penetrate the patient and interact with the orbital electrons of the atoms of which tissue is made up. The interaction processes relevant in diagnostic radiology have been introduced in Chap. II.2.2:

- The photoelectric effect is the most important interaction process with respect to image quality
- The Compton effect or incoherent scattering is responsible for the creation of scattered radiation and therefore interferes with image quality

In the energy range used in diagnostic radiology (see Fig. II.5.1), incoherent scattering predominates over Rayleigh scattering or coherent scattering. This process is

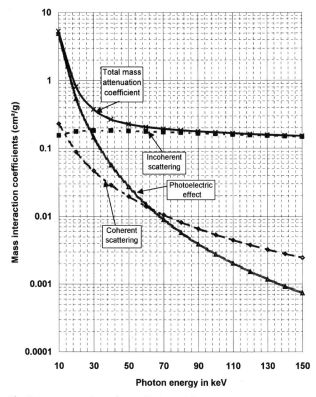

Fig. II.5.1. Interaction of X-radiation with water: Mass attenuation coefficients (photoelectric effect, Rayleigh scattering, Compton scattering, total attenuation) as a function of photon energy (calculation with the help of the XCOM program; see Chap. II.2.3)

mostly ignored when considering the influence of scattered radiation on image quality. Furthermore the scattering angles are small (forward scattering) and therefore the normally employed techniques (see Chap. II.5.3) for the control of scattered radiation in the front of the image receptor, of low efficacy (Johns and Yaffe 1983).

In summary the number of X-ray photons of the primary radiation is reduced by absorption and scattering, when they penetrate the patient. With increasing patient thickness, single and multiple scattered photons appear. However, image information is carried only by the primary radiation, i.e. the spatial distribution of that X-ray photons, which passed the patient without interaction. The simultaneously created scattered radiation (secondary radiation) superimposes this distribution and interferes with the image quality in a serious manner (see Chap. III.2).

Firstly in this chapter, the quantitative evaluation of scattered radiation will be discussed and then the characteristics of scattered radiation, dependent on the imaging parameters, will be shown. Secondly the techniques to reduce its deleterious effect on image quality will be demonstrated.

II.5.1   Measurement of Scattered Radiation

X-ray imaging is based on a central projection of that part of the patient's body on the image receptor, which shall be examined by means of the X-radiation emitted from the focal spot of the X-ray tube (see Fig. II.4.1). Therefore the irradiation field sizes on the entrance and the exit side of the patient, which define the tissue volume, are determined by the collimation of the X-radiation, the focus object distance and the object image receptor distance. The number of X-ray photon interactions depends on the volume of tissue irradiated (see Chap. II.5.2).

The evaluation of the characteristics of scattered radiation in diagnostic radiology as a rule will be realised with suitable phantoms, not with patients. The *percentage of scattered radiation S(a)* (see Chap. II.4.1.1, scatter fraction) emerging from a phantom of thickness d dependent on the irradiation field size a (on the exit side) can be measured with an experimental set-up shown in Fig. II.5.2. It is similar to the measuring arrangement

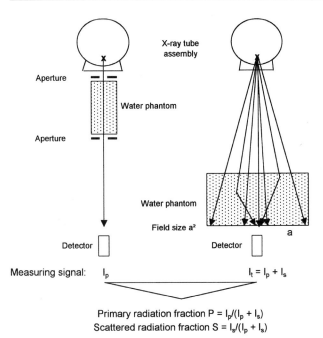

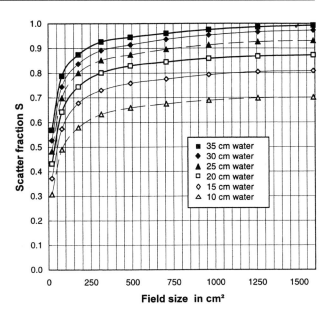

Fig. II.5.2. Measuring arrangement for the evaluation of scattered radiation: measurement of total radiation with phantom in front of the radiation detector, measurement of primary radiation with phantom nearby the focal spot

Fig. II.5.3. The fraction of scattered radiation S dependent on the field size for various object thickness at an X-ray tube voltage of 80 kV (adapted from Reiss and Steinle 1973)

(without grid) given in IEC 60627 (IEC 1978, 2001; compare Figs. II.5.2 and II.5.12). This standard describes the measurement method of the physical characteristics of antiscatter grids, a subject which shall be discussed further in Chap. II.5.4.

For the determination of the scatter fraction $S(a)$, the total radiation intensity $I_t(a)$ is measured first by using patient-equivalent scattering phantoms (e.g. water or polymethylmethacrylate, PMMA) of different thickness d, arranged directly in front of the radiation detector, with increasing field sizes (a; i.e. with respect to Fig. II.5.2, but without the antiscatter grid in place). $I_t(a)$ encloses the primary radiation intensity I_p and that of the scattered radiation $I_s(a)$. Therefore it is:

$$I_t(a) = I_s(a) + I_p \qquad (II.5.1)$$

The measurement of the primary radiation intensity I_p alone is carried out with tight collimation (i.e. similar to Fig. II.5.2, also without the antiscatter grid in place), whereby the patient-equivalent phantoms of thickness d (with reduced outside dimensions) are now arranged near the focal spot.

For each phantom thickness d, the difference between the corresponding measuring results of $I_t(a)$ and I_p divided by the total radiation intensity $I_t(a)$ results in the scatter fraction $S(a)$ (see also Fig. II.4.5):

$$S(a) = \frac{I_t(a) - I_p}{I_t(a)} \qquad (II.5.2)$$

In Fig. II.4.5 the quantities $I_t(a)$ and I_p have been normalised to the intensity of the impinging radiation. Figure II.5.3 shows the scatter fraction $S(a)$ dependent on the field size a in front of the image receptor for various phantom thicknesses d (Reiss and Steinle 1973; Morneburg 1995) at an X-ray tube voltage of 80 kV.

The *scatter fraction S(U)* as a function of the X-ray peak tube voltage U (see Fig. II.5.4) can be determined by an analogous measurement, now always with a fixed phantom thickness d and field size a as parameters, by alteration of the tube voltage U (Reiss and Steinle 1973; Morneburg 1995).

If the scatter fraction S is measured directly when making a patient exposure, as a rule lead discs are arranged at the patient's entrance side (see Fig. II.5.5). The thickness of the lead discs must be calculated in such a way that practically no primary radiation can penetrate them. Within their shadow only scattered radiation is than recorded in the imaging plane.

For obtaining a high accuracy, measurements have to be made with lead discs of decreasing diameter and the results must be extrapolated to a zero disc diameter. If multiple lead discs are spread across the whole field of vision by interpolation between the distributed measuring points, a spatial distribution of the scatter fraction can be obtained. Such investigations have been reported in the literature (Love and Kruger 1987). By subtraction of the spatial profile of the scattered radiation intensity, e.g. from a series of digital radiographs,

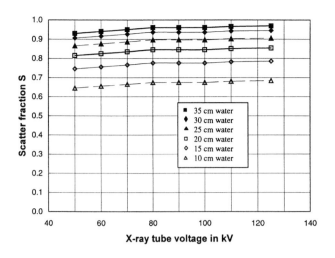

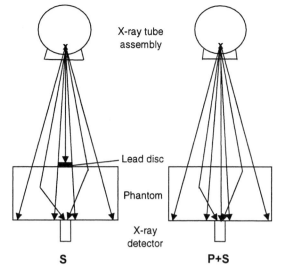

Fig. II.5.4. The fraction of scattered radiation S dependent on the X-ray tube voltage U for various object thickness (radiation field size about 700 cm²; adapted from Reiss and Steinle 1973)

Fig. II.5.5. Determination of the percentage of the primary radiation P and the percentage of the scattered radiation S by the employment of a lead disc

the image quality of these images can be considerably improved.

II.5.2 Properties of Scattered Radiation

When developing techniques to reduce the deleterious effect of scattered radiation on image quality, the dependence of scatter on the imaging parameters must be known. Because the radiation qualities used in general X-ray diagnostics and mammography are very different, it is reasonable to consider these imaging techniques separately.

II.5.2.1 General X-ray Diagnostics

The amount of scattered radiation (scatter fraction) S emerging from a patient depends primarily on the irradi-

ated volume of tissue, i.e. patient thickness and field size, whereas the dependence on tube voltage is very small (see Fig. II.5.4). The scatter fraction S at a given patient thickness and field size is nearly constant in the tube voltage range used in radiography. Furthermore, it is important that, from a field size of about 200 cm² upwards, the scatter fraction (see Fig. II.5.3) reaches a point of saturation (Reiss and Steinle 1973; Morneburg 1995).

Table II.5.1 shows the scatter fraction for a few examples of medical exposure situations. Especially for pelvis exposures, the amount of scatter is higher than 80%, whereas the contribution of the primary radiation is therefore lower than 20% and the scatter degradation factor (SDF) <0.2. The SDF (Morgan 1946), a quantity which describes the contrast reduction in radiographic images (see Fig. II.5.6) by scattered radiation, will be introduced and discussed in detail in Chaps. II.5.5 and III.2.1.

Table II.5.1 Typical scatter fractions for a few examples of medical exposure situations and the corresponding scatter degradation factor SDF (adapted from Morneburg 1995)

Object	X-ray tube voltage (Up)	Scatter fraction (S)	Scatter degradation factor SDF[a]
Cranium p.-a.	70 kV	45%	0.55
Lung (average) p.-a.	120 kV	55%	0.45
Lung (thick) p.-a.	120 kV	65%	0.35
Bulbus p.a.	80 kV	70%	0.3
Pelvis (average) p.-a.	80 kV	80%	0.2
Pelvis (thick) p.-a.	80 kV	85%	0.15
Pelvis from the side	8o kV	90%–95%	0.1–0.05

p.-a. posterior-anterior, [a] See equation II.5.9

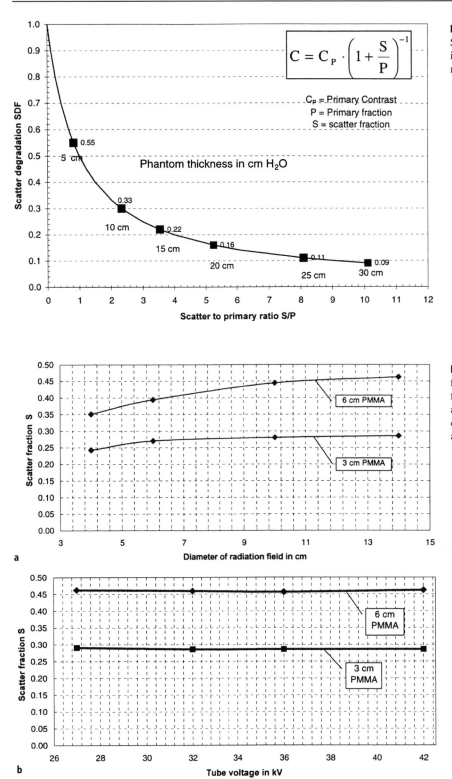

$$C = C_P \cdot \left(1 + \frac{S}{P}\right)^{-1}$$

C_P = Primary Contrast
P = Primary fraction
S = scatter fraction

Phantom thickness in cm H_2O

Fig. II.5.6. Scatter degradation factor SDF (see Eq. III.2.1 and Eq. III.2.4) in dependence on the scatter to primary ratio S/P

Fig. II.5.7a, b. Mammography: Scatter fraction dependent on **a** the radiation field size at a tube voltage of 32 kV and **b** the X-ray tube voltage at a field size diameter of 14 cm (adapted from Barnes and Brezovich 1978)

II.5.2.2   Mammography

In mammography the dependence of the scatter fraction S behind the breast to be imaged on the exposure parameters is very similar to the situation given in general radiography (see Fig. II.5.7). The scatter fraction S is deter-

mined by the tissue volume irradiated and – at a small object thickness, e.g. equivalent to 3 cm PMMA – reaches a point of saturation very fast with increasing field sizes.

S is also nearly independent of the X-ray tube voltage (Barnes and Brezowich 1978; Boone and Cooper 2000; Boone et al. 2000). This behaviour can be understood more easily in mammography than in general radiography: because of the employment of K-edge filters as additional filtration (Aichinger et al. 1994), the mean energy of the X-ray spectrum does not change much with increasing tube voltage. At very large, compressed breast thickness (>6 cm) the percentage of scattered radiation approximates 60%. This smaller "limiting value" of S in comparison with general radiography (see Table II.5.1) results from the lower X-ray energy of the scattered radiation and the consequently higher interaction probability in the tissue. In mammography the scattered X-ray photons reaching the image receptor above all come from the object layers nearby the image receptor.

II.5.3 Techniques to Reduce Scattered Radiation at the Image Receptor

The deleterious effect of scattered radiation requires appropriate techniques for its control. From Figs. II.5.3 and II.5.4 can be deduced that this can be done first by carefully compressing the object to be imaged and by tight collimation. Image quality will be improved and radiation exposure reduced; however, especially in the case of larger field sizes and object thicknesses, additional techniques for reduction of scatter are necessary.

II.5.3.1 Air-Gap Technique

Air-gap technique is based on the application of a certain distance, for example 30 cm, between the patient and the image recording system (Sorenson and Floch 1985). To maintain the geometric scale constant, the focus-patient distance must also be increased to the same extent, otherwise it is possible that – dependent on the focal spot size used – the visual resolution limit gets worse because of the higher geometrical magnification (see Chap. II.4.2.4, Eq. II.4.7 and Fig. II.4.11). The reduction of the scatter fraction S at the plane of the image receptor with increas-

ing distance from the patient's body has been described by various authors (Sorenson and Floch 1985) with the help of an *effective scatter point source*. By taking into consideration that the intensity of the scattered radiation I_S in comparison with the intensity of the primary radiation I_P at the image receptor plane decreases faster with increasing distance from the patient, one can calculate a "selectivity" (see Eqs. II.5.8 and III.2.11), which increases with the extent of the air-gap. By calculation of a Bucky factor and a contrast-improvement factor, in analogy with antiscatter grids (see Fig. II.5.8), one gets information about the radiation exposure of the patient and the improvement in image quality which is attainable in combination with air-gap technique (see Chap. III.2).

These calculations are unfortunately complicated, because the location of the effective scatter point source with respect to the scattering medium depends on the imaging geometry applied (Sorenson and Floch 1985). Very similar considerations with respect to mammography have been made by Krol et al. (1996).

II.5.3.2 Slot Technique

A very effective scatter control system is the slit or slot technique (Barnes et al. 1976). In slot technique the radiation beam is collimated close to the X-ray tube and the image receptor to form a fan beam, e.g. with a width of 1 cm, which is guided over the patient during the exposure (see Fig. II.5.9).

By pivoting the X-ray tube, the angle between the anode disc and the image receptor is constant during the scanning motion, thereby avoiding variations in X-ray fluence due to the heel effect. The extremely high tube load and the mechanical complexity have up to now prevented extensive application of the slot technique. The tube load in comparison with full-field technique – at a given X-ray tube voltage – is approximately increased by the ratio of the image receptor size to the slit width (without grid). With the help of a multiple-slit system, it would be possible to reduce the demands on the X-ray tube as-

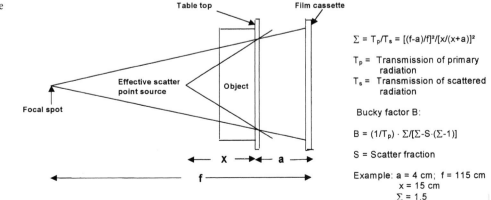

Fig. II.5.8. The air-gap technique

$\Sigma = T_p/T_s = [(f-a)/f]^2/[x/(x+a)]^2$

T_p = Transmission of primary radiation

T_s = Transmission of scattered radiation

Bucky factor B:

$B = (1/T_p) \cdot \Sigma/[\Sigma-S\cdot(\Sigma-1)]$

S = Scatter fraction

Example: a = 4 cm; f = 115 cm
x = 15 cm
Σ = 1.5

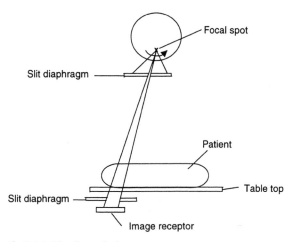

Fig. II.5.9. The slot technique

sembly, but there is very little control of the complex mechanical system. It is very difficult to receive images free of artefacts if the slit aperture in front of the image receptor is not outshone, but this is not allowed with respect to radiation protection.

II.5.3.3   Antiscatter Grid

The use of antiscatter grids in diagnostic X-ray imaging is the most widely applied and accepted method for reducing the amount of scattered radiation at the recording system and improving the contrast of the image. The antiscatter grid consists of a periodic array of radio-opaque

foil strips (usually lead; see Fig. II.5.10), separated by strips of radiolucent spacing material (typically paper or aluminium).

In the case of *focused grids*, the strips are in alignment with the direction to the focal spot, so that the attenuation of the image-forming primary radiation is as low as possible. In contrast, only a small fraction of scattered radiation which emerges from the patient along the strips will pass the grid.

In addition to focused grids, two other types of grids should be mentioned here: the parallel grids and the cross grids.

In *parallel grids* the absorbing strips are adjusted parallel to each other. To avoid lateral dose cut-off (see Chap. II.5.4), the grids are available only for relatively low grid ratios (e.g. r = 6) and they can only be used for larger focus-image receptor distances. Application of parallel grids has the advantage that imaging is insensitive to decentring.

Cross grids consist of two linear grids built together in such a way that the directions of their absorbing strips form an angle of 90°. The advantage of such grids is that very high selectivity values (e.g. ~50) can be obtained. Their drawback is that the primary radiation transmission T_p is reduced to a large extent (e.g. <50%). Cross grids can furthermore only be used with a perpendicular incident radiation beam, whereas with linear focused grids the X-ray tube focal spot can be shifted in the direction of the grid strips, so that oblique beam directions, which are sometimes necessary with respect to medical indications, are possible.

In this connection one grid available in the USA for mammography has to be mentioned, the "cellular grid"

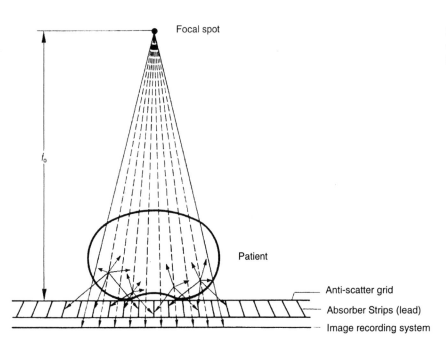

Fig. II.5.10. Principle of a focused antiscatter grid (from Morneburg 1995). fo is the focusing distance of the grid

(Rezentes et al. 1999). Its design has a honeycomb pattern instead of a linear arrangement of its absorbing lamella. Above all it must be pointed out that the additional advantage is that air is used as interspacing medium and therefore beam hardening of the image forming radiation is avoided. A very thin, evaporated Cu-layer is employed as absorbing material instead of lead lamella.

II.5.4 Quantities for the Description of the Characteristics of Grids

The physical characteristics of an antiscatter grid are essentially determined by its geometrical construction (see Fig. II.5.11). The geometric parameters (IEC 1978, 2001) are the *height (h)* and the *thickness (d)* of the *absorbing strips* (mostly lead), moreover the *thickness (D)* of the *interspace material* (e.g. paper, aluminium). With these quantities (see Table II.5.2), the *strip density* N, i.e. the number of grid lines per centimetre and the *grid ratio* (r) are defined as:

$$N = \frac{1}{(D+d)} \tag{II.5.3}$$

and:

$$r = \frac{h}{D} \tag{II.5.4}$$

The physical characteristics (IEC 1978, 2001; see Table II.5.2) which result from the geometrical construction of an antiscatter grid are the *transmission of primary radiation* (T_p) *the transmission of scattered radiation* (T_s) *and the transmission of total radiation* (T_t).

The transmission of primary radiation T_p is defined as the ratio of the intensity I'_p of the primary radiation with the grid in place to the corresponding intensity I_p without the grid in front of the detector plane of the measuring arrangement given in the IEC 60627 standard (IEC 1978, 2001):

Fig. II.5.11. Geometrical design of an antiscatter grid

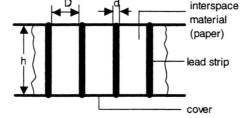

interspace material (paper)

lead strip

cover

Height of lead strips h
Thickness of lead strips d
Thickness of interspace material D
Strip density N = 1/(D+d)
Grid ratio r = h/D

Table II.5.2 Geometrical and physical characteristics of an antiscatter grid according to IEC 60627 (1978, 2001)

Geometrical characteristics		
Height of the lead strips	h	
Thickness of the lead strips	d	
Thickness of interspace material	D	
Strip density	$N = \dfrac{1}{D+d}$	
Grid ratio	$r = \dfrac{h}{D}$	
Physical characteristics		
Transmission of primary radiation	T_p	Measuring conditions:
Transmission of scattered radiation	T_s	Water phantom height 20 cm
Transmission of total radiation	T_t	Radiation field size 30 cm · 30 cm
Selectivity	$\Sigma = \dfrac{T_p}{T_s}$	**IEC 1978:** 100 kV ± 3% Total filtration 4 mm Al **IEC 2001:** Radiation quality RQN6 and RBN6 according to IEC 61267 (80 kV)
Derived physical characteristics		
Contrast improvement factor	$C_{if} = \dfrac{T_p}{T_t} = \dfrac{\Sigma}{\Sigma - S \cdot (\Sigma - 1)}$	
Bucky factor	$B = \dfrac{1}{T_t} = \dfrac{1}{T_p} \cdot \dfrac{\Sigma}{\Sigma - S \cdot (\Sigma - 1)}$	
Improvement of the signal-to-noise ratio	$SRN_{if} = T_p \cdot \sqrt{B}$	

$$T_p = \frac{I'_p}{I_p} \qquad\qquad (II.5.5)$$

The transmission of scattered radiation T_s and the transmission of total radiation T_t are defined by analogy with T_p as:

$$T_s = \frac{I'_s}{I_s} \qquad\qquad (II.5.6)$$

and:

$$T_t = \frac{I'_t}{I_t} \qquad\qquad (II.5.7)$$

The determination of the physical characteristics T_p, T_s and T_t of antiscatter grids are to be made in agreement with IEC 60627 (IEC 1978), as a rule, at a peak tube voltage of 100 kV, with a ripple of less than 10% and in connection with an additional filter of 4 mm Al. The measurements are made with 20-cm-thick water phantoms (see Fig. II.5.12).

In the revised version of the IEC 60627 standard (IEC 2001), some changes are made with respect to the radiation quality and the measuring arrangement used. Measurements are now required mainly at a tube voltage of 80 kV in combination with the normally applied total filtration of 2.5 mm Al, whereby the fine adjustment of the radiation quality (defined as radiation conditions RQN 6 and RBN 6) is made with respect to IEC 61267 (IEC 2001). For grids which are used preferably in the low- or high-voltage region, 60 kV (RQN 3 and RBN 3) and 100 kV (RQN 9 and RBN 9) are recommended. Therefore the 1978 and 2001 versions of the IEC standard differ essentially by the additional filtration at the X-ray tube output, which has only minor influence on T_p and T_s. More critical are the rigorous requirements in the new version on the beam collimation and the adjustment of the smaller lead disc that serves as the beam stop. As shown in Fig. II.5.12, the grid measurements can also be made without the beam stop and the results would than represent the situation with a zero lead disc diameter. In the further discussion, remarks will always be given to the details of the measuring arrangement (IEC 1978, 2001) used for the determination of the grid characteristics under consideration.

Finally the quotient of the transmission of the primary radiation T_p and the transmission of the scattered radiation T_s is defined as the *selectivity* Σ of the antiscatter grid:

$$\Sigma = \frac{T_p}{T_s} \qquad\qquad (II.5.8)$$

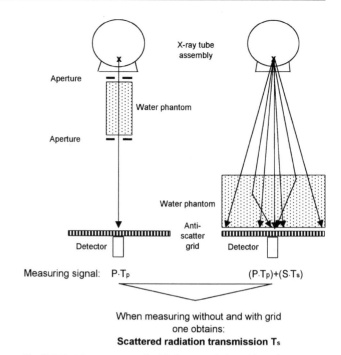

Fig. II.5.12. Measurement of grid characteristics without beam stop

Together with the primary radiation transmission T_p, the selectivity Σ describes clearly the properties of the scatter-reduction devices.

II.5.5   Derived Quantities for the Description of the Efficiency of Antiscatter Grids

The influence of scattered radiation on the primary radiation contrast C_p (see Chaps. II.7.1 and III.2.1) and on the signal-to-noise ratio (SNR) in the radiation image can be evaluated with the scatter degradation factor SDF (Morgan 1946). C_p is that contrast which would be obtained when no scattered radiation was present. The SDF factors without grid and with grid in place can be written as:

$$SDF_o = \frac{1}{(1+\frac{I_s}{I_p})} \qquad\qquad (II.5.9)$$

and:

$$SDF_g = \frac{1}{(1+\frac{I_s \cdot T_s}{I_p \cdot T_p})} = \frac{1}{(1+\frac{1}{\Sigma}\cdot\frac{I_s}{I_p})} \qquad (II.5.10)$$

Two variables which describe the effectiveness of antiscatter grids in conventional X-ray diagnostics, i.e. their influence on image quality and dose, are the *contrast improvement factor* C_{if} and the Bucky factor B.

The contrast improvement factor C_{if} is the ratio of the contrast obtained with grid C_g to the contrast C_o obtained without grid; it can be calculated with the help of the Eqs. II.5.8, II.5.9 and II.5.10:

$$C_{if} = \frac{C_g}{C_o} = \frac{C_p \cdot SDF_g}{C_p \cdot SDF_o} = \frac{T_p}{T_t} \qquad (II.5.11)$$

The contrast improvement factor C_{if} can therefore also be calculated by the ratio of the primary radiation transmission T_p to that of the total radiation transmission T_t.

The Bucky factor B gives the factor by which the entrance dose with grid must be increased in order to compensate the reduction in dose at the image receptor caused by the lower amount of scattered radiation and the small absorption of the primary radiation by the interspacing material:

$$B = \frac{1}{T_t} \qquad (II.5.12)$$

In *digital imaging* the SNR (see Chap. II.7.4) is the physical quantity suitable for describing image quality, since the ability to display poorly contrasted details with the windowing technique is limited only by noise. The effect of a grid on image quality can therefore be evaluated by the *signal to noise improvement factor SNR$_{if}$*. The signal to noise improvement factor is the ratio of the SNR obtained with grid SNR_g to that obtained without grid SNR_o:

$$SNR_{if} = \frac{SNR_g}{SNR_o} \qquad (II.5.13)$$

Considering only quantum noise, e.g. the noise present in the radiation image, it can be derived, with the help of Eqs. II.5.7, II.5.11, II.5.12 and II.7.11, that the improvement which the grid brings to the SNR is equal to the product of the primary radiation transmission T_p and the square root of the Bucky factor B (Chan et al. 1990):

$$SNR_o = C_p \cdot SDF_o \cdot \sqrt{I_p + I_s} \qquad (II.5.14)$$

$$SNR_g = C_p \cdot SDF_g \cdot \sqrt{T_p \cdot I_p + T_s \cdot I_s} \qquad (II.5.15)$$

$$(SNR)_{if} = \frac{T_p}{T_t} \cdot \frac{\sqrt{I_t'}}{\sqrt{I_t}} = \frac{T_p}{T_t} \cdot \sqrt{T_t} = T_p \cdot \sqrt{B} \qquad (II.5.16)$$

Finally, with the help of Eqs. II.5.9 and II.5.10, the contrast improvement factor and the Bucky factor can be convert-

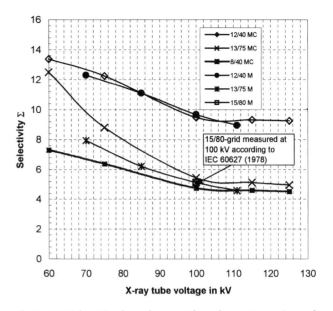

Fig. II.5.13. Selectivity dependent on tube voltage: Comparison of the results of measurements (IEC 1978) and simulation calculations for the grid types Pb 12/40, Pb 8/40, Pb 13/75 and Pb 15/80 (only one measuring point)
(MC = Monte Carlo   M = Measurement)

ed into a form which now shows clearly their dependence on the selectivity Σ and the primary radiation transmission T_p:

$$C_{if} = \frac{\Sigma}{\Sigma - S \cdot (\Sigma - 1)} \qquad (II.5.17)$$

$$B = \frac{1}{T_p} \cdot \frac{\Sigma}{\Sigma - S \cdot (\Sigma - 1)} \qquad (II.5.18)$$

The scatter fraction S behind the scattering medium in front of the antiscatter grid does not depend much on the X-ray tube voltage (see Fig. II.5.4), but on the imaging geometry. The transmission of the primary radiation T_p of a grid changes, in the tube voltage range used in X-ray diagnostic radiology, by less than 10%. The selectivity Σ – dependent on the thickness of the lead strips – decreases with increasing tube voltage, because of the increasing translucence of scattered radiation (see Fig. II.5.13).

In conclusion, for *comparing different grid types* with respect to dose and image quality, *the exposure conditions* (geometry, X-ray tube voltage, filtration) must be known.

Equation II.5.17 shows that, for a given scatter fraction S and a fixed X-ray tube voltage, the contrast improvement factor C_{if} is determined by the selectivity Σ alone. Furthermore, for a high contrast improvement and a Bucky factor as low as possible (see Eq. II.5.18), a high primary radiation transmission T_p is of decisive importance. The selectivity Σ and the primary radiation transmission T_p, and not the grid ratio r, are the physical quantities which give a valuable indication of the image quality and radiation exposure obtainable with an antiscatter grid.

For the estimation of the radiation exposure of the patient (organ dose, effective dose), as a rule the entrance surface air kerma is needed (see Chap. II.3 and III.1). This physical quantity can be calculated – besides other methods – starting from the system dose. In this case the scatter fraction S_g directly in front of the image receptor, i.e. behind the antiscatter grid, must be known (see Chap. III.1.3):

$$S_g = \frac{I_s \cdot T_s}{I_t \cdot T_t} = S \cdot T_s \cdot B \qquad (II.5.19)$$

With the help of Eq. II.5.18, this relation can also be written as:

$$S_g = \frac{S}{S + (1 - S) \cdot \Sigma} \qquad (II.5.20)$$

If the selectivity $\Sigma(U)$ of a grid dependent on the tube voltage – at a fixed filtration – is known (see Table III.1.5) the scatter fraction S_g in the image receptor plane can be calculated by making use of Eq. II.5.20.

The primary radiation fraction of the system dose is therefore:

$$P_g = (1 - S_g) \qquad (II.5.21)$$

By considering only the primary radiation attenuation by the patient and the intermediate layers (ionisation chamber of the automatic exposure control, antiscatter grid, table top) the entrance surface air kerma can be estimated in this way from the primary fraction P_g of the system dose (see Chap. III.1.3).

II.5.6   Application Limits for Focused Grids

A consistently good primary radiation transmission T_p of a grid within the whole field of vision is achieved only if the focusing distance (Fig. II.5.14) is kept and the central radiation beam is perpendicular to the centre line of the grid.

Off-level and lateral decentring results in a uniform loss of primary radiation and an increase in relative intensity of scattered radiation imaged, combined with an increased exposure of the patient. If the grid is centred properly but lies above or below the focusing distance f_0, grid artefacts will be enhanced and grid cut-off will be apparent, which means there is a loss of primary and an increase in scattered radiation at the edges of the X-ray field. Defocusing results in tapering of the optical density, when using film-screen systems, from the centre to the edge of the exposed X-ray film, with the density decrease being perpendicular to the grid lines.

The IEC 60627 (IEC 1978) standard defines the application limits of an antiscatter grid as that focus distance at which the primary radiation transmission is reduced to 60% at the border of the largest film size used in comparison with the central line (see Fig. II.5.15).

This regulation is important for X-ray examination devices which make X-ray exposures possible at different focus image receptor distances. The application limits can be calculated according to formulae developed by Boldingh (1961). For all types of decentring, the loss of primary radiation is proportional to the grid ratio r and it is therefore important that high-ratio grids are carefully adjusted.

Decentring is combined with reduced contrast and increased radiation exposure. Because one works as a rule with an automatic exposure control system, this effect often is discovered purely by chance.

In the following the equations are reproduced from Boldingh (IEC 2001). The application limits without decentring of a focused grid follow from:

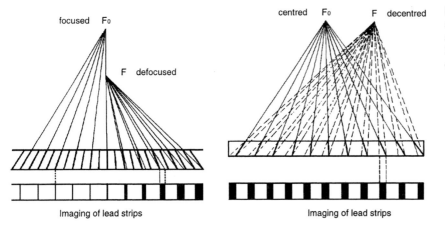

Imaging of lead strips

Imaging of lead strips

Fig. II.5.14. Centring and focusing of an antiscatter grid (adapted from Hoxter and Schenz 1991)
F_0 position of the focal spot, if it is properly centred

Fig. II.5.15. Upper and lower application limits of an antiscatter grid according to IEC 60627 (IEC 1978, 2001)

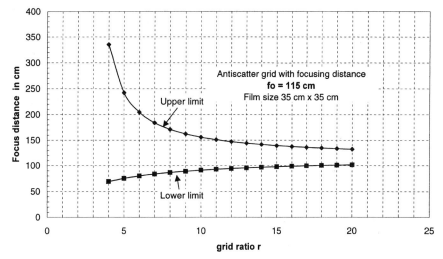

$$f_1 = \frac{f_0}{1 + \dfrac{f_0\, V_1}{r\, c}} \qquad f_2 = \frac{f_0}{1 - \dfrac{f_0\, V_2}{r\, c}} \qquad (II.5.22)$$

The application limits with decentring of a focused grid follow from:

$$f_1 = \frac{c + z}{\dfrac{c}{f_0} + \dfrac{V_1}{r}} \qquad f_2 = \frac{c - z}{\dfrac{c}{f_0} - \dfrac{V_2}{r}} \qquad (II.5.23)$$

where: c is the distance from the true central line to the border of the effective area; f_0 is the focusing distance; f_1 is the lower application limit; f_2 is the upper application limit; r is the grid ratio; V_1 is the loss of transmission of primary radiation at the lower application limit; V_2 is the loss of transmission of primary radiation at the upper application limit; and z is the value of decentring of a focused grid.

References

Aichinger H, Dierker J, Säbel M, Joite-Barfuss S (1994) Bildqualität und Dosis in der Mammographie. Electromedica 62(1): 7–11

Barnes GT, Brezowich IA (1978) The intensity of scattered radiation in mammography. Radiology 126: 243–247

Barnes GT, Cleare HM, Brezowich IA (1976) Reduction of scatter in diagnostic radiology by means of a scanning multiple slit assembly. Radiology 120: 691–694

Boldingh WH (1961) Quality and choice of potter Bucky grids, Parts IV, V. Acta Radiol 55: 225–235

Boone JM, Cooper VN III (2000) Scatter/primary in mammography: Monte Carlo validation. Med Phys 27(8): 1818–1831

Boone JM, Lindfors KK, Cooper VN III, Seibert JA (2000) Scatter/primary in mammography: comprehensive results. Med Phys 27(10): 2408–2416

Chan H-P, Lam KL, Wu Y (1990) Studies of performance of antiscatter grids in digital radiography: Effect on SNR. Med Phys 17(4): 655–664

Hoxter EA, Schenz A (1991) Röntgenaufnahmetechnik und Anwendungen. 14. Überarbeitete und erweiterte Auflage. Siemens, Munich

IEC (International Electrotechnical Commission) (1978) Characteristics of antiscatter grids used in X-ray equipment. Publication 60627. IEC, Geneva

IEC (2001) Diagnostic X-ray imaging equipment – characteristics of general purpose and mammographic antiscatter grids. Publication 60627. IEC, Geneva

IEC (2001) Medical diagnostic X-ray equipment – Radiation conditions for use in the determination of characteristics. Publication 61267. IEC, Geneva

Johns PC, Yaffe MJ (1983) Coherent scatter in diagnostic radiology. Med Phys 10(1): 40–50

Krol A, Bassano DA, Chamberlain CC, Prasad SC (1996) Scatter reduction in mammography with air-gap. Med Phys 23(7): 1263–1270

Love LA, Kruger RA (1987) Scatter estimation for a digital radiographic system using convolution filtering. Med Phys 14(2): 178–185

Morgan RH (1946) An analysis of the physical factors controlling the diagnostic quality of roentgenographic images, Part III. Am J Roentgenol 55: 67–89

Morneburg H (1995) Bildgebende Systeme für die medizinische Diagnostik. Publicis, Erlangen

Reiss KH, Steinle B (1973) Tabellen zur Röntgendiagnostik, Teil II. Bereich Medizinische Technik, Entwicklungsabteilung. Siemens, Erlangen

Rezentes PS, Almeida A de, Barnes GT (1999) Mammography grid performance. Radiology 210: 227–232

Sorenson JA, Floch J (1985) Scatter rejection by air-gaps: an empirical model. Med Phys 12(3): 308–316

II.6   Image Receptors

The X-ray photons that build up the radiation image – possibly after transmission through the antiscatter grid – are partly absorbed by the image receptor. The probability of interaction or the *quantum detection efficiency* (QDE; see Chap. II.7) for photons of energy E is proportional to (Yaffe and Rowlands 1997):

$$\eta = 1 - e^{-\mu(E) \cdot d} \qquad (II.6.1)$$

where $\mu(E)$ is the linear attenuation coefficient of the detector material and d the active thickness of the detector (see also Chap. II.2.1). The main interaction process is the photoelectric effect because of the relatively high atomic number of most detector materials. To get an effective quantum detection efficiency with respect to the tube voltage selected, Eq. II.6.1 must be averaged over the relevant incident X-ray spectrum. The application of the total attenuation coefficient $\mu(E)$ in Eq. II.6.1 is based on the assumption that the highest possible noise equivalent quanta (NEQ) (see Chap. II.7.5) can be obtained with an ideal quantum-counting device, which should therefore be taken as the reference detector (Tapiovaara and Wagner 1985; Zhao et al. 1997). When calculating the sensitivity or the voltage-response of image receptors, we use, however, the energy absorption coefficient $\mu_{en}(E)$ in Eq. II.6.3 (Asai et al. 1998; Stierstorfer and Spahn 1999), because the detector signal is proportional to the energy actually absorbed in the phosphor materials used.

After absorption of the impinging photons in the detector material, the image information is transferred by fast photoelectrons through excitation and ionisation to other information carriers. In the intensifying screens, these are light quanta (Dick and Motz 1981), which are generated by luminescence; in some direct digital image receptors (e.g. amorphous selenium), these are electric charges (electron-hole pairs). The better the absorption of the X-ray photons in the detector is, the more is made use of the image information given by their spatial intensity distribution and extracted to the following imaging chain. The more light quanta (in the intensifying screens) or the more electric charges (in the direct digital detectors) are generated, the more sensitive is the whole imaging system. The detective quantum efficiency DQE (see Chap. II.7.5) has been introduced as the physical quantity to describe the efficiency of this signal transformation.

The DQE(ν) characterises the overall signal and noise performance of imaging detectors dependent on the spatial frequency ν. The DQE is of great importance for the dose which is needed for a radiographic image with good image quality.

In this chapter a short review of the most important detector materials (CsI, Gd_2O_2S, Se, $CaWO_4$) is given and the characteristics which are responsible for their different energy response to X-radiation are discussed. The knowledge of the energy dependence of the image receptor's sensitivity is needed for the production of an optimum image and for an accurate determination of the patient dose (see Chap. III.1).

II.6.1   Characteristics of Phosphor Screens

In general diagnostic radiology, dual-emulsion films and dual-screen cassettes are employed together as *film-screen combinations*. The dual-emulsion film is sandwiched between two X-ray phosphor screens, the so-called intensifying screens. This position of the film reduces the distance between the phosphor screens that light must diffuse through and consequently improves spatial resolution. One exception exists: In mammography film-screen systems, only one intensifying screen (as back screen) is almost exclusively used in combination with a single-emulsion film to reduce the screen blur, albeit at the cost of sensitivity.

In most modern film-screen systems, Gd_2O_2S is used as phosphor. These so-called rare-earth screens have largely removed $CaWO_4$ screens. One important reason for this can be evaluated from Fig. II.6.1, which shows the mass attenuation coefficients of the most common phosphor materials (see also Fig. II.6.5).

The important features of these curves are the positions of the K-absorption edges. The K-edge for $CaWO_4$ is that of tungsten at an energy of 69.5 keV, whereas the K-edge for Gd_2O_2S is that of gadolinium at an energy of 50.2 keV. The two phosphors are roughly equivalent up to 50 keV, but is seen that in the energy range above 50 keV the rare-earth screen shows higher energy absorption than the conventional $CaWO_4$ screen. It is widely assumed that it is an advantage, if the majority of the X-ray photons impinging on the screen have an energy above the K-

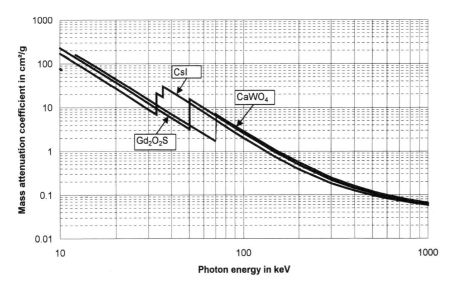

Fig. II.6.1. Mass attenuation coefficients of most common phosphor materials dependent on photon energy

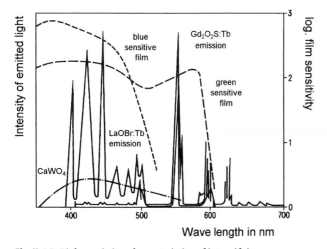

Fig. II.6.2. Light-emitting characteristics of intensifying screens

edge to improve quantum detection efficiency, but re-emission of characteristic radiation by X-ray fluorescence and re-absorption by the scintillator possibly can reduce the performance of the detector by a loss of spatial resolution and an increase in image noise (Boone et al. 1999).

An additional advantage of modern screens is their light-emitting characteristic (see Fig. II.6.2). Depending on the type of phosphor used, intensifying screens emit light of various colours (wavelengths). The rare-earth screen emits predominantly green light (at a wavelength of 544 nm) and the calcium tungstate screen a greater portion of blue light. The *intensification* of the intensifying screens in total is determined by the degree of *absorption* of the X-rays and the degree of *transformation* of the X-ray energy absorbed into visible light (conversion efficiency). The *conversion efficiency* can be expressed in terms of the energy W necessary to release a light photon

in a scintillator (e.g. in Gd_2O_2S, W = 13 eV; and in CsI(Tl), W = 19 eV; see Yaffe and Rowlands 1997). For both processes, i.e. absorption and transformation, the rare-earth screens show more favourable values in comparison with the conventional $CaWO_4$ screens, so that approximately the same spatial resolution can be reached at nearly a third of the patient dose.

The sensitivity of photographic film for X-rays is considerably less than that for visible light. The exposure of the film therefore results mostly from the visible light emitted by the screens and created by the absorption of the imaging X-rays. The sensitive layer of the film is a photographic emulsion, which consists of gelatine in which a large number (10^9–10^{12}/cm²) of microscopically small silver halide crystals are embedded. Depending on the screen type applied, films which are sensitised with regard to green light are employed. During exposure a latent image consisting of silver atoms is formed, which can subsequently be developed. The number of silver atoms present is proportional to the product of the light intensity created in the screen and the exposure time. This behaviour is known as the *law of reciprocity*. At high and low light intensities, often a deviation from linearity is observed, known as *reciprocity law failure* (Wagner et al. 1983; de Almeida et al. 1999).

In conventional film-screen technique, the film serves simultaneously as receptor, storage and display medium for the radiographic image. In addition only a bright viewing box (2000–3000 cd/m²), with masking possibility, and an adequately low ambient light level (below 50 lx) is required for the appropriate display of radiographs. The characteristic curve of the processed film describes the relation between log(exposure) and the resultant optical density (OD) in the image; the gradient curve (see Fig. II.6.3) determines the film contrast, which can be achieved dependent on the optical density. Therefore im-

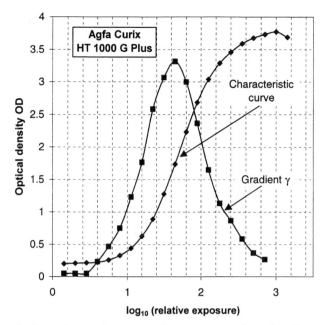

Fig. II.6.3. Characteristic curve and gradient curve of a modern film-screen system: Agfa Curix HT 1000 G Plus

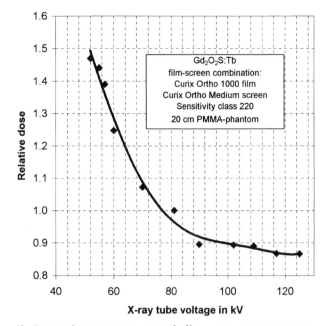

Fig. II.6.4. Voltage-response curve of a film-screen system: Image receptor dose necessary for constant optical density dependent on X-ray tube voltage for a constant object thickness of 20 cm PMMA

portant film properties such as optimum film contrast and maximum density can be derived from the characteristic curve ("Hurter and Driffield curve" or "H&D curve"; James 1977) and the gradient curve (see Fig. II.6.3). The gradient γ is in this context defined by (Barrett and Swindell 1981, p. 209):

$$\gamma = \frac{\Delta(\mathrm{OD})}{\Delta(\log \frac{\Psi}{\Psi_0})} \qquad (\mathrm{II.6.2})$$

where (Ψ/Ψ_0) is the relative exposure, in this case defined as the quotient of photon energy fluence (see Eq. II.3.3), and $\Delta(\mathrm{OD})$ the relevant difference in optical density.

Figure II.6.3 shows as an example the characteristics of a modern screen-film combination. The image receptor dose in the plane of the film cassette necessary to obtain the constant mean optical density (*system dose*) as desired (see Chap. II.6.4) is not a constant (see Fig. II.6.4), but depends on the radiation quality (i.e. the energy distribution of the photons in front of the image receptor).

This behaviour of film-screen systems is sometimes called their voltage response. The responsible dependence of their sensitivity or the relative sensitivity of different systems (Vyborny 1979; Asai et al. 1998) for mono-energetic photons of energy E can be evaluated by calculation of the absorbed energy A(E) in the screens (see Fig. II.6.5):

$$A(E) = E\left\{1 - e^{-\frac{\mu_{en}(E)}{\rho}\rho \cdot d}\right\} \qquad (\mathrm{II.6.3})$$

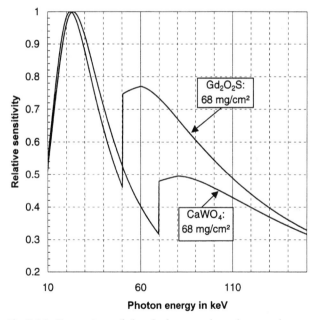

Fig. II.6.5. Comparison of absorbed energy dependent on photon energy of two most common phosphor materials: Gd_2O_2S and $CaWO_4$, both with a mass density of 68 mg/cm^2

where E is the energy of the impinging X-ray photon; μ_{en}/ρ the mass energy absorption coefficient; and ρd the mass density of the phosphor material considered.

For the application to medical radiography, Eq. II.6.3 in each special case of examination (tube voltage, filtration, patient thickness) must be averaged over the relevant whole X-ray spectrum impinging on the image receptor:

$$A(0, E_{max}) = \frac{\displaystyle\int_0^{E_{max}} \Phi_E \cdot E\{1 - e^{-\frac{\mu_{en}(E)}{\rho} \cdot \rho \cdot d}\} \cdot dE}{\displaystyle\int_0^{E_{max}} \Phi_E \cdot dE}$$

where E_{max} (see Eq. II.1.4) is the maximum energy in the X-ray spectrum and Φ_E the fluence of photons in the energy interval from E to E+ΔE.

From Eqs. II.6.3 and II.6.4 can be derived that the course of the screen sensitivity with respect to the X-ray tube voltage is a function of the energy dependence of the mass energy absorption coefficient $\mu_{en}(E)/\rho$ of the phosphor material, the position of the K-edge, the mass density ρd of the screen and also of the energy distribution Φ_E of the impinging photons, i.e. the system dose is dependent also on an additional filtration of the primary beam and on the thickness of the patient.

II.6.2   Digital Image Receptors

In digital radiography detector systems (see Fig. II.6.6), based on a combination of phosphor material with hydrogenated *amorphous silicon* (a-Si:H) in thin-film transistor technology (TFT) or with charged coupled devices (CCD), only a single screen can be used in front of the detector (Yaffe and Rowlands 1997).

With such systems the X-ray photons are incident upon the detector from the front, while the photosensitive a-Si:H layer collects the light behind the screen. The light emitted by the screen is produced preferentially near its front side and must diffuse through the scintillator, possibly resulting in an undesirable unsharpness of the image. It is therefore of advantage that the scintillator CsI(Tl) – most applied together with image intensifiers and digital

image receptors – has a small, column-like crystal structure (Morneburg 1995) that avoids lateral diffusion of the emitted light. For this reason even thick scintillator layers (e.g. of mass density 180 mg/cm^2) can be realised, resulting in a high quantum detection efficiency (QDE) and still a high spatial resolution. The energy dependence of the QDE of an a-Si:H detector is caused by the CsI(Tl) scintillator. Figure II.6.7 shows the voltage response of a 180-mg/cm^2-thick CsI-layer for an additional filtration of 0.1 mm Cu, 0.2 mm Cu and 0.3 mm Cu. The curves have been calculated using Eq. II.6.4.

In the image receptors based on *amorphous selenium* (a-Se) technology (Rowlands et al. 1992; Pang et al. 1998), the absorbed X-ray energy is directly converted to an electrical signal (direct digital detectors). Amorphous selenium can be produced on glass of large surface area by a thermal evaporation process (layer thickness 200–500 µm) and deposited together with a bias electrode on an array of TFT. For the signal generation, the detour around visible light of a scintillator is avoided. This behaviour offers the potential for increased spatial resolution (Yaffe and Rowlands 1997; Pang et al. 1998), which is of interest especially in mammography. Furthermore, in the energy range used in mammography, the QDE is of comparable magnitude with that of CsI(Tl) (see Fig. IV.3.11 and Fig. IV.3.13). The energy dependence of the image signal of direct digital detectors is caused by the X-ray-sensitive detector material, e.g. the amorphous selenium layer.

The effect of the voltage response (see Chap. II.6.1) of image receptors on the resulting image is in practice underexposure at low and high X-ray tube voltages. In combination with film-screen systems, this means that the optical density is too low at low and high X-ray tube voltages and results in the *cut-off dose* of the automatic exposure control (AEC; see Chap. II.6.3) being fixed and adjusted, e.g. at 80 kV, only to the system dose (see Chap. II.6.5) of the image receptor. To obtain a constant optical density, the cut-off dose of the AEC must be adapted to the voltage response by a suitable tube voltage-dependent correction factor for the cut-off dose (see Chap. II.6.3 and Fig. II.6.8).

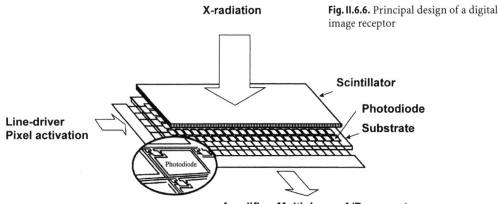

Fig. II.6.6. Principal design of a digital image receptor

Fig. II.6.7. Voltage-response of image intensifier: Relative dose rate for constant brightness of the output screen dependent on X-ray tube voltage for various additional filtration 0 mm Cu, 0.1 mm Cu, 0.2 mm Cu and 0.3 mm Cu. Object thickness values (water) on the abscissa are valid for dose rate regulation via an "anti-isowatt curve" (see Fig. II.6.12)

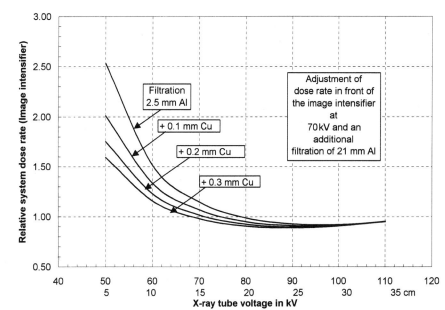

In digital imaging systems, which are provided with an auto-windowing function, underexposure causes increased quantum mottle, which is easily overlooked. In digital imaging the AEC system has therefore also a very important controlling function. Careful adaptation to the voltage-response curve should be made, because there is danger that noise is compensated for too fast by increasing patient exposure.

II.6.3   Automatic Exposure Control

II.6.3.1   Direct Technique

AEC systems of modern X-ray generators have the capability to take the voltage-response of image receptors (see Fig. II.6.4) into account. They relieve the radiographer of this task, e.g. when radiographs are made by using a film-screen system. With past X-ray systems, it is necessary to consider the voltage-response function by means of an exposure table while setting the radiographic parameters (see Chap. II.4). Figure II.6.8 shows as an example a family of correction curves, such as are stored in the memory of an AEC system today. Dependent on the X-ray tube voltage selected, the cut-off dose of the AEC is changed according to these curves.

When the X-ray system is put into operation by the manufacturer, the appropriate correction curve is adjusted. The correction curve selected simultaneously takes into account the *voltage response of the film-screen system*, that will be used, *the processing conditions* of the X-ray film and *the voltage response of the AEC-detector*, which is installed in the examination unit (see Fig. II.6.9).

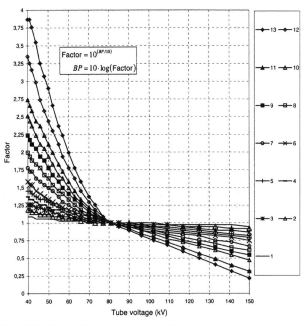

Fig. II.6.8. Correction curves in the AEC

Especially in *mammography*, modern AEC systems consider also the influence of the breast thickness and density on the exposure (Aichinger et al. 1990). This is necessary because the location of the detector behind the film-screen system results in a very pronounced dependence of the exposure on breast thickness.

In *general radiology*, the AEC-detector in most cases is an ionisation chamber. The principal design was introduced in Fig. II.4.10. of Chap. II.4. The sensitivity of the chamber mainly results from the thickness and the atomic number of the metal evaporated onto the measuring

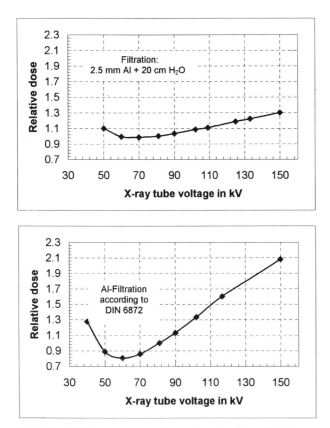

Fig. II.6.9a, b. Voltage-response of ionisation chamber: **a** Relative reciprocal sensitivity normalised at 80 kV; adequate to clinical situation. **b** Relative reciprocal sensitivity normalised at 80 kV; not adequate to clinical situation(DIN 1983)

electrode and the distance of both chamber electrodes (Eberhardt and Jaeger 1954). The number N_{pe} of the photoelectrons released from the electrode during exposure by photons of energy E is proportional to:

$$N_{pe} \propto N_0(E) \cdot \frac{\mu_{en}(E)}{\mu_{en}(E) - \mu(E)} (e^{-\mu(E) \cdot d} - e^{-\mu_{en}(E) \cdot d}) \qquad (II.6.5)$$

whereby: N_{pe} is the number of photoelectrons released; $N_0(E)$ the number of incident X-ray photons of energy E; μ_{en} the corresponding linear energy absorption coefficient; μ the corresponding total linear attenuation coefficient; and d the thickness of the evaporated metal layer. For very thin layers, i.e. if $\mu d \ll 1$ and $\mu_{en}d \ll 1$, Eq. II.6.5 can be approximated by:

$$N_{pe} \propto N_0(E) \cdot \mu_{en}(E) \cdot d \qquad (II.6.6)$$

From Eq. II.6.6 follows that the yield N_{pe} of photoelectrons emitted from the measuring electrode increases proportional to the evaporated metal layer thickness d and its linear energy absorption coefficient μ_{en}. N_{pe} depends therefore especially on the atomic number of the

evaporated metal. The yield of photoelectrons – dependent on the photon-energy – reaches a maximum because of the simultaneously increasing attenuation of the X-radiation by the layer itself (factor $e^{-\mu \cdot d}$). From Eq. II.6.5 an optimal thickness can be calculated as a function of X-ray energy E:

$$d(E)_{opt} = \frac{\ln \dfrac{\mu_{en}(E)}{\mu(E)}}{\mu_{en}(E) - \mu(E)} \qquad (II.6.7)$$

but more decisive to the sensitivity of an ionisation chamber is the range of the released photoelectrons in the metal layer of the measuring electrode (Eberhardt and Jaeger 1954; Aichinger 1968). The number of secondary electrons that can leave the deeper layers of the evaporated metal electrode decreases with increasing layer thickness d. The maximum sensitivity of the ionisation chamber is obtained at photon energies, where the range of the photoelectrons is equal to the metal layer thickness d.

As an example Fig. II.6.9a shows the voltage response (relative reciprocal sensitivity) of an ionisation chamber, which contains a measuring electrode evaporated with 2–3 μm lead and which has an electrode distance of about 6 mm (see Fig. II.4.10), measured in combination with a 20-cm water-phantom. The maximum sensitivity lies in the tube voltage range from 60 to 80 kV. In Fig. II.6.9b the voltage response of the same ionisation chamber is presented, whereby a beam filtration according to the German standard DIN 6872 (DIN 1983) was employed. The effect of the different filtration on the voltage-response curve of the chamber is impressively demonstrated.

II.6.3.2   Fluoroscopy and Indirect Technique

In the indirect technique, the CsI(Tl) input screen of the image intensifier serves as the image receptor. The indirect technique includes fluoroscopy, rarely spot-film operation any more, with the medium-format camera, and, above all, digital radiography, including especially angiography. Here again, the system dose rate or dose required for constant image quality depends on the imaging-beam quality.

Figure II.6.10 shows the relative dose rate required at the image intensifier input with fluoroscopy for constant image brightness as a function of tube voltage and additional filtering. The curve is similar to the film-screen system voltage-response functions, but there are very important differences with respect to the functioning of the AEC system (see Fig. II.6.11):

- The AEC-detector controls the exposure directly via the brightness of the output screen of the image intensifier by regulation of the exposure parameters (X-ray tube voltage, tube current and exposure time). There-

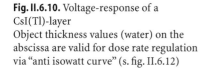

Fig. II.6.10. Voltage-response of a CsI(Tl)-layer
Object thickness values (water) on the abscissa are valid for dose rate regulation via "anti isowatt curve" (s. fig. II.6.12)

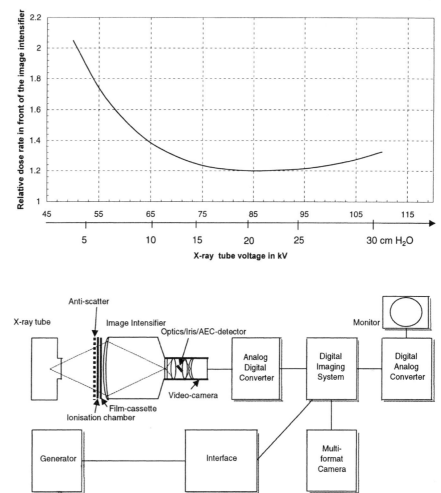

Fig. II.6.11. Principal design of a digital radioscopic system

fore no correction curves are necessary for the compensation of the energy dependence of the input screen
- Image reception, image storage (in digital systems) and display are independent from each other. The dose at the input of the imaging chain can be separately adjusted. The brightness of the image on the display (television monitor) can be set by variation of the aperture (iris diaphragm) located between the output screen of the image intensifier and the video camera or by signal amplification in the video chain. With a large-iris diaphragm aperture, a low dose rate at the image intensifier is sufficient to optimally drive the television camera, but results in a high image noise level. By contrast, with a small-iris diaphragm aperture, a low image noise level results, accompanied by a higher dose at the image intensifier input.
- The signal-to-noise ratio is the decisive quantity for image quality. This fact opens up the possibility to adjust patient exposure to the diagnostic requirements and to apply computer techniques to image processing.

- When applying antiscatter grids in digital imaging, dose-saving is possible – in contrast to the film-screen technique – by further opening of the aperture between image intensifier and television camera. Unacceptably high quantum noise is the limit for this procedure.

II.6.4 Image Receptor Dose/Dose Rate (System Dose/Dose Rate)

The purpose of the AEC is to achieve radiological images with constant mean optical density or brightness independent of the amount of body fat and muscularity of the patients and the selected exposure parameters. The radiologist must be provided with images that have *the best possible diagnostic information* obtainable when the appropriate radiographic technique is employed. The image receptor dose by which this goal is achieved is the system dose (see Chap. II.3.2). In the fluoroscopic mode, the image receptor dose rate must be regulated and is accordingly called the system dose rate.

Table II.6.1 Radiation qualities for the determination of the speed of film-screen combination (ISO 2001)

Equivalent to exposure technique	Approximate x-ray tube voltage (kV)	Half-value layer (mm Al)	Exposure time (ms)	Phantom[a] (mm)	Distance phantom detector (mm)
Extremities	50	3.0	60 ± 30	25 mm ± 1mm (PMMA) 2.0 mm ± 0.1 mm Al 25 mm ± 1mm (PMMA)	60
Skull	70	5.7	60 ± 30	12.0 mm ± 0.1 mm AL	60
Lumbar spine and Colon	90	7.4	60 ± 30	13.0 mm ± 0.1 mm AL	60
Chest	120	8.5	15 ± 5	70 mm ± 0.1 mm (PMMA) 5.0 mm ± 0.1 mm AL	60

[a]Area of the phantoms should be approximately 0.3 m^2
The naming sequence of the phantom components corresponds to the direction from the focal spot to the image receptor

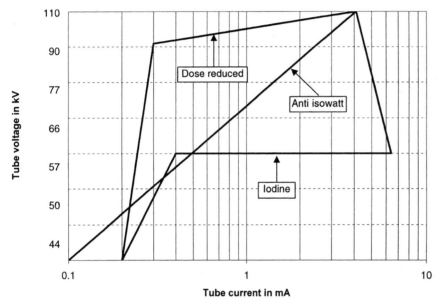

Fig. II.6.12. Regulation of X-ray tube voltage and tube current in modern fluoroscopy systems

Because the sensitivity of each image receptor is dependent on the energy distribution of the impinging X-ray photons (see Chaps. II.6.1 and II.6.2), the system dose/dose rate in general is determined at a specified radiation quality (IEC 2001). Only in this way can different image receptors be compared objectively with respect to their characteristics.

The *sensitivity* of film-screen systems (called "speed" in ISO 2001), is defined as:

$$S = \frac{K_0}{K_S} \tag{II.6.8}$$

where K_0 is equal to 10^{-3} Gy and K_S is the air kerma incident on the film-screen combination behind a specified phantom to produce a *net density of 1.0 above base and fog*. Four exposure conditions are defined in the ISO-document 9236 (ISO 2001; see Table II.6.1).

In addition, in the corresponding German DIN standard (DIN 1995), certain sensitivity value ranges are grouped in so-called sensitivity classes, SC (6, 12, 25, 50, 100, 200, 400, 800, 1600). Systems belonging to the classes SC 12 and 25 are used in mammography, systems belonging to SC 200–800 are used in general radiography. As an example, for a system with sensitivity S = 400, the image receptor dose for a net OD = 1 is about 2.5 µGy. Systems with sensitivity S = 400 are recommended for most of the common X-ray exposures; in paediatrics often combinations with S = 800 (i.e. 1.25 µGy) are preferred (CEC 1996). When using high-speed systems in paediatrics, one should bear in mind that these systems at X-ray tube voltages below 60 kV – because of their voltage-response curves – often need a much higher dose than that which results from Eq. II.6.8.

In the *film-screen technique*, the patient exposure is determined to a high degree by the type of the film-screen combination (high-resolution SC = 100, regular SC = 200,

Table II.6.2 System dose rates for continous fluoroscopy modes (30 frames/s $\triangleq$ 30 P/s) in dependence on image intensifier field size

II-format Ø	system dose rate
25 cm	0.200 µGy/s
27 cm	0.174 µGy/s
40 cm	0.087 µGy/s

Table II.6.3 System dose rates for pulsed fluoroscopy (system dose/pulse = constant) in dependence on image-intensifier field size

II-format Ø	system-dose rate 30 P/s	dose/pulse	mean dose rate at		
			15.0 P/s	7.5 P/s	3 P/s
25 cm	0.200 µGy/s	6.7 nGy	0.100 µGy/s	0.050 µGy/s	0.020 µGy/s
27 cm	0.174 µGy/s	5.8 nGy	0.087 µGy/s	0.044 µGy/s	0.017 µGy/s
40 cm	0.087 µGy/s	2.9 nGy	0.044 µGy/s	0.022 µGy/s	0.009 µGy/s

Remark: With the grey accentuated dose/pulse levels the mean dose rate values at 15.0 P/s, 7.5 P/s and 3 P/s result

Table II.6.4 System dose rates for pulsed fluoroscopy (mean dose rate = constant) in dependence on image-intensifier field size

II-format Ø	system dose rate	dose/pulse 15.0 P/s	dose/pulse 7.5 P/s	dose/pulse 3.0 P/s	mean dose rate
25 cm	0.200 µGy/s	6.7 nGy	13.3 nGy	33.3 nGy	0.100 µGy/s
27 cm	0.174 µGy/s	5.8 nGy	11.6 nGy	29.0 nGy	0.087 µGy/s
40 cm	0.087 µGy/s	2.9 nGy	5.8 nGy	14.5 nGy	0.044 µGy/s

Remark: With the grey accentuated mean dose rate levels the dose/pulse values at 15.0 P/s, 7.5 P/s and 3 P/s result

Table II.6.5 System dose rate values for continous fluoroscopy and corresponding system dose values for digital indirect technique in dependence on the image intensifier field size

II-Image Size (Diameter)	System dose rate Continuous Fluoroscopy	Dose level			
		50	100	200	500
		Dose / Image in Indirect-technique			
40 cm	0.09 µGy/s	0.22 µGy	0.44 µGy	0.88 µGy	2.20 µGy
28 cm	0.16 µGy/s	0.40 µGy	0.80 µGy	1.60 µGy	4.40 µGy
20 cm	0.31 µGy/s	0.80 µGy	1.60 µGy	3.20 µGy	8.00 µGy
14 cm	0.43 µGy/s	1.10 µGy	2.20 µGy	4.40 µGy	11.00 µGy

Remark: The grey accentuated dose levels are allowed to be employed in digital indirect technique only in special medical indications and in DSA
Nominal values at 25 cm Δ II-format: 0.2 µGy/s (fluoroscopy), 1.0 µGy   (indirect exposure)
In comparison: Film-screen system of sensitivity class 400 corresponding to Ks = 2.5 µGy

high-speed SC = 400, SC = 800 very high speed), which has to be selected by the radiologist according to national or international recommendations. In *digital imaging* the signal-to-noise ratio is decisive for image quality, therefore the system dose has to be selected in such a way that noise does not impede diagnosis. Within the limits given by law (FDA 1984, 1997; BMA 1996), the radiologist mainly bears the responsibility for the radiation exposure of the patient, especially the time of a fluoroscopic or interventional examination is of great influence on the resulting patient dose. Available dose-reducing techniques include the use of pulsed fluoroscopy, digital image processing, increasing the optical iris of the video camera, removal of the antiscatter grid, and fluoroscopic tube current and high-voltage control techniques(see Fig. II.6.12) in combination with filter techniques.

Tables II.6.2, II.6.3 and II.6.4 show system dose rates for various fluoroscopy modes and Table II.6.5, system dose rate values for continuous fluoroscopy and system dose values for indirect techniques. In image-intensifier-television systems (II-TV-systems) with a constant-iris diaphragm, the system dose/dose rate changes quadratically with the image field size, to obtain images with constant brightness and noise. This behaviour must be considered during fluoroscopy. The most effective dose-saving possibility is pulsed fluoroscopy with e.g. image frequencies 15 s^{-1}, 7.5 s^{-1}, 3 s^{-1} in connection with additional filtration (see Fig. II.6.13).

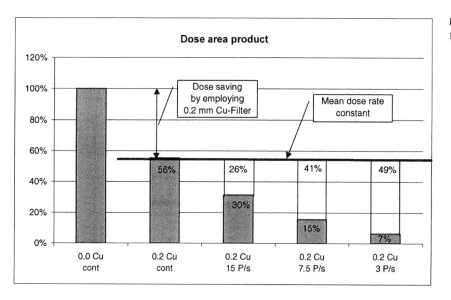

Fig. II.6.13. Dose saving by pulsed fluoroscopy

Two different modes can be applied: pulsed fluoroscopy with a constant mean dose rate, or a constant dose/pulse. In conventional fluoroscopy, the continuous X-ray exposure blurs moving organs; while, in pulsed fluoroscopy, images can be acquired without motion blur. In the last case a sort of stroboscopic effect occurs, which mostly can be tolerated without influencing diagnosis (Xue and Wilson 1998).

References

Aichinger H (1968) Fortschritte in der Technik der Belichtungsautomatik in der Röntgendiagnostik. Radiologe 8(8):233–238

Aichinger H, Joite-Barfuss S, Marhoff P (1990) Automatic exposure control system in mammography. Electromedica 58:61–69

Almeida A de, Sobol WT, Barnes GT (1999) Characterization of the reciprocity law failure in three mammography screen-film systems. Med Phys 26(5):682–688

Asai Y, Tanabe Y, Kubota H, Matsumoto M, Kanamori H (1998) Optimum tube voltage for chest radiographs obtained by psychophysical analysis. Med Phys 25(11):2170–2175

Barrett HH, Swindell W (1981) Radiological imaging. Academic Press, New York

BMA (1996) Richtlinie nach § 16 der Röntgenverordnung zur Durchführung von Prüfungen zur Qualitätssicherung in der Röntgendiagnostik und die Hinweise zur Abnahmeprüfung, 18. Bek. des BMA vom 13.11.1995 (BArBl. 1/96)

Boone JM, Seibert JA, Sabol JM, Tecotzky (1999) A Monte Carlo study of X-ray fluorescence in X-ray detectors. Med Phys 26(6):905–916

CEC (Commission of the European Communities) (1996) European guidelines on quality criteria for diagnostic radiographic images. EUR 16260. CEC, Luxemburg

CEC (Commission of the European Communities) (1996) European Guidelines on Quality Criteria for Diagnostic Radiographic Images in Paediatrics. EUR 16261 (Luxemburg CEC)

Dick CE, Motz JW (1981) Image information transfer properties of X-ray fluorescent screens. Med Phys 8(3):337–346

DIN (Deutsches Institut für Normung) (1983) Strahlenqualitäten für Messungen in der radiologischen Technik: Patientendurchlassstrahlung ohne Streustrahlenanteil in der Röntgendiagnostik. DIN 6872, Part 1. Beuth, Berlin

DIN (Deutsches Institut für Normung) (1995) Sensitometrie an Film-Folien-Systemen für die medizinische Radiographie. Teil 10. Nennwerte der Empfindlichkeit und des mittleren Gradienten. DIN 6867-10. Beuth, Berlin

Eberhardt W, Jaeger R (1954) Eine Folienkammer für empfindliche Röntgenstrahlmessungen. Strahlentherapie 95:641–656

Department of Health and Human Services, Food and Drug Administration (1984) Diagnostic X-ray systems and their major components; amendments to performance standards; final rule, 21 CFR Part 1020. Federal Register 49:171

Food and Drug Administration (1997) Performance standard for diagnostic X-ray systems and their major components, 21 CFR (4-1-97) § 1020.30–1020.33

IEC (International Electrotechnical Commission) (2001) Medical diagnostic equipment – radiation conditions for use in the determination of characteristics. Publication 61267 (Draft). IEC, Geneva

ISO (International Standards Organisation) (2001) Photography-sensitometry of screen/film systems for medical radiography. Part 1. Determination of a sensitometric curve shape, speed and average gradient. Publication 9236-1 (Draft)

James TH (1977) The theory of the photographic process. Macmillan, New York

Morneburg H (1995) Bildgebende Systeme für die medizinische Diagnostik. Publicis, Erlangen

Pang G, Zhao W, Rowlands JA (1998) Digital radiology using active matrix readout of amorphous selenium: geometrical and effective fill factors. Med Phys 25(9):1636–1646

Rowlands JA, DeCrescenzo G, Araj N (1992) X-ray imaging using amorphous selenium: determination of X-ray sensitivity by pulse height spectroscopy. Med Phys 19(4):1065–1069

Stierstorfer K, Spahn M (1999) Self-normalizing method to measure the detective quantum efficiency of a wide range of X-ray detectors. Med Phys 26(7):1312–1319

Tapiovaara M, Wagner RF (1985) SNR and DQE analysis of broad spectrum X-ray imaging. Phys Med Biol 30:519–529

Yaffe MJ, Rowlands JA (1997) Review: X-ray detectors for digital radiography. Phys Med Biol 42:1–39

Vyborny CJ (1979) H and D curves of screen-film systems: factors affecting their dependence on X-ray energy. Med Phys 6(1):39–44

Wagner LK, Barnes GT, Bencomo JA, Haus AG (1983) An examination of errors in characteristic curve measurements of radiographic screen/film systems. Med Phys 10(3):365–369

Xue P, Wilson DL (1998) Effects of motion blurring in X-ray fluoroscopy. Med Phys 25(5):587–599

Zhao W, Blevis I, Germann S, Rowlands JA (1997) Digital radiology using active matrix readout of amorphous selenium: construction and evaluation of a prototype real-time detector. Med Phys 24:1834–1843

II.7 Image Quality and Dose

The radiological image is composed of the spatial variation of some physical quantities, e.g. the X-ray fluence at the input of the imaging chain (*radiation image*), the optical density of the film (*radiograph*) or the grey-scale value on the monitor of the viewing station (*X-ray image*), and represents the spatial distribution of the patient tissue components within the field of view. Visualisation of important details requires separation of the "structures of interest" against the "background" (e.g. in mammography, micro-calcifications in the breast glandular tissue). Loosely speaking, the difference between structures of interest and background is referred to as the *signal*.

Figure II.7.1 shows as an example the case of a very simple signal against a uniform background: the radiograph of a disk-shaped object. The detail can be detected easily according to the difference in optical density between object and background. This has to do with *contrast* in the image. At the boundary of the disk, the optical density does not change step-wise, but the transition has a certain extent. This has to do with *sharpness* of the image. Furthermore there are grainy, *random fluctuations* present in the image which superimpose the image pattern. This has to do with *noise* in the image.

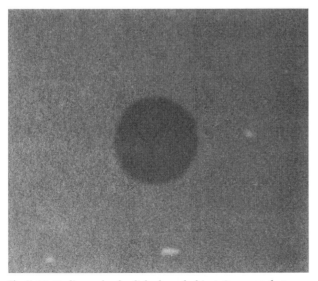

Fig. II.7.1. Radiograph of a disk-shaped object. Some artefacts can also be seen

Depending on its source, noise is also called mottle (quantum number) or granularity (film-screen system) by radiologists and technicians. In the radiograph or the X-ray image on the monitor, only noise resulting from the limited number of X-ray quanta absorbed by the image receptor (quantum noise) can be affected by the system dose (see Chap. II.7.4) applied for imaging. Noise is detrimental to X-ray images because it impairs especially the reliable detection of low-contrast structures. *Noise in a good radiograph should therefore always be quantum-limited.*

The quality of the various components of the imaging chain (focal spot, imaging geometry, image receptor, video camera and amplifier, image-processing software, image display) has also influence on the image signal obtained at the viewing station. These given facts must be taken into account when considering optimisation of image quality and exposure (see Chaps. II.7.6 and II.7.7). In every case image quality and radiation exposure cannot be investigated independently of one another. Figure II.7.2 demonstrates the different components and technical parameters of an imaging system for general radiography which influence absorbed dose and image quality.

The physical characteristics of the X-ray tube assembly and the radiographic device shown must be adapted to the *medical requirements* in such a way that the image receptor generally can produce an optimum X-ray image with respect to the representation of the object details which are needed for diagnosis at the lowest possible level of exposure. The medical requirements on imaging systems in radiology can be derived, e.g. from national and international recommendations and guidelines such as the *European Guidelines on Quality Criteria for Diagnostic Radiographic Images* (CEC 1996a) and the *European Guidelines on Quality Criteria for Diagnostic Radiographic Images in Paediatrics* (CEC 1996b). Since the whole-image information is already included in the radiation image, which exists in front of the imaging chain, it is practical to optimise that image first with respect to the medical indication – assuming theoretically an ideal image receptor – and only then take into consideration the real image receptor used.

The image quality obtained with an imaging system can be described and quantified by the characteristics il-

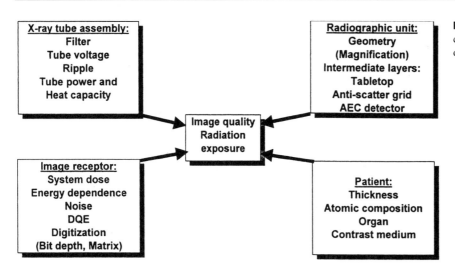

Fig. II.7.2. Characteristics of the imaging chain, which affect image quality and dose

lustrated by Fig. II.7.1: *contrast*, *sharpness* and *noise* (see Chaps. II.7.1–II.7.3). Especially in digital imaging the derived quantity *signal-to-noise ratio* (see Chap. II.7.4) has been used as a comprehensive image quality parameter . Finally the *detective quantum efficiency DQE* will be introduced (see Chap. II.7.5), a quantity which is accepted today as the fundamentally significant performance parameter that characterises an image-detection system. Its usefulness lies in the fact that it can be interpreted as the efficiency of a system to transfer the information that it receives. Some other image-quality figures are discussed in Chap. II.7.6, followed by a short discussion of the dependence of image quality on exposure parameters in Chap. II.7.7. The different image-quality characteristics will be used in Part III for calculations in relation to clinical applications

II.7.1   Contrast

Contrast is generated by the different attenuation of X-radiation by the tissue (e.g. adipose and glandular tissue in mammography). Furthermore the contrast in the image is affected by the *X-ray spectrum*, which in its turn is determined by the tube voltage applied, the anode material and the filtration. As already mentioned above, the contrast of the radiation image directly behind the patient contains already all the information about tissue composition obtainable with the selected X-ray spectrum. With the help of the image receptor, the *radiation contrast* is transformed into differences in optical density in the radiograph (*image contrast*) or differences in brightness on the monitor (image contrast). The capability to convert subtle density differences in the patient's tissue into image information is called *contrast resolution*. The image contrast is influenced by the *gradient* (γ) of the film (see Chap. II.6.1 and Fig. II.6.3) or by the *window level* and

width setting of the digital image. The higher the contrast the stronger the bright and dark image areas will emerge. Sometimes this effect creates the impression of a sharper image (see Chap. II.7.2).

The possibility to discriminate between the varying internal structures of the human body (lungs, tissue, tissue fluids, bone) in the radiographic image is caused by their different linear attenuation coefficients for X-radiation (see ICRU 1989; and Fig. II.7.3).

In the literature (Biedermann et al. 1967; Herz 1969; Koedooder 1986; Christensen 1990; Yaffe 1994), a number of different definitions for the resulting radiation contrast is given (see Fig. II.7.4).

The most often used definitions are (see also Fig. II.7.5):

$$C_1 = \frac{I_1}{I_2} = e^{(\mu_1 - \mu_2) \cdot d} \approx 1 + \Delta\mu \cdot d \qquad (II.7.1)$$

$$C_2 = \frac{(I_1 - I_2)}{I_1} = 1 - e^{(\mu_1 - \mu_2) \cdot d} \approx \Delta\mu \cdot d \qquad (II.7.2)$$

$$C_3 = \frac{(I_1 - I_2)}{(I_1 + I_2)} = \frac{(1 - e^{(\mu_1 - \mu_2) \cdot d})}{(1 + e^{(\mu_1 - \mu_2) \cdot d})} \approx \frac{1}{2} \cdot \Delta\mu \cdot d \qquad (II.7.3)$$

The equations make use of the exponential law of attenuation (see Chap. II.2.1). I_1 is the intensity of the radiation which penetrates the phantom as well as the detail to be imaged and I_2 the intensity which penetrates the phantom in the region of the detail of thickness d. The results on the right side of Eqs. II.7.1, II.7.2 and II.7.3 are valid for low-contrast details under the condition $\Delta\mu \cdot d \ll 1$. The

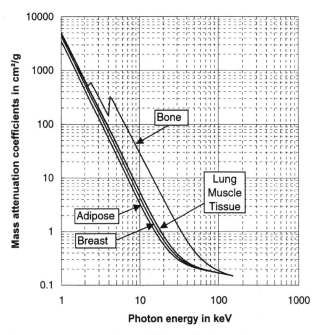

Fig. II.7.3. Mass attenuation coefficients of various tissues

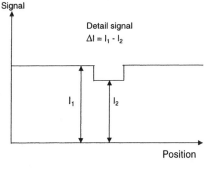

Fig. II.7.4. Contrast definitions (radiation contrast)

Name	Contrast definition	Data range ($I_2=I_1$ to $I_2=0$)
Contrast ratio	$C_1 = \dfrac{I_1}{I_2}$	1 - ∞
Contrast	$C_2 = \dfrac{I_1 - I_2}{I_1}$	0 - 1
Modulation	$C_3 = \dfrac{I_1 - I_2}{I_1 + I_2}$	0 - 1

Fig. II.7.5. Characteristics of image quality in analogue and digital imaging

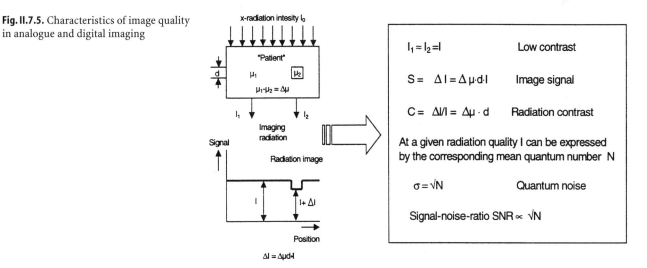

formulation of C_3 is also named *modulation* (Barrett and Swindell 1981, p. 41). This contrast definition is used very often instead of C_2, because its form is in harmony with the definition of the modulation transfer function. C_2 and C_3 differ from each other only by a factor of 2. The radiation contrast of a low-absorbing detail is proportional to the product of the detail thickness d times the difference $\Delta\mu$ of the linear attenuation coefficients of the detail μ_2 and its surrounding medium μ_1 (see Fig. II.7.5).

The derivation of Eqs. II.7.1, II.7.2 and II.7.3 is carried out by assuming mono-energetic radiation (see Chap. II.2.1). For low-contrast details, if $\mu_1 \cdot d \approx \mu_2 \cdot d \ll 1$ is valid,

the exponential attenuation law can be applied also to polychromatic X-radiation (bremsstrahlung). Instead of $\mu(E)/\rho$, the effective attenuation coefficient $(\mu/\rho)_{\text{eff}}$ must be used in the exponential attenuation equation:

$$I = I_0 \cdot e^{-\left(\frac{\mu}{\rho}\right)_{\text{eff}} \cdot d \cdot \rho} \qquad (II.7.4)$$

where $(\mu/\rho)_{\text{eff}}$ is equal to the mean value $<\mu/\rho>$ of the mass attenuation coefficient $\mu(E)/\rho$, weighted according to the spectral distribution of the X-ray energy spectrum under consideration. The contrast C_3 of a low-absorbing detail can therefore be written as:

$$C_3 = \frac{1}{2} \cdot (\mu_{1,\text{eff}} - \mu_{2,\text{eff}}) \cdot d = \frac{1}{2} \cdot \Delta\mu_{\text{eff}} \cdot d \qquad (\text{II.7.5})$$

Equations II.7.3 or II.7.5 for the radiation contrast do not imply any information about the system dose, i.e. the dose, which is needed for an optimal image (see Chap. II.6.5). When using a film-screen system as image receptor, the system dose is determined by the speed or sensitivity class of the system (DIN 1995; ISO 2001) itself and the image contrast ΔOD is equal to (see also Eqs. II.7.2; Barnes 1979)

$$\Delta\text{OD} = \gamma \cdot \log\frac{\Delta I}{I} = \frac{1}{\log_e 10} \cdot \gamma \cdot \Delta\mu_{\text{eff}} \cdot d = 0{,}43 \cdot \gamma \cdot \Delta\mu_{\text{eff}} \cdot d$$

$$(\text{II.7.6})$$

where ΔOD is the difference in the optical density arising from the difference in the radiation exposure to the two areas of the film-screen system under consideration; γ is the gradient (see Eq. II.6.2) of the film-screen system's sensitometric curve between the two different exposure levels I_1 and I_2 as defined in Figs. II.7.4 and II.7.5.

II.7.2   Sharpness

The capability of an imaging system to display separation between small, juxtaposed details and furthermore large details clearly defined against the background can be described by the sharpness of the image. What subjectively is found as sharpness can be quantitatively defined by the

concept of the *modulation transfer function* [MTF(ν); see Fig. II.7.6].

In the analysis of imaging systems, MTF(ν) is used to express the transfer of the image signal amplitude at each spatial frequency ν. The example in Fig. II.7.6 shows the MTF(ν) of a focal spot with a size according to 0.3 IEC (IEC 1993), whereby an ideal Gaussian radiation intensity distribution within the focal spot is assumed. The MTF(ν) indicates the contrast which can be transmitted at best by the imaging system to the viewing station

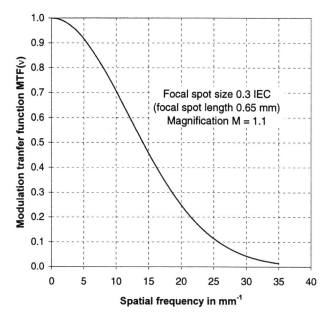

Fig. II.7.6. Modulation transfer function, MTF(ν), of a focal spot 0.3 IEC with a Gaussian-shaped distribution of the radiation intensity

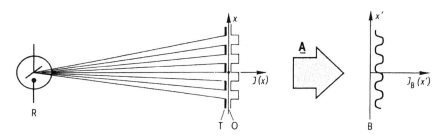

A = Image transfer, B = Image plane, J(x), J$_B$ (x) = Intensities, O = Object plane, R = X-ray tube, T = Test grid

Fig. II.7.7. Lead bar pattern; a suitable tool for the determination of the MTF

dependent on the spatial frequency ν. The better the contrast transfer at high spatial frequencies is, the smaller are the details that can be recognised separately in the image. In practice, resolution is often determined by the imaging of a lead bar pattern (see Fig. II.7.7). The highest number of line pairs per millimetre (lp/mm) that are displayed separated from each other is called the *visual resolution limit* of the whole imaging device. The reproduction of high spatial frequencies is also of great importance for a realistic and sharp visualisation of large details.

When measuring the MTF(ν) – from its original definition – the insertion of a pattern in the X-ray beam is demanded that *modulates the radiation intensity in a sinusoidal way* (Morneburg 1995). But in X-ray-technique it is very hard to realise this requirement. For the measurement of the MTF, one uses therefore most often the lead bar pattern, which is – as mentioned above – also used for the determination of the visual resolution limit. In the object plane, a rectangular radiation intensity distribution is therefore produced by this test pattern, which in the image plane is displayed more or less smoothed down according to the quality of the imaging system (see Fig. II.7.7). Coltman (1954) has given a *correction formula* which makes it possible to determine the "correct" MTF(ν) which would be obtained by a sinusoidal modulated radiation intensity. These connections are the reason that on the x-axis of MTF-diagrams the spatial frequency should be given per millimetre. Only if the "contrast-transfer" of a lead bar pattern is shown directly without correction in a diagram can line pairs per millimetre be used as unity.

An often-used measuring method which results directly in the correct MTF(ν) is the determination of the line spread function (LSF-function) by imaging a narrow jaw slit (see ICRU 1986). In numerous studies the difficulties involved with the alignment of the parallel jaw slit have been reported (Bradford et al. 1999) and therefore the method shall not discussed in detail in this book.

From the Fourier theory, it is known that each continual function can be described as the sum of sine functions of defined frequency and amplitude. Figure II.7.8 shows this by the approximation of a rectangular signal pulse by superposition of few sine functions.

The entirety of the sine functions, which are characteristic of this signal, is called its *spatial frequency spectrum*. During the imaging process, the spatial frequency spectrum A(ν) is filtered by the MTF_{ch} of the whole imaging chain. On this occasion the high frequencies of the signal pulse are often suppressed. The imaging of large and small details in principle is illustrated in Fig. II.7.9.

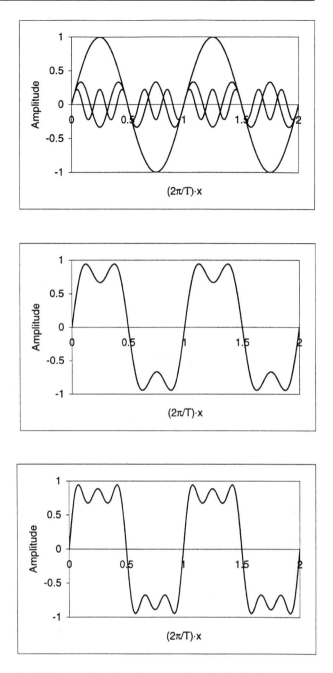

Fig. II.7.8. Approximation of a rectangular signal pulse by superposition of sine functions

The absence of the high frequencies in the images results, for the large detail, in blurred edges. In addition the image of the small detail shows a reduced amplitude. In summary one can say that the MTF of the whole imaging chain, which means from the focal spot to the viewing station, is not only responsible for the visual resolution limit of the system but also for a sharp image display.

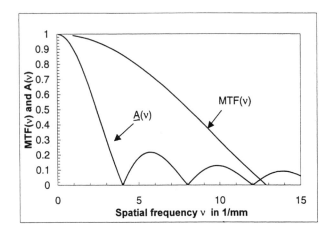

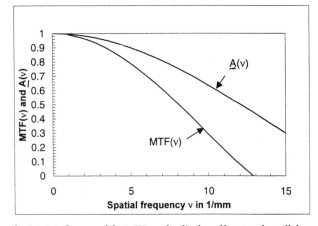

Fig. II.7.9. Influence of the MTF on the display of large and small details: In imaging of large details, edges are blurred (top). Imaging of small details results also in a reduced signal amplitude (bottom) A(ν) is the spatial frequency spectrum

II.7.3   Noise

Noise overlays the image information. It can be recognised, e.g. within homogeneous tissue areas by fluctuations in the optical density or brightness (see also Fig. II.7.1). An X-ray image is built up by the individual absorbed X-ray photons within the area of the radiological image. The contribution of each single X-ray photon is added up to the overall picture. The more X-ray photons per image area are absorbed, the lower are the fluctuations due to noise. The number of X-ray quanta absorbed by the image receptor determines the quantum noise.

The number of absorbed X-ray photons as well as the efficiency of the transformation of the absorbed energy in the image receptor to other information carriers (e.g. visible light or charge carriers, i.e. electrons or holes; see Fig. II.7.10 and Chap. II.6) contribute also to the noise, because these processes limit the number of light photons that build up the visible image on the viewing station.

In addition, the total noise is still affected and increased by the system noise, i.e. the statistical fluctuations within the imaging system, be that a film-screen system (graininess) or a digital system (electrical noise, quantisation noise).

The methods of Poisson statistics can be used to characterise and quantify noise (Barrett and Swindell 1981, p. 82). The noise normally will be defined by the variance σ^2 of the image signal, i.e the mean square deviation of the signal about the mean $\bar{N}$:

$$\sigma^2 = \bar{N} \tag{II.7.7}$$

where $\bar{N}$ is the mean number of the detected photons. The variance σ^2 or the standard deviation σ describes the magnitude of the noise fluctuations but says nothing about their spatial or temporal characteristics. Figure II.7.11 shows the *probability density function* (PDF) of a noisy uniform image (Yaffe and Nishikawa 1991).

The spatial or temporal characteristics of the noise fluctuations can be described by the *noise power spectrum* (NPS) or *Wiener spectrum* (W). According to ICRU 54 (ICRU 1996) it can be expressed (for simplicity in one dimension) as:

$$W(\nu) = \lim_{x \to \infty} \frac{1}{X} \left\langle \left| \int_{-\frac{x}{2}}^{\frac{x}{2}} dx \cdot e^{-2\pi i \nu x} \cdot n(x) \right|^2 \right\rangle \tag{II.7.8}$$

where $n(x)$ is the noise at the location x. For stationary noise, the Wiener spectrum is the Fourier transform of the noise autocorrelation function (ICRU 1996).

Since the photons emitted from the focal spot of the X-ray tube are uncorrelated, the input signal – assuming an ideal imaging geometry without any unsharpness – shows *white noise*, i.e. noise which is independent of the spatial frequency [n(x) is constant]. The noise in the output signal of an imaging system is, however, dependent on the transfer characteristics of the system, modulated in accordance with the input modulation, but additional noise originating from the multiple imaging stages is obtained. In the analysis of imaging systems, the MTF(ν) is used to express the transfer of the signal amplitude at each spatial frequency (see also Chap. II.7.2) but says nothing about the influence of noise on the detectability of important image details. In comparing signal *and* noise properties of imaging systems, it is therefore useful to analyse the noise versus spatial frequency as well. As a good first approximation, it can be assumed that the propagation of noise (power) behaves as MTF[2] (Lubberts 1968; Williams et al. 1999; see also Eq. II.7.26). Lubberts showed that this is not strictly correct and in fact the NPS

Fig. II.7.10. Information transfer within the image receptor. Example image intensifier: 1 photon number in front of the II-input, 2 absorbed quanta in the input screen, 3 created light quanta in the input screen, 4 number of photo electrons, 5 light quanta in the output screen (Morneburg 1995)

Fig. II.7.11. Noise: mean-square deviation of the signal about its mean value

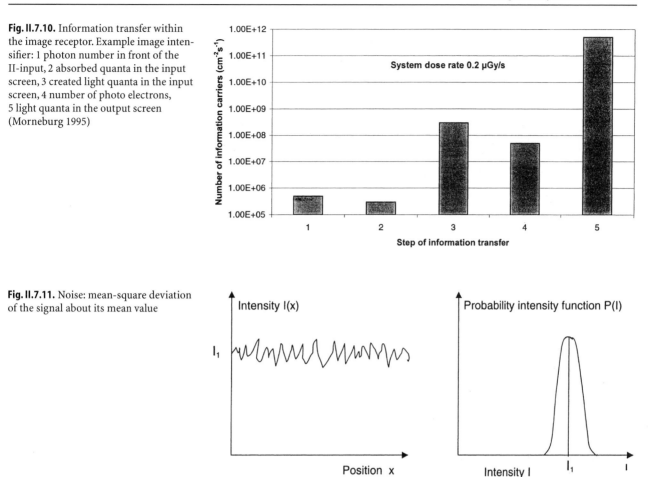

generally falls off more slowly than MTF^2 at high frequencies (Lubberts 1968).

The NPS or W(ν) [without the dc ($\nu = 0$) component; see Yaffe and Nishikawa 1991] describes the noise fluctuations in the image with respect to their spatial or temporal characteristics in terms of Fourier components, whereby:

$$\int W(\nu)d\nu = \sigma^2 \tag{II.7.9}$$

For independent noise sources, the standard deviation is calculated by addition of the individual σs in quadrature, i.e.:

$$\sigma_{tot} = (\sigma_1^2 + \sigma_2^2 + ... + \sigma_n^2)^{1/2} \tag{II.7.10}$$

II.7.4 Signal-to-Noise Ratio

Especially in *digital radiography*, image quality can be better described by the *signal-to-noise ratio* (SNR), because – when taking advantage of the possibilities of image processing (e.g. windowing techniques) – the display of low-contrast details is limited only by noise. The SNR is that physical quantity with which a more flexible opti-

misation of the imaging process with respect to *image quality and dose* can be carried out. The signal-to-noise ratio shall be therefore discussed in the following.

In radiography an image signal which is relevant for diagnosis is created by the "important image details" as they are described, e.g. in the *European Guidelines on Quality Criteria for Diagnostic Radiographic Images* (CEC 1996a). It must be perceptible for diagnosis against the background of the surrounding tissue. The perceptibility within the image is complicated by noise (Hasegawa 1991), which is caused, as explained above, by statistical fluctuations of the X-ray photons (quantum mottle), which contribute to the image composition. The difference between signal and noise must therefore be as high as possible. In theoretical considerations one assumes that the SNR shall be at least 3–5, so that an image detail can be recognised by the eye with sufficient reliability (Rose 1948, 1973). The selection of an imaging system, which has as low an inherent system noise as possible, is therefore also of great significance.

The signal ΔI for a low-contrast detail $I_1 \approx I_2 = I$ can be written in accordance with Eq. II.7.2:

$$\Delta I = (I_1 - I_2) = \Delta\mu \cdot d \cdot I \tag{II.7.11}$$

For a given beam quality, the signal ΔI in the plane of the image receptor can be also expressed by the corresponding difference in the mean quantum numbers $\Delta \bar{N}$. When considering only quantum noise for the SNR in the radiation image (or in an ideal image receptor) it follows that

$$SNR = \frac{\Delta \bar{N}}{\sqrt{\bar{N}}} = \Delta \mu \cdot d \cdot \sqrt{\bar{N}} \qquad (II.7.12)$$

because the signal (ΔI, respectively $\Delta \bar{N}$) is proportional to the number of quanta $\bar{N}$; which escape from the phantom, and the quantum noise (σ) is proportional to the square root of $\bar{N}$ (see Eq. II.7.7). Furthermore the SNR is proportional to the product of the detail thickness d times the difference $\Delta \mu$ of the linear attenuation coefficients of the detail and its surrounding medium. For a given beam quality, doubling of the dose at the image receptor would result in an improvement of the SNR by a factor of $\sqrt{2}=1.4$.

Until now the influence of the detail size on the perceptibility of small details has not yet been considered in the discussion about contrast and SNR. For the calculation of the *SNR of small details in the radiation image* (e.g. micro-calcifications), in Eq. II.7.12 the modulation transfer function MTF(v) of the imaging geometry must be supplemented and the noise power spectrum must be introduced. For this general case, Eq. II.7.12 can be written as (Yaffe and Nishikawa 1991):

$$SNR(v) = \frac{I(v) \cdot MTF(v)}{\sqrt{W(v)}} \qquad (II.7.13)$$

whereby ΔI is now replaced by the spatial frequency spectrum I(v) of the incident X-ray pattern and $\sqrt{\bar{N}}$ by W(v). The SNR(v) is dependent on the spatial frequency v. The MTF(v) takes into account the influence of the focal spot characteristics and organ movement on the radiation image in front of the image receptor.

II.7.5   Detective Quantum Efficiency

Imaging systems can be characterised by how well they transmit information between their input and output. The *DQE* is suitable to quantify this behaviour. The DQE is defined as the ratio of the square of SNR $(SNR_{out})^2$ at the output of the system to that at the input $(SNR_{in})^2$:

$$DQE = \frac{(SNR_{out})^2}{(SNR_{in})^2} \qquad (II.7.14)$$

Equation II.7.14 can be made more clear with the special case of a simple *photon-counting detector* which has a quantum-detection efficiency (see Eq. II.6.1):

$$QDE = \eta. \qquad (II.7.15)$$

The SNR of the input signal SNR_{in} follows in this case from Eq. II.7.12:

$$SNR_{in} \propto \sqrt{\bar{N}} \qquad (II.7.16)$$

where $\bar{N}$ is the mean number of incident quanta. The SNR of the output signal SNR_{out} is determined by the really absorbed photons $\eta \cdot \bar{N}$:

$$SNR_{out} \propto \frac{\eta \cdot \bar{N}}{\sqrt{\eta \cdot N}} = \sqrt{\eta \cdot \bar{N}} \qquad (II.7.17)$$

whereby the standard deviation σ of the output noise inserted in the denominator of Eq. II.7.17 is:

$$\sigma_{out} = \sqrt{\eta \cdot \bar{N}} \qquad (II.7.18)$$

From Eqs. II.7.16 and II.7.17 it follows that the DQE in this simple case is identical to the quantum detection efficiency QDE $= \eta$ and one says that the noise in the imaging system is quantum-limited:

$$DQE = \eta \qquad (II.7.19)$$

The noise in the output signal is equal to that which would result from $\eta \cdot \bar{N}$; quanta, which in this example is the number of X-ray quanta which actually interacted in the detector. From that reason one defines as noise equivalent quanta (NEQ):

$$NEQ = (SNR_{out})^2 \qquad (II.7.20)$$

From Eq. II.7.16 one sees that the DQE in addition to the fundamental definition in Eq. II.7.14 can also be expressed as:

$$DQE = \frac{NEQ}{\bar{N}} \qquad (II.7.21)$$

The DQE is the ratio of the NEQ to the actual mean number of exposure quanta $\bar{N}$ from which the image was made up.

The only way to improve the SNR in a quantum-limited system (see Eq. II.7.19) is to use more X-ray photons, i.e. to increase radiation dose or to increase η by using a thicker detector or a detector material which shows a higher absorption of X-radiation.

Another interesting example can be given with regard to the imaging with antiscatter grids: The concept of the DQE was also applied to the grid performance by Wagner (1977) to rank the *relative effectiveness of scatter reduction techniques*. With the help of Eqs. II.5.10, III.2.2, III.2.3 and the total radiation intensity I' behind the grid:

$$I' = I_p \cdot T_p + I_s \cdot T_s = I_p \cdot T_p \cdot \frac{1}{SDF_g} \qquad (II.7.22)$$

it follows for the SNR_g in front of the image receptor:

$$SNR_g = C_p \cdot \sqrt{(I_p \cdot T_p)} \cdot \sqrt{(SDF_g)} \qquad (II.7.23)$$

Furthermore the SNR_{ig} that would be obtained with an ideal grid (i.e. $T_p = 1$ and $T_s = 0$) is:

$$SNR_{ig} = C_p \cdot \sqrt{I_p} \qquad (II.7.24)$$

From Eqs. II.7.23 and II.7.24, the following expression can be derived for the DQE:

$$DQE = \left(\frac{SNR_g}{SNR_{ig}}\right)^2 = T_p \cdot SDF_g \qquad (II.7.25)$$

For a given patient thickness (i.e. a given S/P-ratio; see Eq. II.5.10), the selectivity Σ and the primary radiation transmission T_p are decisive for the improvement of the image quality by the application of an antiscatter grid, not the grid ratio r (see Chap. III.2).

In real imaging systems, especially those with multiple stages, there are also noise contributions to the output due to sources other than quantum detection, such as electronic noise. In a film-screen system, often the granular structure of the film emulsion and the phosphor screen is an important noise source. These noise contributions are independent from the quantum fluctuations and increase the output noise, as explained in Eq. II.7.10. This effect reduces SNR_{out}, causing the DQE to a lower value than the quantum detection efficiency QDE and furthermore the NEQ is less than the number of X-ray quanta actually used for imaging. With regard to Eq. II.7.14, the frequency dependence of the DQE can be derived (Yaffe and Nishikawa 1991):

$$DQE(\nu) = \frac{k \cdot [MTF(\nu)]^2}{W(\nu)} \qquad (II.7.26)$$

where $W(\nu)$ is the Wiener spectrum, dependent on the spatial frequency ν and k is a constant. For a given beam

quality and mean number of exposure quanta $\bar{N}$, it follows from Eqs. II.7.14, II.7.20 and II.7.21:

$$k = \frac{(I(\nu))^2}{\bar{N}} \qquad (II.7.27)$$

where $I(\nu)$ is the signal which is included in the *incident X-ray pattern*.

II.7.6   Other Image-Quality Figures

The assessment of image quality, including the aspect of minimising radiation exposure, is often carried out with the help of the concept of an image-quality figure (Gajewski and Reiss 1974; Thijssen et al. 1989; Jennings et al. 1993; Aichinger et al. 1994; Tapiovaara et al. 1999), which is sometimes derived from the SNR. Some examples are given in the following paragraphs.

Hessler et al. (1985) have defined an image quality index (IQI) for film-screen systems as the diameter of the smallest visible object with a high probability of correct decision. This index depends on the three quantities, which have been described in Chaps. II.7.1 to II.7.3, that is, contrast, sharpness and noise:

$$IQI \propto (s^2 + \frac{G}{c \cdot \gamma \cdot C})^{\frac{1}{2}} \qquad (II.7.28)$$

Contrast is represented by the term $c\gamma C$, where c is a proportinality constant, γ is the gradient of the film-screen system and C is the radiation contrast. Sharpness is represented by s, which is a measure of the width of the line spread function. Finally noise is expressed in terms of the Selwyn coefficient G (Barrett and Swindell 1981, p. 215), which specifies the first-order statistics of the image noise. The image quality index has been applied to film-screen mammography by Desponds (Desponds et al. 1991).

Another image-quality figure (IQF), which sometimes is also called "figure of merit (FOM)", is defined as the quotient of the *square of the SNR and the system dose*:

$$IQF = \frac{(SNR)^2}{dose} \qquad (II.7.29)$$

In an optimised imaging system, most of the image noise is caused by the limited number of X-ray photons detected. Consequently the quotient $(SNR)^2$/dose is essentially independent of the dose. This conclusion follows from Eq. II.7.12. The IQF can be called also a dose-to-information conversion factor and is measured per dose (Tapi-

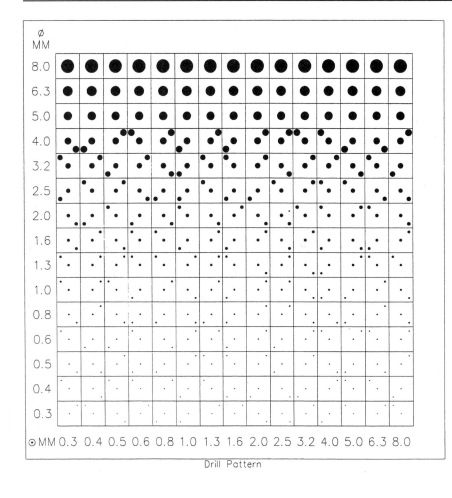

Fig. II.7.12. Principal design of the CDRAD-Phantom

Drill Pattern

ovaara et al. 1999). The optimum imaging technique is found at exposure conditions where SNR^2/dose is at its maximum. The final decision on the required image quality and patient dose is dependent on the diagnostic task, because the radiologist has to define the noise, which he can tolerate in the image for diagnosis. Examples for this optimisation process will be discussed in Chap. III.3.

Another image-quality figure which can be used for optimisation of image quality with regard to the medical indication and minimising radiation exposure has been presented by Thijssen et al. (1988, 1989). A *contrast-detail curve* is determined whereby a special phantom (CDRAD; see Fig. II.7.12) is needed to visualise details dependent on their size and contrast at the threshold level of visibility.

Holes of different diameters and depth are drilled in a polymethylmethylate (PMMA) plate with dimensions of 26,5 cm · 26.5 cm and of 1 cm thickness. With the radiological equipment under evaluation, radiographic images of the phantom must be made with optimised imaging parameters. To simulate different patient thicknesses, the phantom must be positioned between additional PMMA layers (e.g. 5–25 cm in total). The phantom is divided into 225 squares of equal size, in each of which two holes of identical dimensions are drilled, one in the centre and an-

other in one of the four corners. The location of the eccentric hole in the square is chosen in a random manner. In every row, the depth of the holes is changed logarithmically in 15 steps from 0.3–8.0 mm, in each column the diameter of the holes is varied similarly. In this way, in the horizontal direction contrast is varied, in the vertical direction detail diameter. During exposure the entrance dose used to generate the image shall be measured.

The radiographic images of the phantom can be examined by several observers, whereby the threshold level of visibility is determined by indicating on a form in which of the corners of each square the just-visible second eccentric disk is seen. In this way a contrast-detail curve can be drawn (see Fig. II.7.13).

Low-contrast details like masses need a minimum contrast to the background to be visualised. That threshold contrast is dependent on the size of the lesion and the noise in the imaging system. In other words, the system must have a good contrast transfer. The fact that small details need high contrast and that large details can do with lower contrast is expressed with regard to Rose (1973) by the formula:

$$C \cdot D = k \qquad (II.7.30)$$

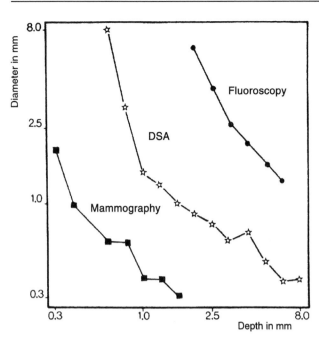

Fig. II.7.13. Contrast detail curve of various imaging techniques (Thijssen 1989)

i.e. the product of contrast C and detail size D at the visibility threshold is constant (see Fig. II.7.13; Thijssen et al. 1989). To quantify the image quality, the parameter k is important: When image quality is better, smaller details and/or lower contrasts are visualised in the image. So, the lower k is, the better the imaging system is. Limits to that are given by system properties and dose constraints.

To compare the image quality obtained with various radiological equipment or exposure techniques, a single image-quality figure (IQF$_{th}$) from the phantom image can be calculated (Thijssen 1989). By taking, at every contrast step, the product of the depth and the diameter of the hole that generates the radiation contrast of the just-visible disc and adding up all 15 products, a single figure is produced:

$$IQF_{th} = \sum_{i=1}^{N} (C_i \cdot D_i) \qquad (II.7.31)$$

The better the image quality is, the lower the IQF$_{th}$ is. In Chap. III.3 an example will be given for optimising fluoroscopy.

II.7.7   Dependence of Image Quality on Exposure Parameters

The contrast is the physical quantity used in imaging to indicate the detectability of a detail of interest and is given by the difference in image signals at two locations in the radiograph (see, e.g. Fig. II.7.5). The dependence of the contrast of low-contrast details within a phantom or a patient on the X-ray tube voltage or the radiation quality and the thickness of the phantom or the patient itself can

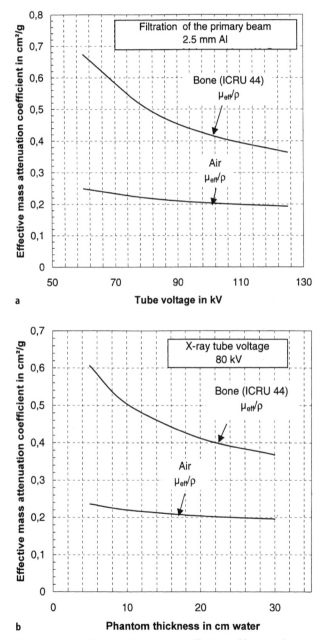

Fig. II.7.14a, b. Effective attenuation coefficients of bone and air: **a** embedded in 10 cm water dependent on X-ray tube voltage; **b** embedded in water of various thickness at an X-ray tube voltage of 80 kV

be calculated with the help of Eq. II.7.5, which was derived in Chap. II.7.1. By using the effective mass attenuation coefficients μ_{eff}/ρ instead of the mass attenuation coefficients $\mu(E)/\rho$, the radiation quality used and the beam hardening within the patient are taken into account. As an example, Fig. II.7.14a shows the effective mass attenuation coefficients of details consisting of bone and air, when embedded in a 10-cm-thick water phantom dependent on the X-ray tube voltage, and Fig. II.7.14b shows the effective mass attenuation of the details, when embedded in water phantoms of various thickness at a tube voltage of 80 kV.

From the course of the μ_{eff}/ρ-curves in Fig. II.7.14a, it can be derived that the employment of X-ray tube voltages above 100 kV in chest radiography results in an improvement of the detectability of the lung structures behind the ribs (see Chap. III.3.1). Whereas the X-ray attenuation by bone decreases very much with increasing tube voltage, the change in attenuation by air is low. The contrast of the lung structures is not much affected by the tube voltage.

The dependence of the SNR in the *radiation image* on radiation quality agrees with that of the contrast, but in addition it is affected by the system dose used for imaging. From Eqs. II.7.12 and II.7.13, respectively, it follows with regard to the example of Fig. II.7.5:

$$\text{SNR} = \Delta\mu_{eff} \cdot d \cdot \sqrt{N} \cdot \text{MTF}(\nu) \qquad (\text{II.7.32})$$

The noise power spectrum of the impinging X-rays is white, therefore the dependence of the SNR on spatial frequency is given by the MTF(ν) of the imaging geometry and the focal spot.

The considerations in Chap. II.7 deal especially with the image quality of the radiation image. If one considers the whole imaging system, one must bear in mind that its output signal is merely the input signal multiplied by the MTF(ν)s of the various components of the system. The determination of the output noise is more complex (Moy 2000).

References

Aichinger H, Dierker J, Säbel M, Joite-Barfuss S (1994) Image quality and dose in mammography. Electromedica 62:7–11

Barnes GT (1979) Characteristics of scatter. In: Logan WW, Muntz EP (eds) Reduced dose mammography. Masson, New York

Barrett HH, Swindell W (1981) Radiological imaging. Academic, New York

Biedermann K, Borcke E, Buchmann F, Frieser H, Munker H, Schober H, Schott, Stieve FE, Widenmann L (1967) Über Kontrastbegriffe in der Radiologie und ihre Definitionen. Röntgenblätter 20(3):131–147

Bradford CD, Peppler WW, Waidelich JM (1999) Use of a slit camera for MTF measurements. Med Phys 26(11):2286–2294

CEC (Commission of the European Communities) (1996a) European guidelines on quality criteria for diagnostic radiographic images. EUR 16260. CEC, Luxemburg

CEC (1996b) European guidelines on quality criteria for diagnostic radiographic images in paediatrics. EUR 16261. CEC, Luxemburg

Christensen (1990) Introduction to the physicsof diagnostic radiology. In: Curry TS, Dowdey JE, Murray RC (eds) 4th edn. Lea and Febiger, Philadelphia

Coltman JW (1954) The specification of imaging properties by response to sine wave input. J Opt Soc Am 44:468–471

Desponds L, Depeursinge C, Grecescu M, Hessler C, Samiri A, Valley JF (1991) Image quality index (IQI) for screen-film mammography. Phys Med Biol 36:19–33

DIN (Deutsches Institut für Normung) (1995) Sensitometrie an Film-Folien-Systemen für die medizinische Radiographie. Teil10. Nennwerte der Empfindlichkeit und des mittleren Gradienten. DIN 6867-10. Beuth, Berlin

Gajewski H, Reiss KH (1974) Physik und Technik der Weichstrahltechnik. Radiologe 10:438–46

Hasegawa BH (1991) The physics of medical X-ray imaging, 2. Medical Physics, Madison, WI

Herz RH (1969) The photographic action of ionizing radiation. Wiley, London

Hessler C, Depeursinge C, Grecescu M, Pochon Y, Raimondi S, Valley JF (1985) Objective assessment of mammography systems. Radiology 156:215–219

ICRU (International Commission on Radiation Units and Measurements) (1986) Modulation transfer function of screen-film systems. Report 41. ICRU, Bethesda, MD

ICRU (1989) Tissue substitutes in radiation dosimetry and measurements. Report 44. ICRU, Bethesda, MD

ICRU (1996) Medical imaging – the assessment of image quality. Report 54. ICRU, Bethesda, MD

IEC (International Electrotechnical Commission) (1993) X-ray tube assemblies for medical diagnosis – characteristics of focal spots. Publication 60636. IEC, Geneva

ISO (International Standards Organisation) (2001) Photography-sensitometry of screen/film systems for medical radiography. Part 1. Determination of a sensitometrc curve shape, speed and average gradient. Publication 9236-1 (Draft)

Koedooder K (1986) Dose reduction by X-ray beam filtration in screen-film radiography. Thesis, University of Amsterdam

James TH (1977) The theory of the photographic process. Macmillan, New York

Jennings RJ, Quinn PW, Gagne RM, Fewell TR (1993) Evaluation of X-ray sources for mammography. (Physics of medical imaging, SPIE Vol 1896) pp 259–268

Lubberts G (1968) Random noise produced by fluorescent screens. J Opt Soc Am 58:1475–1483

Morneburg H (Herausgeber) (1995) Bildgebende Systeme für die medizinische Diagnostik. Publicis, Erlangen

Moy J-P (2000) Signal-to-noise ratio and spatial resolution in X-ray electronic imagers: Is the MTF a relevant parameter? Med Phys 27 (1):86–93

Rose A (1948) The sensitivity performance of the human eye on an absolute scale. J Opt Soc Am 38 (2):196–208

Rose A (1973) Vision:human and electronic. Plenum, New York

Tapiovaara MJ, Sandborg M, Dance DR (1999) A search for improved technique factors in paediatric fluoroscopy. Phys Med Biol 44:537–559

Thijssen MAO, Rosenbusch G, Gerlach H-J (1988) Reduktion der Strahlenexposition bei Durchleuchtung. Electromedica 56 (4):126–133

Thijssen MAO, Thijssen HOM, Merx JL, Lindeijer JM, Bijkerk KR (1989) A definition of image quality: the image-quality figure. In: Moores BM, Wall BF, Eriskat H, Schibilla H (eds) Optimization of image quality and patient exposure in diagnostic radiology. BIR Report 20. British Institute of Radiology, London, pp 29–34

Wagner RF (1977) Noise equivalent parameters in gerneral medical radiography: the present and future pictures. Photo Sci Eng 21:252–262

Williams MB, Mangiafico PA, Simoni PU (1999) Noise power spectra of images from digital mamography detectors. Med Phys 26 (7):1279–1293

Yaffe MJ (1994) X-ray spectral considerations for mammography. In: Haus AG, Yaffe MJ (eds) A categorical course in physics: technical aspects of breast imaging. RSNA, Oak Brook, pp 63–74

Yaffe MJ, Nishikawa RM (1991) X-ray imaging concepts: Noise, SNR and DQE. In: Specification acceptance testing and quality control of diagnostic X-ray imaging equipment, Vol I. Proceedings Summer School University of California, Santa Cruz, CA pp 123–154

Part III   Clinical Applications

In X-ray diagnostic radiology there are essentially two reasons for determining radiation doses to patients. Firstly, knowledge of the absorbed doses to tissues and organs in the patient is needed to estimate the associated *radiation risk*. Secondly, this knowledge plays a significant role in the optimisation of image quality versus radiation exposure and therefore in the process of setting and checking *standards of good practice*.

Most of the specific dosimetric quantities used for patient-dose evaluation have already been presented in Chap. II.3.2. Additionally the computed tomography (CT) dose index (CTDI) is introduced in Chap. III.1.1. This is followed by a short discussion of quantities influencing patient dose. For the estimation of organ doses, clearly defined initial dosimetric quantities are required which can be measured easily with readily available instruments of sufficient precision and accuracy. These fixed quantities are given in Chap. III.1.3 and their determination is discussed in Chap. III.1.4. Finally the most important methods for the estimation of organ doses from the initial dosimetric quantities are described. The structure of the presentation corresponds to a large extent to the structure of the draft of the German standard DIN 6809-7 (DIN 2002).

III.1.1 Specific Dosimetric Quantities Used for Patient-Dose Evaluation

Most of the specific dosimetric quantities used for patient-dose evaluation are defined by using the air kerma (see Eq. II.3.4). They have been introduced in Chap. II.3.2.

An exception is the CTDI, which is the line integral of the absorbed dose profile D(z) of a single scan (360° rotation) over ±7 slices and parallel to the axis of rotation in a solid-state phantom, divided by the product of nominal slice thickness s and the number N of slices per rotation of the X-ray tube assembly:

$$ \text{CTDI} = \frac{1}{N \cdot s} \int_{-7s}^{+7s} D(z) dz \qquad (\text{III.1.1}) $$

This definition can be traced back to the US Department of Health and Human Services/Food and Drug Adminis-

tration rules (DHHS/FDA 1984). There the axial absorbed dose profile D(z), including the contribution of scattered radiation, is determined from measurements with thermoluminescent dosimeters (TLD) in the solid-state phantom. These phantoms consist of polymethylmethacrylate (PMMA). They are cylindrical, with diameters of 16 cm or 32 cm (representing head and trunk, respectively) and a length of 16 cm. In addition to a central bore, four peripheral bores at a depth of 1 cm are provided for taking up the ionisation chamber or the TLD arrays. If the CTDI is estimated from a measurement of the dose-length product the different lengths of integration must be considered.

In the literature (CEC 1997), more definitions of the CTDI can be found, which differ mainly in their integration length. Examples are the CTDI_∞ (integration more than ±∞) and the $\text{CTDI}_{10\,\text{cm}}$ (integration more than 10 cm, corresponding to the length of an ionisation chamber for the measurement of the dose-length product). Another dosimetric quantity is the weighted CT dose index:

$$ \text{CTDI}_\text{W} = \frac{1}{3} \text{CTDI}_{10\,\text{cm,A}} + \frac{2}{3} \text{CTDI}_{10\,\text{cm,P}} \qquad (\text{III.1.2}) $$

where the indices A and P indicate the point of measurement "axis of rotation" and "periphery" (i.e. at a depth of 1 cm),, respectively. The CTDI_w is used to define another dosimetric quantity, which is also called "dose-length product" (DLP) and is used to characterise the patient exposure of a complete CT examination with N slices of thickness s:

$$ \text{DLP} = \text{CTDI}_\text{W} \cdot s \cdot N \qquad (\text{III.1.3}) $$

III.1.2 Quantities Influencing Patient Dose

There are a lot of quantities that influence patient exposure. Some of them, which are of practical importance for patient-dose estimation, are briefly reviewed in the following.

For the characterisation of *radiation quality*, as a rule the statement of tube voltage and total filtration or tube

Table III.1.1 Backscatter factors for water and PMMA*) in dependence on radiation quality and field size adapted from (Petoussi et al 1998)

Radiation quality				Backscatter factors for various field sizes					
X-ray Tube voltage kV	Total filtration	HVL	Mean Energy	10 x 10 cm²		20 x 20 cm²		25 x 25 cm²	
		mm Al	keV	Water	PMMA	Water	PMMA	Water	PMMA
50	2.5 mm Al	1.74	32.0	1.24	1.33	1.26	1.36	1.26	1.36
60	2.5 mm Al	2.08	35.8	1.28	1.36	1.31	1.41	1.31	1.42
70	2.5 mm Al	2.41	39.3	1.30	1.39	1.34	1.45	1.35	1.46
70	3.0 mm Al	2.64	40.0	1.32	1.40	1.36	1.47	1.36	1.48
70	3.0 mm Al + 0.1 mm Cu	3.96	44.0	1.38	1.48	1.45	1.58	1.46	1.59
80	2.5 mm Al	2.78	42.9	1.32	1.41	1.37	1.48	1.38	1.50
80	3.0 mm Al	3.04	43.7	1.34	1.42	·1.39	1.51	1.40	1.52
80	3.0 mm Al + 0.1 mm Cu	4.55	48.2	1.40	1.49	1.48	1.61	1.49	1.63
90	2.5 mm Al	3.17	46.3	1.34	1.43	1.40	1.51	1.41	1.53
90	3.0 mm Al	3.45	47.0	1.35	1.44	1.42	1.53	1.42	1.55
90	3.0 mm Al + 0.1 mm Cu	5.12	51.7	1.41	1.50	1.50	1.62	1.51	1.65
100	2.5 mm Al	3.24	48.1	1.34	1.42	1.40	1.51	1.41	1.53
100	3.0 mm Al	3.89	50.0	1.36	1.45	1.44	1.55	1.45	1.57
100	3.0 mm Al + 0.1 mm Cu	5.65	54.8	1.41	1.50	1.51	1.64	1.53	1.66
110	2.5 mm Al	3.59	50.8	1.35	1.43	1.42	1.53	1.43	1.55
120	3.0 mm Al	4.73	55.4	1.37	1.46	1.46	1.58	1.48	1.60
120	3.0 mm Al + 0.1 mm Cu	6.62	60.1	1.41	1.50	1.53	1.64	1.54	1.67
130	2.5 mm Al	4.32	55.6	1.36	1.44	1.44	1.55	1.45	1.57
150	2.5 mm Al	4.79	59.1	1.36	1.44	1.45	1.55	1.46	1.58
150	3.0 mm Al	6.80	64.9	1.39	1.47	1.50	1.61	1.52	1.63
150	3.0 mm Al + 0.1 mm Cu	8.50	69.2	1.40	1.48	1.53	1.64	1.55	1.67

PMMA Polymethylmethacrylate (acrylic glass, i.e. Plexiglas)

Table III.1.2 Backscatter factor as a function of half value layer HVL in mammography (adapted from CEC 1996).

HVL (mm Al)	0.25	0.30	0.35	0.40	0.45	0.50	0.55	0.60	0.65
Backscatter factor	1.07	1.07	1.08	1.09	1.10	1.11	1.12	1.12	1.13

voltage and half-value layer (HVL) is sufficient. Additionally the form of the tube voltage and the anode material (if different from tungsten) can be given (see also Chap. II.1.1).

The *radiation beam geometry* is characterised by the distance of the entrance surface of the patient r_E and of the image receptor r_B to the focus, the field size at the entrance surface of the patient A_E or at the image receptor A_B, and the patient thickness d_P.

The *backscatter factor* (B) is the quotient of the entrance surface dose K_O and the entrance surface air kerma K_E (also sometimes called "entrance dose", see Chap. II.3.1) at the entrance surface of the patient or the phantom. The backscatter factor is dependent on radiation quality, field size and patient (or phantom) thickness. Examples for the backscatter factor are given for general X-ray diagnostic radiology in Table III.1.1 and for mammography in Table III.1.2.

Attenuation factors characterise the attenuation of the radiation beam by the penetrated material for a defined radiation quality. As a rule they are expressed as the quo-

tient of two air kerma values (without and with the attenuating medium). The reciprocal value of the attenuation factor is often called radiation transmission (see Chap. II.5.4). Two attenuation factors of practical importance are the equipment and the patient attenuation factor:

The *equipment attenuation factor* characterises the attenuation of the radiation beam by materials, arranged between the exit surface of the patient and the image receptor (e.g. patient table, AEC chamber, scattered radiation grid; see also Chap. II.4).

The *total equipment attenuation factor* (m) is the product of the attenuation factors of the particular components, including also the geometrical attenuation of the radiation between the exit surface of the patient and the image receptor:

$$m = m_{absorber} \cdot m_{grid} \cdot m_{geometry} \qquad (III.1.4)$$

Some typical values of equipment attenuation factors are given in Table III.1.3. The *patient attenuation factor*

Table III.1.3 Typical values for the attenuation factor of components of the X-ray examination equipment between the patient output plane and the image receptor. Measurement is made with 80 kV and an additional filter of 25 mm Al arranged nearby the focal spot (DIN 6815 1992)

Components of X-ray examination equipment	Attenuation factor
Patient table	1.25
Front cover of the film changer	1.25
Anti-scatter grid	1.4 – 1.6
Radiation detector of the automatic exposure control system	1.10

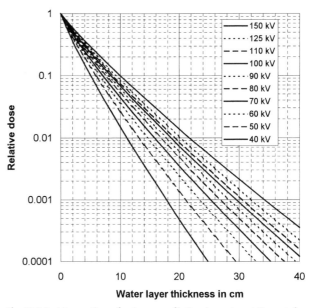

Fig. III.1.1. Attenuation of primary radiation in water at X-ray tube voltages (ripple <1%) from 40 to 150 kV (*from left to right*); total filtration 2.5 mm Al

$(m_{patient})$ characterises the attenuation of the primary beam by the patient (or water as a substitute). It is determined under scatter-free conditions in narrow-beam geometry and for radiation qualities used for imaging. It can also be calculated by using Eqs. II.2.1 and II.3.6 and the corresponding X-ray spectra. Examples for the patient attenuation factor are given in Figs. III.1.1 and III.1.2.

Finally two dosimetric quantities are given, which describe the dose distribution on the central ray inside the patient (or the phantom):

The *relative depth dose* $D_{rel}(z)$ is defined by:

$$D_{rel} = \frac{D(z)}{D_0} \qquad (III.1.5)$$

where $D(z)$ is the absorbed dose on the central ray inside the patient (or the phantom) at depth z and D_0 is the absorbed dose reference value. As a rule D_0 is the maximum of the absorbed dose on the central ray inside the patient (or the phantom). For diagnostic radiation qualities, the maximum is situated at the entrance surface of the patient (or the phantom).

The *tissue-air ratio* $T_a(z)$ is defined by:

$$T_a(z) = \frac{D(z)}{D_{\Delta m}} \qquad (III.1.6)$$

where $D(z)$ is the absorbed dose on the central ray inside the patient (or the phantom) at depth z and $D_{\Delta m}$ is the absorbed dose in a mass element Δm of the same material at this point free-in-air (i.e. in the absence of the patient or phantom).

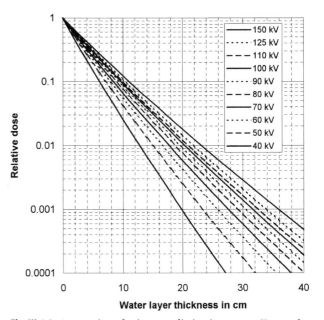

Fig. III.1.2. Attenuation of primary radiation in water at X-ray tube voltages (ripple <1%) from 40 to 150 kV; total filtration 2.5 mm Al, added filtration 0.1 mm Cu

Table III.1.4 Initial dosimetric quantities for the estimation of organ doses

Branch of X-ray diagnostic radiology	Initial dosimetric quantity
Radiography	entrance surface air kerma K_E
	air kerma-area product P_F
Fluoroscopy	entrance surface air kerma K_E
	air kerma-area product P_F
Mammography	entrance surface air kerma K_E
Computed tomography	slice averaged axial air kerma K_{CT}

III.1.3 Initial Dosimetric Quantities for the Estimation of Organ Doses

The selection of dosimetric quantities as initial quantities for estimation of organ doses is oriented on their availability in radiological routine and on the existence of conversion procedures in the literature. In this connection, conversion procedures, which allow only the estimation of the effective dose, are also taken into consideration. The selected initial dosimetric quantities are given in Table III.1.4.

III.1.4 Determination of the Initial Dosimetric Quantities

The value of the initial dosimetric quantity in question can be determined from measurable or *equipment-specific dosimetric quantities* as well as from *investigation-specific quantities*. The quantities, which are measurable during an X-ray investigation, are – apart from radiographic parameters – the air kerma-area product (P_F) and the entrance surface air kerma K_E at a fixed focus distance. In particular situations dose measurements at the surface of the patient are also possible; in this case the entrance surface dose K_O is measured. Equipment-specific dosimetric quantities can have been received during acceptance testing or can be determined from radiographic parameters. Equipment-specific dosimetric quantities are, e.g. the nominal air kerma rate $\dot{K}_{a,100}$, the image receptor dose K_B and the dose-length product P_L. Investigation-specific quantities are, e.g. radiographic parameters such as tube voltage U, tube current-exposure time product $I_R \cdot t$, total filtration, focus distance, field size and the body region under investigation.

For the determination of the initial dosimetric quantities, all radiological investigations have to be separated into sections, where neither the radiographic parameters nor the region under investigation changes significantly. Therefore, a set of dose values is received for a single investigation, which has to be converted separately into or-

gan doses. In the following the determination of the initial dosimetric quantities is briefly discussed for the different branches of X-ray diagnostic radiology.

III.1.4.1 Radiography

Depending on availability, the entrance surface air kerma K_E can be determined from measured quantities as the air kerma-area product P_F (or in particular situations the entrance surface dose K_O) or calculated from equipment-specific dosimetric quantities. In most cases additional knowledge about quantities influencing patient dose is necessary. Examples for these quantities are (see Chap. III.1.2) radiation quality, the distance of the entrance surface of the patient r_E and of the image receptor r_B to the focus, the field size at the entrance surface of the patient A_E or at the image receptor A_B, and the patient thickness d_P.

In the following the different methods for the determination of the entrance surface air kerma K_E are arranged according to their priority, which depends on the inherent uncertainty of the method in question. The priority decreases, starting with the measurement of the entrance surface air kerma and finishing with its determination from the image receptor dose.

III.1.4.1.1 Determination of the Entrance Surface Air Kerma K_E from the Value K_m, Measured at Focus Distance r_m

The entrance surface air kerma K_E at focus distance r_E is calculated from the value K_m, measured at focus distance r_m, according to:

$$K_E = K_m \cdot \left(\frac{r_m}{r_E}\right)^2 \tag{III.1.7}$$

Special ionisation chambers, which are integrated into the large-area, parallel-plate ionisation chamber for the measurement of the air kerma-area product make the measurement of K_m at focus distance $r_m < r_E$ possible (Gfirtner et al. 1997).

III.1.4.1.2 Determination of the Entrance Surface Air Kerma K_E from the Air Kerma-Area Product P_F

According to Eq. II.3.12, the entrance surface air kerma K_E is approximately equal to the quotient of the air kerma-area product P_F and the field size at the entrance surface of the patient A_E:

$$K_E = \frac{P_F}{A_E} \tag{III.1.8}$$

The field size A_E can be calculated from:

$$A_E = A_B \cdot \left(\frac{r_E}{r_B}\right)^2 \qquad \text{(III.1.9)}$$

where A_B is the field size at the image receptor with focus distance r_B.

III.1.4.1.3 Determination of the Entrance Surface Air Kerma K_E from the Entrance Surface Dose K_O

The entrance surface air kerma K_E is the quotient of the entrance surface dose K_O and the backscatter factor B (see Chap. III.1.2):

$$K_E = \frac{K_O}{B} \qquad \text{(III.1.10)}$$

The backscatter factor is essentially dependent on radiation quality and field size (see Tables III.1.1 and III.1.2). Therefore these investigation-specific quantities must be known.

III.1.4.1.4 Determination of the Entrance Surface Air Kerma K_E from the Radiation Output of an X-ray Tube Assembly Y_{100} and the Radiographic Parameters

The entrance surface air kerma K_E at focus distance r_E is calculated from the radiation output of an X-ray tube assembly Y_{100} at focus distance $r_Y = 100$ cm and the tube current-exposure time product $I_R \cdot t$ according to:

$$K_E = Y_{100} \cdot I_R \cdot t \cdot \left(\frac{r_Y}{r_E}\right)^2 \qquad \text{(III.1.11)}$$

Guidance levels for Y_{100} are given in Fig. III.1.3.

III.1.4.1.5 Determination of the Entrance Surface Air Kerma K_E from the Image Receptor Dose K_B or the Sensitivity of the Image Receptor S

If radiation quality, patient thickness, radiation beam geometry and equipment attenuation factors are known, the entrance surface air kerma K_E can be determined from the image receptor dose K_B according to:

$$K_E = K_B \cdot m_{patient} \cdot m_{absorber} \cdot m_{grid} \cdot m_{geometry} \cdot (1 - S_g) \qquad \text{(III.1.12)}$$

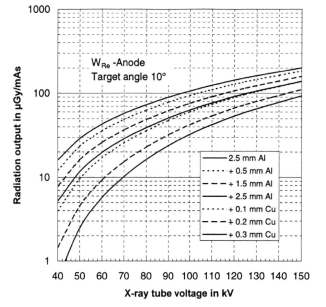

Fig. III.1.3. Dose output Y_{100} of X-ray tube assemblies with W_{Re}-anode at a target angle of 10° and various additional filtration; focus distance of 100 cm

The different attenuation factors (m) are have already been explained in Chap. III.1.2. S_g is the scatter fraction behind the grid, defined already by Eqs. II.5.19 or II.5.20. As also already mentioned, the attenuation factor of the patient (or water as a substitute) $m_{patient}$ is determined under scatter-free conditions. For $m_{geometry}$:

$$m_{geometry} = \left(\frac{r_B}{r_E}\right)^2 \qquad \text{(III.1.13)}$$

In Table III.1.5, m_{grid} together with some values for the selectivity Σ is given for four different antiscatter grids.

If only the sensitivity S of the image receptor is known (see Eq. II.6.8), it can be assumed that K_B is approximately equal to K_S, which can be calculated according to Eq. II.6.8 by:

$$K_s = 1000 \ \mu Gy / S \qquad \text{(III.1.14)}$$

Table III.1.5 Typical values of the selectivity of anti-scatter grids in dependence on the X-ray tube voltage and for the reciprocal primary radiation transmission (1/Tp) at 100 kV according to IEC 60627 (1978)

Type Anti-scatter grid	Selectivity Σ [a]			1/Tp[a]
	60 kV	75 kV	100 kV	100 kV
Pb 8/40	8.0	7.1	5.2	1.56
Pb 12/40	13.4	12.3	9.5	1.56
Pb 15/80	11.0	8.7	5.4	1.35
Pb 17/70	19.4	15.4	9.7	1.58

[a] Characteristics are valid for anti-scatter grids with Al-cover and paper interspace medium. The reciprocal primary radiation transmission (1/Tp) agrees with the attenuation factor m_{grid}, which is affected not much by X-ray tube voltage (< 10%) in the tube voltage range from 60 kV to 125 kV.

III.1.4.2 Fluoroscopy

Fluoroscopy investigations have to be separated into sections where neither the fluoroscopic parameters nor the region under investigation changes significantly. For standardised investigations, as a rule the air kerma-area product P_F is the initial dosimetric quantity for the estimation of organ doses. Alternatively the methods described in Chap. III.1.4.1. for the determination of the entrance surface air kerma K_E can be applied accordingly. In this context dose values can also be calculated from measured dose rates by multiplication with the corresponding fluoroscopy time.

III.1.4.3 Mammography

For mammography the entrance surface air kerma K_E can be determined according to Chap. III.1.4.1. from the value K_m, measured at focus distance r_m or from the entrance surface dose K_O. If the radiographic parameters are known, the determination of the entrance surface air kerma from the radiation output of the X-ray tube assembly corresponding to Eq. III.1.11 is also possible. A detailed description of these methods is given in the *European Protocol on Dosimetry in Mammography* (EPDM), published by the European Commission (CEC 1996), and in a paper by Zoetelief et al. (1998), published in the proceedings (Bauer et al. 1998) of a workshop on *Reference Doses and Quality in Medical Imaging*, organised by the European Commission and the German Institute of Radiation Hygiene. As an example, guidance levels for the radiation output Y_{60} are given in Fig. III.1.4.

III.1.4.4 Computed Tomography

For computed tomography the slice-averaged axial air kerma K_{CT} can be calculated from the measured dose-length product P_L according to Eq. II.3.14 or II.3.15. A comprehensive compilation of equipment typical K_{CT} values is given by Nagel (1999).

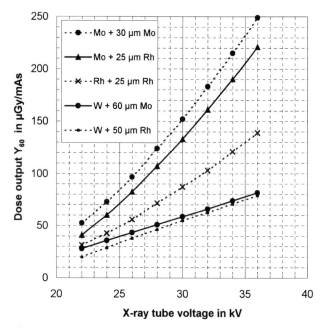

Fig. III.1.4. Typical values for the dose output Y_{60} of mammography X-ray tube assemblies at a emission angle of 16° and various anode-filter combinations; focus distance 60 cm; X-ray tube voltages from 22 to 36 kV; 2 mm polycarbonate compression plate within the X-ray beam

III.1.5 Estimation of Organ Doses from the Initial Dosimetric Quantities

The estimation of organ doses H_T by the use of conversion factors is described in Chap. III.1.5.1. In this case the mean dose, i.e. the dose averaged over the mass of the organ, is determined. In Chap. III.1.5.2. the estimation of the absorbed dose to an organ D_{org} by the use of tissue-air ratios $T_a(z)$ or relative depth doses $D_{rel}(z)$ is presented. In this case a point dose is calculated (see Chap. II.3.3) and averaging over the organ must follow subsequently if necessary.

III.1.5.1 Estimation of Organ Doses H_T by the Use of Conversion Factors

Conversion factors for the estimation of organ doses are based on mathematical models of the human body with various modifications to account for differences between male and female, adult and child. Monte Carlo computation techniques are used to model the X-ray interaction for commonly used beam spectra, projections and radiation field sizes. The result are tables of data which can be used to estimate organ doses given the exposure conditions.

III.1.5.1.1 Radiography and Fluoroscopy

If the organ dose conversion factor f_T is related to the entrance surface air kerma K_E, the organ dose H_T is given by:

$$H_T = f_T \text{ (exposure conditions)} \cdot K_E \qquad \text{(III.1.15)}$$

If the organ dose conversion factor f_T is related to the air kerma-area product P_F, the organ dose H_T is given by:

$$H_T = f_T \text{ (exposure conditions)} \cdot P_F \qquad \text{(III.1.16)}$$

If the entrance surface air kerma K_E is known, organ-dose conversion factors f_T from Drexler et al. (1993) and from Hart et al. (1998) can be used. Organ-dose conversion factors f_T related to the air kerma-area product P_F can also be found in Hart et al. 1998. For selected projections in fluoroscopy of the coronary arteries organ-dose conversion factors f_T are tabulated by Stern et al. (1995).

III.1.5.1.2 Mammography

Presently it is generally assumed that the glandular tissue is the most vulnerable in the breast, as compared for example with adipose or skin tissue. Therefore, in the meantime it is widely accepted that the average dose to the glandular tissue most usefully characterises the risk of carcinogenesis. Consequently, countries that have established mammography screening are using average glandular dose (AGD) as a risk-relevant dose quantity. As for most organ doses, AGD cannot be measured directly, but is calculated for a specified thickness of the compressed breast under certain assumptions (mainly concerning the tissue composition) from entrance surface air kerma K_E or entrance surface dose K_O. For the assessment of AGD, phantoms are often used, which are considered to be representative of the average-sized female breast of average tissue composition.

The term "average glandular dose" was introduced by the International Commission on Radiological Protec-

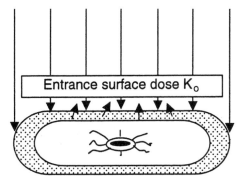

Cross-section of a compressed breast

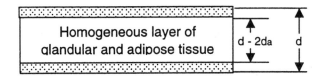

$$K_O = B \times K_E$$
$$AGD = K_E \times g$$

Homogeneous layer of glandular and adipose tissue

Layer model of the breast for the determination of the average glandular dose AGD

Fig. III.1.5. Standard breast model for the evaluation of the average glandular dose (AGD). d is the tickness of the compressed breast and d_a is the thickness of the surrounding adipose tissue

tion in 1987 (ICRP 1987). At that time the ICRP recommended, for radiation dose estimation from X-ray mammography, the use of the average absorbed dose in the glandular tissue (excluding skin) in a uniformly compressed breast. A simplified breast model according to a proposal of Stanton et al. (1984) is presented in Fig. III.1.5. The ICRP (1987) further recommended that the tissue composition should be 50% adipose/50% glandular tissue by weight and that the breast thickness should be specified. Consequently, in the EPDM (CEC 1996) the "standard breast" has been defined as "a model used for calculations of glandular dose consisting of a 40-mm-thick central region comprising a 50%/50% mixture by weight of adipose tissue and glandular tissue surrounded by a 5-mm-thick superficial layer of adipose tissue". AGD calculated by application of this model is called "standard AGD" (sAGD; CEC 1996).

AGD is then calculated from:

$$AGD = K_E \cdot g \qquad \text{(III.1.17)}$$

where g is the conversion factor to be used to calculate the AGD from K_E for a specified breast thickness. If the en-

Table III.1.6 Conversion factor g in mGy/mGy, for conversion of entrance surface air kerma K_E of the breast into average glandular dose AGD, for various breast compositions and thicknesses, anode-filter combinations and tube voltages. For the mix the breast composition is 50 % adipose and 50 % glandular tissue by mass (adapted from Klein et al 1997)

Composition: 100 % adipose

Anode material	Filter thickness (μm)	Filter material	Tube voltage (kV)	g in mGy/mGy for breast thickness in mm							
				20	30	40	50	60	70	80	90
Mo	30	Mo	25	0.458	0.340	0.264	0.213				
Mo	30	Mo	28		0.371	0.292	0.236	0.197	0.168		
Mo	30	Mo	30				0.248	0.207	0.177	0.153	
Mo	30	Mo	32					0.215	0.184	0.160	0.140
W	60	Mo	25	0.497	0.373	0.292	0.236				
W	60	Mo	28		0.384	0.303	0.245	0.205	0.174		
W	60	Mo	30				0.258	0.216	0.184	0.160	
Mo	25	Rh	28				0.269	0.226	0.193	0.168	
Mo	25	Rh	30					0.234	0.201	0.175	0.154
Mo	25	Rh	32					0.242	0.207	0.181	0.159
Rh	25	Rh	28				0.282	0.238	0.204	0.178	
Rh	25	Rh	30					0.255	0.220	0.192	0.170
Rh	25	Rh	32					0.270	0.233	0.204	0.181
W	50	Rh	28				0.328	0.277	0.239	0.209	
W	50	Rh	30				0.339	0.288	0.248	0.218	
W	50	Rh	32					0.293	0.254	0.223	0.197
W	50	Rh	34					0.302	0.262	0.230	0.205

Composition: 100 % gland

Anode material	Filter thickness (μm)	Filter material	Tube voltage (kV)	20	30	40	50	60	70	80	90
Mo	30	Mo	25	0.365	0.241	0.176	0.136				
Mo	30	Mo	28		0.267	0.197	0.153	0.124	0.104		
Mo	30	Mo	30				0.162	0.132	0.111	0.095	
Mo	30	Mo	32					0.138	0.116	0.100	0.087
W	60	Mo	25	0.399	0.267	0.192	0.151				
W	60	Mo	28		0.278	0.205	0.159	0.130	0.109		
W	60	Mo	30				0.170	0.139	0.117	0.100	
Mo	25	Rh	28				0.178	0.145	0.122	0.105	
Mo	25	Rh	30					0.152	0.128	0.110	0.096
Mo	25	Rh	32					0.157	0.133	0.114	0.100
Rh	25	Rh	28				0.190	0.156	0.132	0.114	
Rh	25	Rh	30					0.170	0.143	0.124	0.109
Rh	25	Rh	32					0.181	0.153	0.132	0.116
W	50	Rh	28				0.223	0.183	0.155	0.134	
W	50	Rh	30				0.232	0.192	0.162	0.140	
W	50	Rh	32					0.196	0.166	0.144	0.126
W	50	Rh	34					0.204	0.174	0.150	0.132

Composition: mix with 50 % adipose and 50 % gland

Anode material	Filter thickness (μm)	Filter material	Tube voltage (kV)	20	30	40	50	60	70	80	90
Mo	30	Mo	25	0.407	0.284	0.213	0.168				
Mo	30	Mo	28		0.313	0.237	0.187	0.154	0.130		
Mo	30	Mo	30				0.198	0.163	0.137	0.118	
Mo	30	Mo	32					0.170	0.143	0.124	0.109
W	60	Mo	25	0.444	0.314	0.236	0.186				
W	60	Mo	28		0.325	0.246	0.195	0.160	0.135		
W	60	Mo	30				0.206	0.170	0.144	0.124	
Mo	25	Rh	28				0.216	0.178	0.151	0.130	
Mo	25	Rh	30					0.186	0.158	0.136	0.120
Mo	25	Rh	32					0.192	0.163	0.141	0.124
Rh	25	Rh	28				0.229	0.190	0.161	0.140	
Rh	25	Rh	30					0.205	0.175	0.152	0.134
Rh	25	Rh	32					0.218	0.186	0.162	0.142
W	50	Rh	28				0.267	0.222	0.189	0.164	
W	50	Rh	30				0.278	0.232	0.198	0.172	
W	50	Rh	32					0.237	0.202	0.176	0.155
W	50	Rh	34					0.246	0.210	0.183	0.162

Table III.1.7 Conversion factors g to calculate the average glandular dose AGD for different breast thicknesses from the entrance surface air kerma K_E (from CEC 1996)

HVL	Compressed breast thickness in mm					
(mm Al)	30	40	50	60	70	80
0.25	0.234	0.174	0.137	0.112	0.094	0.081
0.30	0.274	0.207	0.164	0.135	0.114	0.098
0.35	0.309	0.235	0.187	0.154	0.130	0.112
0.40	0.342	0.261	0.209	0.172	0.145	0.126
0.45	0.374	0.289	0.232	0.192	0.163	0.140
0.50	0.406	0.318	0.258	0.214	0.177	0.154
0.55	0.437	0.346	0.287	0.236	0.202	0.175
0.60	0.466	0.374	0.310	0.261	0.224	0.195
0.65	0.491	0.399	0.332	0.282	0.244	0.212

trance surface dose K_O is given, K_E can be calculated from Eq. III.1.10.

Values of g are obtained through Monte Carlo simulations of AGD as a function of entrance surface air kerma K_E or entrance surface dose K_O and are available for various radiation qualities, breast thicknesses and breast compositions (Boone 1999; Dance 1990; Klein et al. 1997; NCS 1993; Rosenstein et al. 1985; Wu et al. 1991, 1994). As an example the data from Klein et al. (1997) are given in Table III.1.6 (see Chap. IV.6.2)

For the standard composition of 50% adipose/50% glandular tissue, the conversion factor g is essentially dependent on the HVL of the incident radiation and the compressed breast thickness. Although differences exist in g values for various anode/filter combinations, their magnitude (up to 5%) is small compared with the estimated accuracy (approximately ±30%) in the determination of AGD. It is therefore justified to use only one set of conversion factors as a function of HVL (Table III.1.7) for the anode/filter combinations currently employed (CEC 1996; Dance 1990; Zoetelief 1998).

Standard AGD (sAGD) for the 50-mm-thick standard breast can be calculated from the entrance surface air kerma K_E measured at the tube loading recorded for exposure of a 45-mm-thick PMMA phantom, which is defined as the "standard phantom" in the EPDM (CEC 1996):

$$sAGD \text{ (Standard breast)} = g_{PB} \cdot K_E \text{ (standard phantom)}$$
$$(III.1.18)$$

Values of g_{PB} are presented in Table III.1.8 as a function of HVL. Methods for measurement of the HVL are presented, e.g. in the EPDM (CEC 1996). For assessment of the results of HVL measurements, a comparison can be made with the values presented in Table III.1.9. For compression plates thinner than 3 mm PMMA, the HVL can be derived from the table by interpolation.

Table III.1.8 Conversion factors g_{PB} for calculating the AGD for a 50 mm thick standard breast from the entrance surface air kerma K_E measured at the tube loading recorded for exposure of the 45 mm thick PMMA (polymethylmethacrylate) standard phantom (from CEC 1996)

HVL (mm Al)	g_{PB} in mGy/mGy
0.25	0.149
0.30	0.177
0.35	0.202
0.40	0.223
0.45	0.248
0.50	0.276
0.55	0.304
0.60	0.326
0.65	0.349

For the calculation of AGD of patients, it has been assumed in the EPDM (CEC 1996) that all breasts have a standard composition (50% adipose/50% glandular tissue). However, evidence is accumulating that breast glandularity decreases with increasing breast thickness and also with increasing age (Beckett and Klotre 2000; Geise and Palchevsky 1996; Heggie 1996; Klein et al. 1997; Young et al. 1998). The assumption of a standard composition will result in overestimation of the AGD to breasts with a higher glandular content and underestimation of the AGD to breasts with a lower glandular content. To a certain extent, these effects will tend to cancel each other out when calculating the mean value of AGD for a sample of patients with a range of breast compositions. However, individual doses may be in error by as much as 25% (Heggie 1996; Klein et al. 1997). In particular, the highest radiation exposures, which occur for the largest breasts, will be underestimated by something of this order. Methods of estimating and correcting for breast glandularity are proposed by Beckett and Klotre (2000), Geise and Palchevsky (1996), Heggie (1996), Klein et al. (1997), and Young et al. (1998).

Table III.1.9 Typical half-value layers HVL in mm Al for mammography units with different anode-filter combinations operated at various tube voltages (from CEC 1996)

Anode- and filter-material	Tube voltage (kV)	HVL (mm Al) Without compression-plate	With compression-plate (3 mm PMMA)
Mo+30 μm Mo	25	0.28	0.34
	28	0.32	0.37
	30	0.34	0.38
	31	0.35	0.39
	34	0.36	0.40
Mo+25 μm Rh	22	0.30	0.34
	25	0.36	0.40
	28	0.40	0.44
	34	0.41	0.46
W+60 μm Mo	22	0.33	0.37
	25	0.35	0.39
	28	0.37	0.41
	30	0.38	0.42
W+50 μm Rh	22	0.41	0.43
	25	0.48	0.51
	28	0.51	0.54
	30	0.53	0.56
W+40 μm Pd	22	0.36	0.40
	25	0.44	0.48
	28	0.48	0.53
	30	0.50	0.55
Rh+25 μm Rh	23	0.31	0.36
	25	0.34	0.40
	28	0.39	0.45
	30	0.42	0.48

III.1.5.1.3 Computed Tomography

For computed tomography, the organ dose H_T is calculated from the slice-averaged axial air kerma K_{CT} according to:

$$H_T = K_{CT} \cdot \sum_{z_u}^{z_o} f_T(z) \qquad (III.1.19)$$

where $f_T(z)$ are the conversion factors for the organ dose contributions of subsequent 1 cm thick phantom slices at position z, z_u and z_o are the lower and upper limit of the exposed part of the body. $f_T(z)$ data for three radiation qualities have been published by Zankl et al. (1991).

III.1.5.2 Estimation of the Absorbed Dose to an Organ D_{org} by the Use of Tissue-Air Ratios or Relative Depth Doses

For special questions it is often possible to calculate the point dose quantity D_{org}, which then represents the organ-averaged dose H_T with sufficient accuracy. This can be done by the use of tissue-air ratios or relative depth doses.

If tissue-air ratios $T_a(z)$ are used, the absorbed dose to an organ D_{org} is given by:

$$D_{Org} = K_E \cdot \left(\frac{r_E}{(r_E + d)}\right)^2 \cdot t_{w/a} \cdot T_a \qquad (III.1.20)$$

where r_E is the distance of the entrance surface of the patient to the focus, d is the depth of the organ and $t_{w/a}$ is the quotient of the mass energy absorption coefficients of water and air. For the X-ray diagnostic energy range, $t_{w/a} = 1.05 \pm 0.03$ is valid. Data of tissue-air ratios $T_a(z)$ have been tabulated by Säbel et al. (1980) and Harrison (1983). As an example, the data of Säbel et al. (1980) are given in Tables III.1.10, III.1.11, III.1.12, III.1.13, III.1.14, III.1.15.

If relative depth doses $D_{rel}(z)$ are used, the absorbed dose to an organ D_{org} is given by:

$$D_{Org} = K_E \cdot B \cdot t_{w/a} \cdot D_{rel} \qquad (III.1.21)$$

where B is the back-scatter factor (see Chap. III.1.2). Data of relative depth doses $D_{rel}(z)$ for the X-ray diagnostic energy range have been published by Harrison (1981).

Table III.1.10 Tissue-air ratios T_a (from Säbel et al. 1980). Total filtration 2.6 mm Al; tube voltage: 60 kV; half-value layer: 2.2 mm Al

Depth (cm)	Field size (cm²)			
	10×10	15×15	20×20	30×30
0	1.269	1.280	1.280	1.280
1	1.120	1.166	1.166	1.200
2	0.917	0.960	0.965	0.982
3	0.723	0.763	0.770	0.797
4	0.563	0.611	0.623	0.642
5	0.442	0.490	0.502	0.525
6	0.349	0.393	0.405	0.429
7	0.273	0.315	0.326	0.349
8	0.215	0.253	0.263	0.285
9	0.169	0.203	0.213	0.233
10	0.133	0.162	0.170	0.190
12	0.082	0.105	0.110	0.126
14	0.051	0.067	0.072	0.085
16	0.031	0.043	0.046	0.056
18	0.019	0.029	0.030	0.037
20	0.013	0.018	0.021	0.025

Table III.1.11 Tissue-air ratios T_a (from Säbel et al. 1980). Total filtration 2.6 mm Al; tube voltage: 70 kV; half-value layer: 2.6 mm Al

Depth (cm)	Field size (cm²)			
	10×10	15×15	20×20	30×30
0	1.257	1.303	1.314	1.314
1	1.200	1.246	1.246	1.269
2	1.030	1.061	1.061	1.090
3	0.831	0.875	0.878	0.906
4	0.672	0.711	0.723	0.755
5	0.541	0.584	0.600	0.632
6	0.434	0.480	0.498	0.529
7	0.349	0.394	0.413	0.443
8	0.280	0.325	0.342	0.371
9	0.225	0.266	0.285	0.311
10	0.181	0.218	0.237	0.261
12	0.117	0.147	0.162	0.182
14	0.075	0.099	0.112	0.128
16	0.049	0.067	0.077	0.090
18	0.031	0.045	0.053	0.063
20	0.021	0.031	0.037	0.045

Table III.1.12 Tissue-air ratios T_a (from Säbel et al. 1980). Total filtration 2.6 mm Al; tube voltage: 80 kV; half-value layer: 3.0 mm Al

Depth (cm)	Field size (cm²)			
	10×10	15×15	20×20	30×30
0	1.303	1.360	1.360	1.360
1	1.246	1.314	1.314	1.314
2	1.070	1.177	1.166	1.166
3	0.885	0.994	0.993	0.989
4	0.722	0.825	0.833	0.840
5	0.591	0.682	0.697	0.715
6	0.483	0.565	0.584	0.609
7	0.395	0.466	0.489	0.518
8	0.323	0.386	0.409	0.441
9	0.265	0.319	0.342	0.375
10	0.216	0.264	0.287	0.319
12	0.145	0.181	0.201	0.231
14	0.097	0.123	0.141	0.167
16	0.065	0.085	0.099	0.121
18	0.043	0.058	0.069	0.088
20	0.029	0.039	0.048	0.064

Table III.1.13 Tissue-air ratios T_a (from Säbel et al. 1980). Total filtration 2.6 mm Al; tube voltage: 90 kV; half-value layer: 3.5 mm Al

Depth (cm)	Field size (cm²)			
	10×10	15×15	20×20	30×30
0	1.291	1.337	1.371	1.371
1	1.269	1.314	1.349	1.349
2	1.110	1.166	1.211	1.189
3	0.925	0.989	1.040	1.030
4	0.763	0.837	0.887	0.880
5	0.629	0.703	0.753	0.757
6	0.517	0.591	0.640	0.651
7	0.425	0.497	0.544	0.560
8	0.350	0.418	0.462	0.481
9	0.288	0.352	0.392	0.414
10	0.237	0.296	0.333	0.357
12	0.161	0.209	0.241	0.264
14	0.109	0.149	0.174	0.195
16	0.074	0.105	0.125	0.144
18	0.050	0.074	0.090	0.107
20	0.034	0.053	0.065	0.079

Table III.1.14 Tissue-air ratios T_a (from Säbel et al. 1980). Total filtration 2.6 mm Al; tube voltage: 100 kV; half-value layer: 3.9 mm Al

Depth (cm)	Field size (cm²)			
	10×10	15×15	20×20	30×30
0	1.314	1.371	1.383	1.383
1	1.269	1.349	1.360	1.360
2	1.141	1.246	1.246	1.246
3	0.962	1.061	1.080	1.080
4	0.802	0.907	0.928	0.949
5	0.667	0.770	0.795	0.823
6	0.555	0.654	0.681	0.715
7	0.462	0.555	0.584	0.621
8	0.384	0.471	0.501	0.539
9	0.319	0.400	0.429	0.469
10	0.266	0.341	0.368	0.406
12	0.184	0.245	0.270	0.306
14	0.127	0.176	0.198	0.231
16	0.088	0.127	0.146	0.174
18	0.061	0.093	0.107	0.131
20	0.042	0.066	0.079	0.099

Table III.1.15 Tissue-air ratios T_a (from Säbel et al. 1980). Total filtration 2.6 mm Al; tube voltage: 120 kV; half-value layer: 4.7 mm Al

Depth (cm)	Field size (cm²)			
	10×10	15×15	20×20	30×30
0	1.326	1.406	1.406	1.429
1	1.326	1.406	1.406	1.474
2	1.166	1.280	1.280	1.349
3	1.021	1.166	1.166	1.246
4	0.869	1.010	1.021	1.090
5	0.729	0.864	0.893	0.955
6	0.614	0.741	0.774	0.835
7	0.517	0.634	0.672	0.730
8	0.434	0.543	0.583	0.638
9	0.365	0.465	0.506	0.558
10	0.307	0.398	0.439	0.488
12	0.218	0.293	0.330	0.373
14	0.154	0.214	0.249	0.285
16	0.109	0.157	0.187	0.218
18	0.077	0.115	0.141	0.167
20	0.055	0.085	0.106	0.128

III.1.6 Estimation of Effective Dose

As already mentioned in Chap. II.3.3, in order to calculate effective dose, it is first necessary to estimate the radiation dose to the individual organs, included in Table II.3.1. Some authors have carried out Monte Carlo simulations for the determination of conversion factors, which allow the calculation of the effective dose directly from initial dosimetric quantities:

Le Heron (1992) has calculated conversion factors for the estimation of effective dose to the patient during medical X-ray examinations from measurements of the dose-area product. Hart et al. (1994) have determined corresponding factors for the conversion of entrance surface dose and dose-area product, respectively. Details of the problems concerning the calculation of effective dose are discussed by McCollough and Schueler (2000).

Table III.1.16 The maximum relative deviations of the relevant exposure parameters

Parameters which influence the determination of K_E and K_{CT}	Radiography	Fluoroscopy	Mammography	Computed tomography
X-ray tube voltage U_p[a]	5%	5%[b]	5%	5%
Total filtration in Al or Cu equivalent thickness[a]	10%	10%	10%	10%
Tube current-time-product $I_R t$	5%	–[b]	5%	5%
Dose output Y_{100} (Fig. III.1.3)c	25%	25%	25%	–
System dose K_B[d]	+100%/–25%	+100%/–25%	+50%/–10%	–
Focus distance r_m	2%	–	2%	–
Focus image receptor distance r_B	2%	5–15%[e]	2%	–
Focus entrance surface distance r_E	5%	5–20%[e]	5%	–
Field size A_B	10%	30%[f]	–	–
Slice thickness s or table feed (pitch) v	–	–	–	5%
Patient thickness d_p	10%	10%	10%	10%
Patient attenuation ratio $m_{patient}$	42%[g]	42%[g]	–	–
Attenuation ratio of materials between patient and image receptor $m_{absorber}$	10%	10%	–	–
Attenuation ratio of antiscatter grid m_{grid}	10%	10%	–	–
Back-scatter factor B	ca. 3%[h]	–	3%[h]	–
Percentage of scattered radiation S_g in the plane of the image receptor	15–30%[i]	15–30%[i]	–	–

[a] Today from manufacturers the accuracy of the peak tube voltage U of X-ray generators is stated as ±5%, but the entrance surface air kerma K_E is also dependent on the waveform of the tube voltage. In the uncertainty, only the accuracy of the peak voltage is included. Furthermore the total filtration of the X-ray tube assembly (inherent filtration and additional filtration) effect the entrance surface kerma, as shown by the examples in Table III.1.17

[b] During fluoroscopy often the X-ray projection is changed, resulting in a corresponding change of the exposure parameters. In modern X-ray equipment therefore, the tube voltage applied in the last fluoroscopy sequence is indicated on the generator console.

[c] The dose output Y_{100} depends – besides the waveform of the X-ray tube voltage – also on the anode target angle. The uncertainty given is related to the reference axis focal spot-image receptor

[d] The system dose K_B of the AEC as a rule will be adjusted, e.g. with respect to IEC 61223-3-1. Because of the energy dependence of the image receptor's sensitivity, the high uncertainty of the system dose shown in the table results. In modern X-ray equipment with integrated automatic correction of the voltage response (see Chap. II.6), the uncertainty of the system dose K_B can be reduced to about ±25%. In mammography a reduction to about ±10% is possible if an automatic transparency control (Aichinger et al. 1990) is available

[e] The determination of the *focus image receptor distance* r_B and of the *focus entrance surface distance* r_E in the fluoroscopy mode is complicated by varying projections. Furthermore the different geometrical circumstances of over- and under-table and C-arm units must be taken into account

[f] In the fluoroscopy mode, an average field size A_B (on the exit side of the patient) must be determined; therefore an uncertainty of 30% is assumed

[g] Especially for the determination of K_E, with the help of the system dose, the patient thickness must be estimated and contributes to the resulting uncertainty

[h] Uncertainty can be estimated from Petoussi et al. (1998; A_E field size on the entrance side of the patient)

[i] If the selectivity Σ of the antiscatter grid used is known, the percentage of the primary radiation in the plane of the image receptor can be estimated with the help of Fig. II.5.3 (example 80 kV) and Eqs. II.5.20 and II.5.21

III.1.7 Uncertainties in Patient-Dose Estimation

When estimating the dose to a patient, as a rule one starts with the entrance surface air kerma K_E as *initial dosimetric quantity* (see Table III.1.4). The possibilities for the direct measurement of K_E or the determination from other quantities has been discussed in Chap. III.1.4. To get an idea of the accuracy of the dose results which can be achieved with the various methods of measurement or calculation, the uncertainties of the physical parameters used for the determination of K_E must be known. Tables III.1.16 and III.1.17 give an overview of the uncertainties of the exposure parameters which have influence on the patient-dose estimation.

From the uncertainties of the exposure parameters follow finally the uncertainties of the initial dosimetric quantities (see Table III.1.18).

Table III.1.17 Influence of the tube voltage and filtration uncertainties on entrance surface air kerma K_E

Examples of exposure parameters (tube voltage and total filtration)		Uncertainty of K_E resulting from	
		X-ray tube voltage±5%	Filter thickness±10%
Radiography and fluoroscopy	80 kV: 3.0 mm Al	12%	8%
	80 kV: 2.5 mm Al+0.1 mm Cu	15%	6%[a]
	80 kV: 2.5 mm Al+0.2 mm Cu	18%	8%[a]
	80 kV: 2.5 mm Al+0.3 mm Cu	21%	9%[a]
Mammography	28 kV: Mo-anode/30 μm Mo-Filter	15%	11%
	28 kV: Mo-anode/25 μm Rh-Filter	16%	11%
	28 kV: W-anode/60 μm Mo-Filter	11%	18%
	28 kV: W-anode/50 μm Rh-Filter	13%	17%
	28 kV: Rh-anode/25 μm Rh-filter	16%	12%

[a] Only Cu-filter is taken into consideration

Table III.1.18 Relative uncertainties[a] of the measured or estimated initial dose quantities K_E or K_O, K_m, P_F, P_L and K_{CT}

Methods for the determination of the initial dosimetric quantities	Radiography	Fluoroscopy	Mammography	Computed Tomography
Direct measurement of K_E or K_O	10%	10%	10%	–
Measurement of K_m at focus distance r_m[b]	25%	–	–	–
Measurement of dose-area product P_F[b]	25%	25%	–	–
Measurement of P_L	–	–	–	10%
K_E determination from measurement of K_m	27%	–	–	–
K_E determination from measurement of P_F	29%	49%	–	–
K_E determination from measurement of K_O	11%	–	11%	–
K_E determination from dose output Y_{100}	35%	38%	38%	–
K_E determination from cut-off dose K_B	67%	70%	–	–
K_{CT} determination from measurement of P_L (P_L-tables)	–	–	–	13 (25)%

[a] The uncertainties ε of the measured or estimated quantities are given as relative uncertainties, whereby ε is given by ε = 2S. The variance S^2 of the resultant initial dose quantity is equal to $S^2 = \Sigma(s_i)^2$, s_i is the variance of the individual quantities, which are of influence to the initial dose quantity

[b] The relative uncertainties given are the permissible maximum uncertainties. They are valid with respect to routine measurements of the dose-area product or of the entrance surface air kerma in combination with a measuring device mounted in a focus distance r_m. Included are the uncertainties created by an additional filtration until 0.3 mm Cu, neglect of corrections with respect to the air pressure and neglect of the differences between over- and under-table examination units (see Chap. II.4.3)

References

Aichinger H, Joite-Barfuss S, Marhoff P (1990) Die Belichtungsautomatik in der Mammographie. Electromedica 58:68–69

Bauer B, Corbett RH, Moores BM, Schibilla H, Teunen D (1998) Reference doses and quality in medical imaging. Radiat Prot Dosimetry 80:1–3

Beckett JR, Klotre CJ (2000) Dosimetric implications of age related glandular changes in screening mammography. Phys Med Biol 45:801–813

Boone JM (1999) Glandular breast dose for monoenergetic and high energy X-ray beams: Monte-Carlo assessment. Radiology 213:23–37

CEC (Commission of the European Communities) (1996) European protocol on dosimetry in mammography. Report EUR 16263. CEC, Luxembourg

CEC (1997) Quality criteria for computed tomography. Working document EUR 16262. CEC, Brussels

Dance DR (1990) Monte Carlo calculation of conversion factors for the estimation of mean glandular breast dose. Phys Med Biol 35:1211

Department of Health and Human Services, Food and Drug Administration (1984) 21 CFR, Part 1020: Diagnostic X-ray systems and their major components; amendments to performance standards; final rule. Federal Register 49:171

DIN (Deutsches Institut für Normung) (1992), Medizinische Röntgenanlagen bis 300 kV – Regeln für die Prüfung des Strahlenschutzes nach Errichtung, Instandsetzung und Änderung. DIN 6815, Berlin

DIN (Deutsches Institut für Normung) (2002) Klinische Dosimetrie:Verfahren zur Ermittlung der Patientendosis in der Röntgendiagnostik. DIN 6809, Part 7 (Draft). Beuth, Berlin

Drexler G, Panzer W, Stieve F-E, Widenmann L, Zankl M (1993) Die Bestimmung von Organdosen in der Röntgendiagnostik. Hoffmann, Berlin

Geise RA, Palchevsky A (1996) Composition of mammographic phantom materials. Radiology 198:347–350

Gfirtner H, Stieve F-E, Wild J (1997) A new Diamentor for measuring kerma-area product and air-kerma simultaneously. Med Phys 24 (12):1954–1959

Harrison RM (1981) Central-axis depth-dose data for diagnostic radiology. Phys Med Biol 26:657–670

Harrison RM (1983) Tissue-air ratios and scatter-air ratios for diagnostic radiology (1–4 mm Al HVL). Phys Med Biol 28:1–18

Hart D, Jones DJ, Wall BF (1994) Estimation of effective dose in diagnostic radiology from entrance surface dose and dose-area product measurements. NRPB-R262. National Radiological Protection Board

Hart D, Jones DJ, Wall BF (1998) Normalised organ doses for medical X-ray examinations calculated using Monte Carlo techniques. NRPB-SR262 (Software Report). National Radiological Protection Board

Heggie JCP (1996) Survey of doses in screening mammography Australas Phys Eng Sci Med 19:207–216

ICRP (International Commission on Radiological Protection) (1987) Statement from the 1987 Como meeting of the ICRP. ICRP Publication 52. Ann ICRP 17(4)

IEC (International Electrotechnical Commission (1978) Characteristics of Anti-Scatter Grids used in X-ray Equipment. Publication 60627 (Geneva: IEC)

IEC (International Electrotechnical Commission (1999) Evaluation and routine testing in medical imaging departments – Part 3–1: Acceptance tests – Imaging performance of X-ray equipment for radiographic and radioscopic systems. Publication 61223–3–1 (Geneva: IEC)

IEC (International Electrotechnical Commission) (2001) Diagnostic X-ray imaging equipment – Characteristics of general purpose and mammographic antiscatter grids. Publication 60627 (Geneva: IEC)

Klein R, Aichinger H, Dierker J, Jansen JTM, Joite-Barfuss S, Säbel M, Schulz-Wendtland R, Zoetelief J (1997) Determination of average glandular dose with modern mammography units for two large groups of patients. Phys Med Biol 42:651–671

Le Heron JC (1992) Estimation of effective dose to the patient during medical X-ray examinations from measurements of the dose-area product. Phys Med Biol 37:2117–2126

McCollough CH, Schueler BA (2000) Educational treatise: calculation of effective dose. Med Phys 27 (5):828–837

Nagel HD (ed) (1999) Strahlenexposition in der Computertomographie. ZVEI-Fachverband Elektromedizinische Technik, Frankfurt

NCS (Netherlands Commission on Radiation Dosimetry) (1993) Dosimetric aspects of mammography. Report 6. NCS, Delft

Petoussi N, Zankl M, Drexler G, Panzer W, Regulla D (1998) Calculation of backscatter factors for diagnostic radiology using Monte Carlo methods. Phys Med Biol 43:2237–2250

Rosenstein M, Andersen LW, Warner GG (1985) Handbook of glandular tissue doses in mammography. FDA 85-8239. US Department of Health and Human Services, Rockville, MD

Säbel M, Bednar W, Weishaar J (1980) Untersuchungen zur Strahlenexposition der Leibesfrucht bei Röntgenuntersuchungen während der Schwangerschaft. 1. Mitteilung: Gewebe-Luft-Verhältnisse für Röntgenstrahlen mit Röhrenspannungen zwischen 60 kV und 120 kV. Strahlentherapie 156:502–508

Stanton L, Villafana T, Day JL, Lightfoot DA (1984) Dosage evaluation in mammography. Radiology 150:577–584

Stern SH, Rosenstein M, Renauld L, Zankl M (1995) Handbook of selected tissue doses for fluoroscopic and cineangiographic examination of coronary arteries. HHS Publication FDA 95–8289

Wu X, Barnes GT, Tucker DM (1991) Spectral dependence of glandular tissue dose in screen-film mammography. Radiology 179:143–148

Wu X, Gingold EL, Barnes GT, Tucker DM (1994) Normalized average glandular dose in molybdenum target – rhodium filter and rhodium target – rhodium filter mammography. Radiology 193:83–89

Young KC, Ramsdale ML, Bignell F (1998) Review of dosimetric methods for mammo-graphy. Radiat Prot Dosimetry 80:183–186

Zankl M, Panzer W, Drexler G (1991) The calculation of dose from external photon exposures using reference human phantoms and Monte Carlo methods. Part VI. Organ doses from computed tomographic examinations. GSF-Bericht 30/91

Zoetelief J, Fitzgerald M, Leitz W, Säbel M (1998) Dosimetric methods for and influence of exposure parameters on the establishment of reference doses in mammography. Radiat Prot Dosimetry 80:175–180

The influence of scattered radiation on image quality has been discussed in the previous chapters over and over again, because the knowledge of its effect on the visibility of image details is decisive for diagnosis. In the following it will be explained that it is of great importance for image quality and dose which method of scatter reduction technique is selected. In comparing different grid types, for example, one must take into consideration not only the geometrical characteristics of the grid design (e.g. strip density N, grid ratio r; see Chap. II.5.4), but also the materials used for their cover and interspaces (aluminium, paper, carbon fibre). Various publications (Boldingh 1961; Chan and Doi 1982; Aichinger et al. 1992, 1995; Sandborg et al. 1993; Wamser et al. 1997, 2001) have shown, that the characteristics of antiscatter grids should be adapted to the medical requirements. Moreover the introduction of digital radiography necessitates the reassessment of the grid design (Chan et al. 1990), because in digital imaging it is possibly not always necessary to increase the incident exposure, when a grid is employed and, furthermore, image-processing techniques can be used to improve image quality (see Chap. II.5.1). Besides a short general discussion about grid design, examples for these statements will be given, especially for scatter reduction in paediatrics and mammography. Especially in mammography, the air-gap technique will be of interest when digital imaging is applied.

III.2.1   Influence on Image Quality and Dose

X-Radiation leaving the patient consists of primary radiation, which contains all information about the tissue composition in its spatial intensity distribution, and scattered radiation, which impedes detail visibility. No new information can be gathered from this point until the image representation, it can only be selectively emphasised or suppressed, lost or deliberately thrown away (see Chaps. II.5 and II.7). Techniques to reduce the deleterious effect of scattered radiation on image quality are therefore decisive for a successful diagnosis (see Fig. III.2.1).

For imaging of low-contrast details, the radiation contrast C can be expressed as the product of the primary-beam radiation contrast C_p and the scatter degradation factor SDF (see Chap. II.5, Fig. II.5.6 and Table II.5.1):

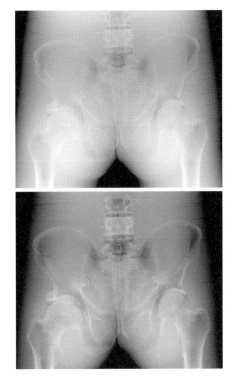

Fig. III.2.1a, b. Radiograph of a pelvis phantom: **a** without employing an antiscatter grid; **b** employing an antiscatter grid. Because of the improved contrast, the perceptibility is noticeably raised

$$C = C_p \cdot SDF \qquad (III.2.1)$$

This relation follows from (see also Eq. II.7.2):

$$C_p = \frac{\Delta I_P}{I_P} \qquad (III.2.2)$$

and:

$$C = \frac{\Delta I_P}{I_P + I_S} = \frac{\Delta I_P}{I_P \cdot (1 + \frac{I_S}{I_P})} = C_p \cdot SDF. \qquad (III.2.3)$$

where ΔI_P is the difference in the primary radiation intensity between the image detail and its surroundings; I_S is the scattered radiation intensity; and:

$$SDF = \frac{1}{(1+\frac{I_S}{I_P})} = \frac{1}{(1+\frac{S}{P})} \qquad (III.2.4)$$

whereby:

$$S = \frac{I_S}{I_P + I_S} \qquad (III.2.5)$$

is the percentage of scattered radiation and:

$$P = \frac{I_P}{I_P + I_S} \qquad (III.2.6)$$

the percentage of primary radiation in the radiation image.

The primary-beam radiation contrast C_p is the contrast that would result if no scattered radiation was imaged, i.e. the maximum possible radiation contrast. The SDF is the fraction of contrast that is imaged in the presence of scatter. With no scatter, the SDF factor is 1; with substantial scatter (S >90%), it ranges to approximately zero (see Chap. II.5).

Scattered radiation affects especially low-contrast details, so that radiographs taken without applying any method to reduce scattered radiation are often unsuitable for diagnosis (see Fig. III.2.1). This fact is understandable if one bears in mind that, for example, in chest radiographs in the region of the mediastinum (see Fig. III.2.2), more than 90% of the dose is contributed by scattered radiation (Niklason et al. 1981; Baydush et al. 2000).

The use of antiscatter grids is the most important and widely applied tool to reduce the amount of scattered radiation at the recording system. In chest examinations, high X-ray tube voltages are used (see Chap. II.7.7). Because of the high scatter fraction, grids which show a high selectivity (Σ) at tube voltages above 100 kV (e.g. a 12/40- or 17/70-grid) should be employed. In favour of an improved diagnosis, the increase in the patient dose by the grid (see Fig. II.4.9) must be accepted. In contrast to this, the employment of a *high-selectivity grid in paediatrics makes no sense* (see Chap. III.2.2.1).

In *digital imaging* the signal-to-noise ratio can be improved to some extent only by increasing the system dose (see Chap. II.7.4) or by using a grid above all with a *high primary radiation transmission T_p* (see Chap. III.2.2.3). If employing a grid, the dose in front of the grid, i.e. the patient exposure, should be increased only if diagnosis of the X-ray images would be otherwise hampered by a too-high image noise. From this statements follows – as already mentioned above – that the scatter reduction technique to be used must be adapted to the medical requirements in order to meet the demands of the ALARA-principle.

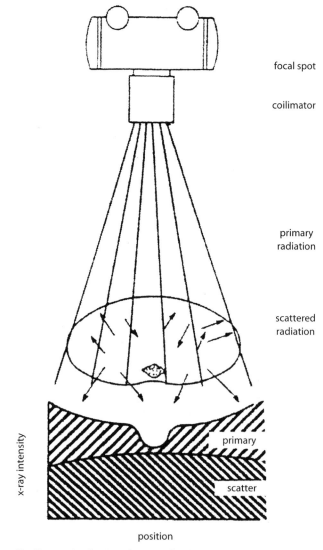

Fig. III.2.2. Distribution of scattered radiation in the plane of the image receptor (Chest; adapted from Barnes 1991)

III.2.2 The Various Types of Antiscatter Grids

The various focused antiscatter grid types differ from each other in the strip density N, the grid ratio r and in the materials used for the interspaces and the cover:

- The strip density of the grid decides whether the grid can be used as a stationary grid or if it must be moved during the exposure. The so-called high-strip density grids with more than 60 strips/cm no longer require a mechanical moving device and thus decentring of the grid during the exposure is eliminated. The use of such stationary, high-strip density grids leads to images free of grid lines in patient radiographs under normal viewing conditions. This has been shown in previous investigations of the imaging performance of high-strip density grids (Doi et al. 1983; Aichinger et al. 1992).

- An interspace medium, as a rule paper, carbon fibre, low-absorbing plastic or aluminium, is used. The choice of the material is especially decisive for the primary radiation transmission value T_p that can be obtained. Air would be an ideal interspace medium, but is restricted to a few applications (Rezentes et al. 1999; see Chap. II.5)

- In most cases lead is used for the strips. The height of the strips divided by the thickness of the interspace medium is called the grid ratio r (see Eq. II.5.4). The grid ratio is a simple geometrical characteristic which describes only approximately the performance of an antiscatter grid, although it serves often as a characteristic of grid design in national and international recommendations

- As cover, very often aluminium is employed. In fluoroscopy, mammography and paediatrics preference is given to carbon fibre, because the patient exposure can be further reduced by about 5% at low X-ray tube voltages.

III.2.2.1 Employment of Grids in Paediatrics

Especially in paediatrics, antiscatter grids should be used only if diagnosis can be improved, and if grids are employed they should be adapted with their geometrical design to the special exposure situation given (i.e. lower X-ray tube voltages and scatter fractions, higher radiation risk). Chapters II.5 and II.7.5 showed that the primary radiation transmission T_p and the selectivity Σ are decisive for optimal performance of antiscatter grids (i.e. low Bucky-factor B and high-contrast improvement factor C_{if}). Contrary to current opinion, an antiscatter grid with a *high grid ratio*, e.g. r = 15, if properly designed, can be much better suited for paediatrics than grids with a low ratio, e.g. r = 8 (as recommended in the European Guidelines; CEC 1996). The characteristics of two examples of such grid types measured according to IEC 60627 (IEC 1978) are shown in Table III.2.1. The corresponding detective quantum efficiency (DQE)-curves (Wagner 1977; see Chap. II.7.5) as a function of the scatter fraction S at a tube voltage of 60 kV and calculated with Eq. II.7.25, are drawn in Fig. III.2.3.

In general the DQE of the 15/80-grid is superior to that of the 8/40-grid. Only if the scatter fraction approximates 100% is the performance of the 8/40-grid (i.e. r = 8 and N = 40 cm^{-1}) nearly equivalent to that of the 15/80-grid (i.e.

r = 15 and N = 80 cm^{-1}). The reason for this behaviour can be explained with the example shown in Fig. III.2.4.

Represented is the interconnection between the height h of the lead strips and their thickness d (see Chap. II.5.4) for grids with a fixed strip density of N = 40 cm^{-1} and N = 80 cm^{-1}, and a ratio r = 8 and r = 15, respectively. With decreasing strip thickness d, the strip height h must be increased to obtain a constant grid ratio r:

$$h = r \cdot \left(\frac{1}{N} - d\right) \qquad (III.2.7)$$

Therefore the thickness D of the interspace medium increases simultaneously according to:

$$D = \frac{h}{r} \qquad (III.2.8)$$

and the primary radiation transmission T_p is improved. T_p is proportional to the ratio:

$$T_p \propto \frac{D}{D+d} = 1 - d \cdot N \qquad (III.2.9)$$

and this varies in Fig. III.2.4 from 52 to 71% for the 8/40-grid and from 4 to 92% for the 15/80-grid. Lead strips with a thickness of 20 µm are used for the 15/80-grid instead of 72 µm for the 8/40-grid, a thickness which just can be handled in the manufacturing process. With Eq. III.2.9 a primary radiation transmission of 84% results, a value which still must be corrected with respect to the attenuation of the imaging radiation by the interspace material and by the cover of the grid.

From these considerations it is clear that with the grid ratio r alone it is not possible to describe the grid performance completely. Especially in paediatrics, one should take into consideration that for small scatter fractions S → zero the Bucky-factor approximates the value B → $1/T_p$. The primary radiation transmission decides whether an antiscatter grid is recommended for paediatrics and also for that reason aluminium is unsuited as interspace material, because it would make the primary radiation transmission T_p worse.

At a given strip density N, the height h and the thickness d of the lead strips determine the selectivity Σ of an antiscatter grid; the thickness of the lead strips influences especially the dependence of the selectivity Σ on the X-ray tube voltage, because their thickness determines their translucence of scattered radiation.

To achieve, at a tube voltage of 100 kV (regarding IEC 60627, 1978; see Fig. III.2.5), the same selectivity for the 80-line grid in comparison with the 8/40-grid, a grid ratio

Table III.2.1 Typical values of the characteristics of antiscatter grids used especially in paediatrics at 60 kV, 70 kV and 100 kV

Characteristic	Pb 8/40-grid			Pb 15/80-grid		
	60 kV	75 kV	100 kV	60 kV	75 kV	100 kV
	2 mm Al	2 mm Al	4 mm Al	2 mm Al	2 mm Al	4 mm Al
T_p	0.61	0.63	0.65	0.68	0.70	0.74
T_s	0.076	0.088	0.125	0.062	0.080	0.137
T_t	0.17	0.18	0.22	0.17	0.19	0.24
Σ	8.0	7.1	5.2	11.0	8.7	5.4
C_{if}	3.54	3.38	2.96	3.93	3.65	3.01
B	5.80	5.37	4.56	5.77	5.21	4.07
SNR_{if}	1.50	1.46	1.39	1.63	1.60	1.49

Measuring arrangement according to IEC 60627 (1978), i.e. phantom (H₂O): field size 30 cm×30 cm, height 20 cm; additional filtration: 2 mm Al at 60 kV, 2 mm Al at 75 kV and 4 mm Al at 100 kV

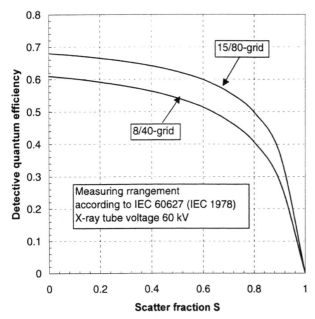

Fig. III.2.3. Detective quantum efficiency of the moving 8/40-grid in comparison with the stationary high-strip density 15/80-grid

of r = 15 is needed, because of the thinner lead strips used. In the direction of tube voltages ≤100 kV, the increase in the selectivity of the 80-line grid is steeper. By using Eqs. II.5.17, II.5.18 and II.7.25, it can be derived that the 15/80-grid delivers at least the same image quality at a lower patient dose than the 8/40-grid, as shown in Fig. III.2.6. The curves drawn in Figs. III.2.5 and III.2.6 have been determined by Monte Carlo simulation calculation and partly confirmed by measurements within ±10% (Aichinger et al. 1995; Aichinger and Säbel 1999).

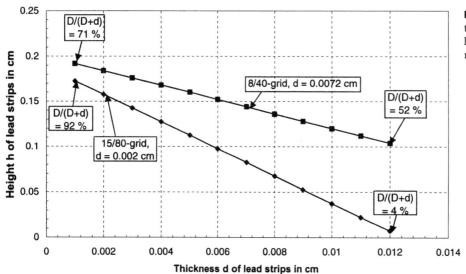

Fig. III.2.4. Interconnection between the height h and thickness d of the lead strips for grids with a fixed grid ratio and strip density

Fig. III.2.5. Dependence of the selectivity on X-ray tube voltage (MC = Monte Carlo, M = Measurement)

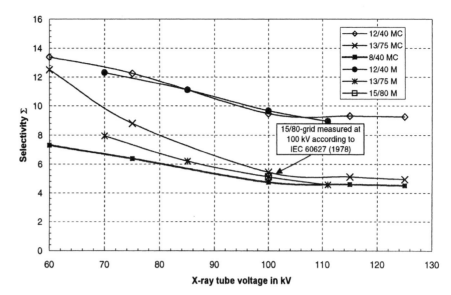

Fig. III.2.6a, b. Contrast improvement factor (**a**) and Bucky-factor (**b**) of the 8/40-grid, the 12/40-grid and the 15/80-grid as a function of the scatter fraction

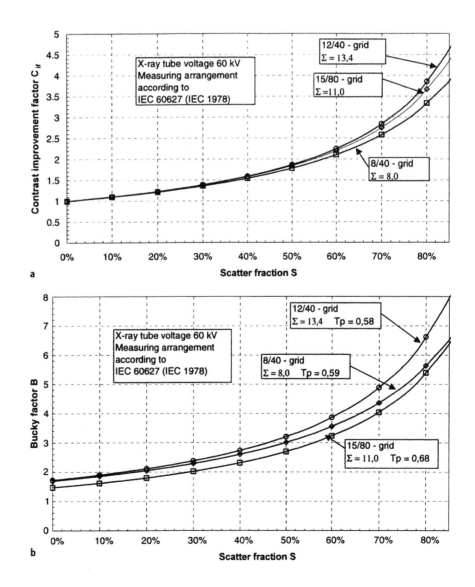

III.2.2.2 Employment of Grids in Mammography

Also in *film-screen mammography*, it is possible to improve the image contrast most effectively by the employment of antiscatter grids. They are made of an arrangement of exceptionally thin lead strips (16 μm) and a low-absorbing medium between them (see Table III.2.2 and Fig. III.2.7).

As a rule their cover is made from carbon fibre. Usually the mammography grid moves during the exposure, but occasionally a stationary grid arranged either inside or outside the cassette may be found. Moving grids normally have a grid ratio of 4:1 or 5:1, with a strip density of approximately 30 strips/cm. Stationary grids are high-strip density grids with a strip density of typically 80 strips/cm (Chan et al.1985; IPSM 1994); most often these grids have aluminium as interspace medium and cover.

For the employment of antiscatter grids, always a compromise between image quality and dose increase must be found. Therefore it makes no sense to increase the ratio of a grid in mammography above r = 4 to 5 (Friedrich 1975). This statement can be proven with the help of the physical characteristics of the grids (see Chap. II.5, Eqs. II.5.17 and II.5.18), which follow from their geometrical design. In contrast to the IEC standard 60627 of 1978 (IEC 1978), in the new version of 2001 (IEC 2001) the definitions for the measurement of the physical characteristics of mammography grids are included.

Analogous to the considerations in the previous chapter, Chap. III.2.2.1, the primary radiation transmission T_p and the selectivity Σ are decisive for the effectiveness of mammography grids. The most important difference between mammography and general radiography is the percentage of scattered radiation that can be reached. In mammography the maximum possible scatter fraction lies at about 60% (see Chap. II.5.2.2), whereas in general radiography nearly 100% are found for obese patients. As a result the selectivity (i.e. the grid ratio) in favour of a lower patient exposure should not be chosen too high. The primary radiation transmission, however, should be as high as possible (see Chap. III.2.2.3).

Table III.2.2 Typical values of the geometrical and physical characteristics of mammography antiscatter grids

Grid	Grid ratio	Strip density	Primary radiation transmission	Scattered radiation transmission	Total radiation transmission	Selectivity	Contrast improvement factor	Bucky factor
	r	$N\,(\mathrm{cm}^{-1})$	T_p (%)	T_s (%)	T_t (%)	Σ	C_{if}	B
Pb 4/27	4	27	72	23	49	3.10	1.48	2.05
Pb 5/31	5	31	72	22	47	3.35	1.55	2.14
Pb 3/80	3	80	55	18	40	3.06	1.37	2.5

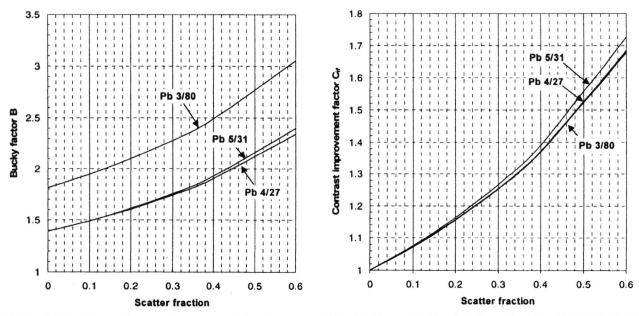

Fig. III.2.7. Bucky-factor and contrast improvement factor of mammographic grids: Mo-anode / 30 μm Mo-filter; tube voltage 28 kV. (IEC 2001)

Whereas mammography in connection with a moving grid is well accepted and considered as mandatory for thicker breasts, the application of stationary, high-strip density grids is controversial in mammography (Dershaw et al. 1985; Friedrich 1986). Possibly the recognisability of the shape of micro calcifications is negatively influenced by the grid strips, which can be seen during evaluation of the mammograms with the help of a magnifying glass. Furthermore high-strip density grids, in which aluminium is used as interspace medium, result in 25–35% more radiation exposure to the patient (Säbel and Aichinger 1996).

III.2.2.3 Employment of Grids in Digital Radiography

In digital radiography the *effect of an antiscatter grid on image quality* can be described by the evaluation of the *improvement of the signal-to-noise ratio* SNR_{if} (Chan et al. 1990; see Chap. II.5.4). In contrast to the film-screen technique, the dose must not be increased when a grid is inserted behind the breast. Only if the noise in the image cannot be tolerated must the dose be increased. Figure III.2.8 shows SNR_{if} as a function of the scatter fraction S at a tube voltage of 60 kV for different grids which are used in general radiography, and Fig. III.2.9, for the mammography grids mentioned above, when using a Mo-anode/Mo-filter-system at a tube voltage of 28 kV. The curves are calculated with the help of Eq. II.5.16.

The primary radiation transmission values T_p of the grids determine the point of intersection $T_{p,i}$ with the line where the SNR improvement factor $SNR_{if} = 1$. From Eq. II.5.16 follows, with regard to the intersection point:

$$T_{p,i} = \frac{1}{\sqrt{B}} \qquad\qquad (III.2.10)$$

On the left of the intersection point, i.e. at lower scatter fractions, the SNR_{if} becomes less than 1. The image quality is not improved by the grid, therefore the grid should be removed during imaging. In general radiography this point of intersection is especially important in paediatrics.

Figure III.2.9 supports investigations which suggest doubts that grids should be used in digital mammography (Baydush et al. 2000). Especially grids made with aluminium as interspace medium should not be employed in digital mammography. In the examples shown (see Fig. III.2.9), the image quality is improved by the standard mammography grids only for very thick breasts because of their low primary radiation transmission (see Table III.2.2). Only an ideal mammography grid (i.e. with $T_p = 1$ and $\Sigma = \infty$) would improve the signal-to-noise ratio also for small breast thicknesses.

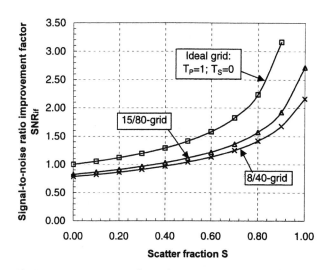

Fig. III.2.8. Improvement of signal-to-noise ratio (SNR_{if}) for 8/40-grid and 15/80-grid as a function of the scatter fraction at a tube voltage of 60 kV in comparison to an ideal grid

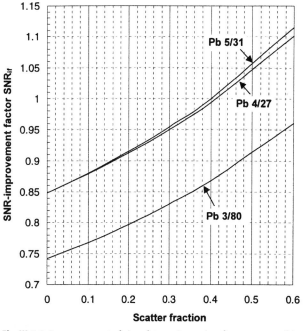

Fig. III.2.9. Improvement of signal-to-noise ratio of mammographic grids as a function of scatter fraction for the Mo/Mo-anode-filter system at a tube voltage of 28 kV

The scanning slot technique is a powerful tool to remove scattered radiation (see Chap. II.5). Such imaging systems have been evaluated for use in mammography in combination with the film-screen technique (Friedrich 1984) and digital line-detectors (e.g. CCD charge-coupled devices; Yaffe 1992). The high tube load and the mechanical complexity have up to now prevented application in routine diagnostics. However, if one combines the advantages of the air-gap technique and the slot technique (e.g.

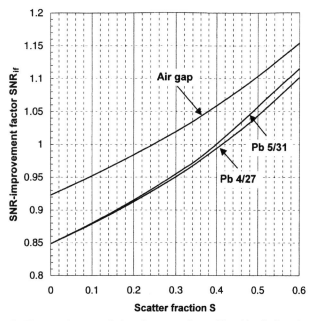

Fig. III.2.10. Air-gap technique in comparison with grid technique in mammography

Fig. III.2.10, where the SNR improvement by antiscatter grids and a 5-cm air gap is compared.

By application of Eq. II.4.6, a primary radiation transmission $T_p = 0.852$ results for the air-gap technique and, with the help of the Eqs. II.5.16 and II.5.18, one can calculate the curves shown in Fig. III.2.10 (see also Chap. IV.4.2). The image quality of a digital mammogram is – with the assumptions made – in general more improved (see also Krol et al. 1996) by the air-gap technique.

In conclusion, because in digital imaging the improvement of the signal-to-noise ratio SNR_{if} by an antiscatter grid is proportional to its primary radiation transmission T_p (see Eq. II.5.12), only low-absorbing material (e.g. paper, carbon fibre or air) should be used as interspace medium. Very impressively can this statement verified, if one replaces the low-absorbing medium of a grid design by aluminium without changing the strip and interspace thickness and consequently also the strip density.

In Fig. III.2.11, as an example the resulting improvement of the signal-to-noise ratio SNR_{if} dependent on the Bucky-factor is shown for a grid type which has a strip density of 70 cm^{-1} and a lead strip thickness of 36 μm. Only the strip height, i.e. the grid ratio r has been varied. From the diagram it follows that a grid which has an interspace medium made from paper and a grid ratio of r = 13 offers the same improvement of the signal-to-noise ratio as a grid which has an aluminium-interspace medium and a grid ratio of r = 17. The patient exposure, however, is reduced by about 25% when using the grid with r = 13. The broken borderline in Fig. III.2.11 agrees with the behaviour of a grid which in the ideal case would have a strip thickness of d = 0 (i.e. $T_p = 1$), whereby the transmission of scattered radiation T_s is varied from 1 to zero. In the limiting value $SNR_{if} = 2.36$, the ideal grid ($T_p = 1$ and

3 cm air gap and 10 mm slot width; Jing et al. 1998), an imaging system will be possible which delivers a scatter reduction comparable with grid technique and a restrained tube load. In addition it is of advantage in this technique that no beam hardening takes place by an interspace medium.

Furthermore the combination of a digital full-field image receptor with air-gap technique possibly makes the employment of antiscatter grids restricted only to very thick breasts. This conclusion can be drawn from

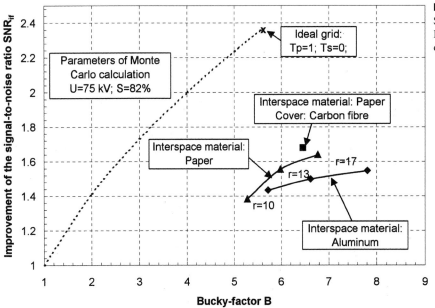

Fig. III.2.11. Interconnection between the SNR-improvement factor SNR_{if} and the Bucky-factor B for antiscatter grids with different material as interspace medium

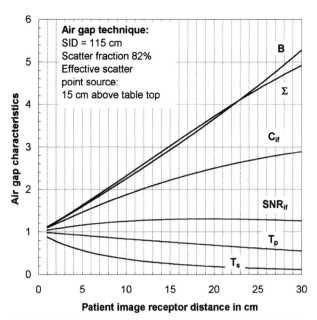

Fig. III.2.12. Air-gap characteristics (T_p, T_s, Σ, C_{if}, B and SNR_{if}) dependent on the distance between the patient output plane and the image receptor (SID source-to-image distance)

$T_s = 0$) would be obtained. It is therefore not surprising that, if the grid were to be provided with a cover made from carbon fibre instead of aluminium, the measuring point would be shifted in the direction of the broken line. Grids with the same grid ratio and strip density but with different interspace material show different grid characteristics. It is not allowed to compare different grids types simply and solely with respect to their grid ratio r.

In general, with respect to digital imaging, the advantages of grids and air gaps must be reconsidered. The properties of the scatter-reduction devices can be described by the primary radiation transmission T_p and the selectivity Σ. The obtained SNR_{if} can be used as performance indicator (Neitzel 1992). For all grid types, the SNR_{if} factor stays below a value of 2 (see Chap. IV.4.1 air-gap characteristics). Only for very high patient thicknesses are values >2 reached. For low and medium scatter conditions, optimised air-gap technique yields SNR_{if} factors of similar magnitude. This statement can be evaluated by using the concept of the *effective scatter point source* (Sorensen and Floch 1985) as discussed in Chap. II.5.3.1. The contrast-improvement factor and the Bucky–factor in this case will be calculated by using Eqs. II.5.17 and II.5.18, whereby the selectivity now is given by:

$$\Sigma = \frac{T_p}{T_s} = \frac{\dfrac{(f-a)^2}{f^2}}{\dfrac{x^2}{(x+a)^2}} \qquad (III.2.11)$$

where x is the distance of the effective scatter point source from the patient output plane (see Fig. II.5.8) and a the distance of the patient output plane from the location of the image receptor. f is the SID. In the resulting diagram (see Fig. III.2.12), the distance a is varied from 0 to 30 cm.

As mentioned in Chap. II.5.3.1, the most important uncertainty in this consideration is the location of the effective scatter point source. Nevertheless one comes to the conclusion that, in several medical examinations (e.g. paediatrics, mammography), one should do radiographs possibly without a grid.

References

Aichinger H, Säbel M (1999) Techniken der Streustrahlenreduktion bei der Röntgendiagnostik. Röntgenpraxis 52:251–264

Aichinger H, Staudt F, Kuhn H (1992) Multiline grids for imaging in diagnostic radiology – a physical and clinical assessment. Electromedica 60:74–81

Aichinger H, Seyferth W, Staudt F (1995) The specific effectiveness of high-strip-density grids in diagnostic radiology (abstract). Eur Radiol (Suppl) 5:70

Barnes GT (1991) Contrast and scatter in X-ray imaging. Radiographics 11:307–323

Baydush AH, Floyd CE Jr (2000) Improved image quality in digital mammography with image processing. Med Phys 27(7): 1503–1508

Baydush AH, Ghem WC, Floyd CE Jr (2000) Anthropomorphic versus geometric chest phantoms: a comparison of scatter properties. Med Phys 27(5):894–897

Boldingh WH (1961) Quality and choice of Potter Bucky grids. Parts IV, V. Acta Radiol 55:225–235

CEC (Commission of the European Communities) (1996) European guidelines on quality criteria for diagnostic radiographic images in paediatrics. EUR 16261EN (ISBN 92-827-7843-6) Brussels

Chan H-P, Doi K (1982) Investigation of the performance of antiscatter grids: Monte Carlo simulation studies. Phys Med Biol 27:785–803

Chan HP, Frank PH, Doi K, Iida N, Higashida Y (1985) Ultra-high-strip-density radiographic grids: a new antiscatter technique for mammography. Radiology 154:807–815

Chan H-P, Lam KL, Wu Y (1990) Studies of performance of antiscatter grids in digital radiography: effect on signal-to-noise ratio. Med Phys 17 (4):655–664

Dershaw DD, Masterson ME, Malik S, Cruz NM (1985) Mammography using an ultrahigh-strip-density, stationary, focused grid. Radiology 156:541–544

Doi K, Frank PH, Chan H-P, Vyborny CJ, Makino S, Iida N, Carlin M (1983) Physical and clinical evaluation of new high-strip-density radiographic grids. Radiology 147:575–582

Friedrich M (1975) Der Einfluss der Streustrahlung auf die Abbildungsqualität bei der Mammographie. Fortschr Rontgenstr 123:556–566

Friedrich M (1984) Schlitzblendentechnik für die Mammographie. Fortschr Rontgenstr 141 574–582

Friedrich M (1986) Ultra-Hoch-Linien-Dichte-Raster für die Mammographie: Eine Alternative zur Rastermammographie? Fortschr Rontgenstr 144:566–571

IEC (International Electrotechnical Commission) (1978) Characteristics of antiscatter grids used in X-ray equipment. Publication 60627. IEC, Geneva

IEC (International Electrotechnical Commission) (2001) Diagnostic X-ray imaging equipment – characteristics of general purpose and mammographic antiscatter grids. Publication 60627. IEC, Geneva

IPSM (Institute of Physical Sciences in Medicine) (1994) The commissioning and routine testing of mammographic X-ray systems, 2nd edn. Topic Group Report 59. IPSM, York, UK

Jing Z, Huda W, Walker JK (1998) Scattered radiation in scanning slot mammography. Med Phys 25(7):1111–1117

Krol A, Bassano DA, Chamberlain CC, Prasad SC (1996) Scatter reduction in mammography with air gap. Med Phys 23(7):1263–1270

Neitzel U (1992) Grids or air gaps for scatter reduction in digital radiography: a model calculation. Med Phys 19(2):475–481

Niklason LT, Sorenson JA, Nelson JA (1981) Scattered radiation in chest radiography. Med Phys 8(5):677–681

Rezentes PS, Almeida A de, Barnes GT (1999) Mammography Grid Performance. Radiology 210:227–232

Säbel M, Aichinger H (1996) Recent developments in breast imaging. Phys Med Biol. 41:315–368

Sandborg M, Dance DR, Alm Carlsson G, Persliden J (1993) Selection of antiscatter grids for different imaging tasks: the advantage of low atomic number cover and interspace materials. Br J Radiol 66:1151–1163

Sorenson JA, Floch J (1985) Scatter rejection by air gaps: an empirical model. Med Phys 12(3):308–316

Wagner RF (1977) Noise equivalent parameters in general medical radiography: the present and future pictures. Photo Sci Eng 21:252–262

Wamser G, Maier W, Aichinger H, Bohndorf K (1997) Clinical evaluation of a new stationary grid in comparison with a conventional grid: influence on image quality and patient dose (abstract). Radiology 205:617

Wamser G, Aichinger H, Maier W, Bohndorf K (2001) Clinical evaluation of a new stationary high-strip density antiscatter grid in comparison with a conventional moving grid: influence on image quality and patient radiation dose. Eur Radiol 11:1710–1719

Yaffe MJ (1992) Digital mammography. In: Haus AG, Yaffe MJ (eds) Syllabus: a categorical course in physics: technical aspects of breast imaging. RSNA, Oak Brook, IL, pp 245–256

The knowledge of the relationship that links image quality and radiation dose is a prerequisite to any optimisation of medical diagnostic radiology, because – according to the ALARA concept – the dose received by the patient during a radiological examination should be kept "as low as reasonably achievable" (see preface). The image quality and dose required for a successful and reliable diagnosis depends on physical parameters such as contrast, resolution and noise, the constitution of the patient, the viewing conditions (Brandt et al. 1983) and also on the characteristics of the observer that assesses the image. In Chap. II.7 (see Fig. II.7.2) the importance of the coordination of these influencing quantities has been pointed out. Furthermore, one should take into consideration that reducing the system noise by increasing the dose will not always improve task performance. This observation indicates that the imaging process might be optimised by accepting a higher system noise (e.g. in paediatrics). The following shows how exposure parameters can be adapted to the medical indication.

III.3.1 Image-Quality Figure
for General X-ray Diagnostics

Image quality is mainly evaluated by subjective methods, e.g. by visual comparison of patient images or images obtained with test objects. Due to the lack of reproducibility and the possibility of bias in these subjective methods, other techniques which are more objective and reproducible have been developed (ICRU 1996). Receiver operating characteristic analysis (ROC) is a frequently used method for quantifying image quality, but it needs interpretation of a large number of radiographic images with and without "the signal" to determine the detectability of an object. In addition, for optimisation problems the number of parameters which influence image quality is large and ROC techniques on the basis of "trial and error" are therefore very time-consuming and cumbersome.

One technique that needs less effort and can be used for optimisation of image quality with regard to the medical indication, including the aspect of minimising radiation exposure, was introduced in Chap. II.7.6 (Thijssen et al. 1988, 1989):

The threshold contrast of details with different contrast and size (contrast-detail curve) can be determined with the help of the CDRAD-phantom (see Fig. II.7.12) and the calculation of a single image-quality figure IQF_{th} (see Eq. II.7.31). The better the image quality is, the lower is the IQF_{th}. For some readers the inverse of this definition would be intuitively a better definition, therefore in the following example the inverse definition will be used:

$$IQF_{inv} = c \cdot \frac{N}{\sum\limits_{i=1}^{N} D_i \cdot C_i} \qquad (III.3.1)$$

where i is the number of the contrast step, N the total number of steps and c is a constant.

The CDRAD-phantom can be used to evaluate the radiation exposure and image quality also in the fluoroscopy mode, if the equipment enables the last image hold (LIH)-function. The LIH-function is an operation mode used in X-ray fluoroscopy systems which opens up the possibility to reduce fluoroscopy time and therefore to reduce patient dose. After each fluoroscopic scene, the last video frame is permanently shown on the monitor. In the case of the fluoroscopic image sequence, "temporal filtering" in the human visual system reduces perceived noise. This phenomenon is not present in a single video frame, but a stationary detail detected in the LIH-image in every case can be seen also in the fluoroscopic image sequence.

In the following example, an investigation (Supervision from Siemens) will be reported, which was carried out at the University Hospital in Nijmegen (Thijssen et al. 1988, 1997)

If the system dose rate at the image receptor during fluoroscopy and simultaneously the scanning speed of the television camera are cut in half, then also only half the exposure of the patient results. By integrating always two images on the television camera target (scanning time per image 80 ms instead of 40 ms), the signal-to-noise ratio remains the same in comparison with the normal fluoroscopy mode. However, each video image must then be reproduced twice on the monitor (see Fig. III.3.1).

In the case of fast organ movement, possibly sharpness is lost by a type of stroboscopic effect, which in most cases in routine diagnostics can be tolerated. If one combines

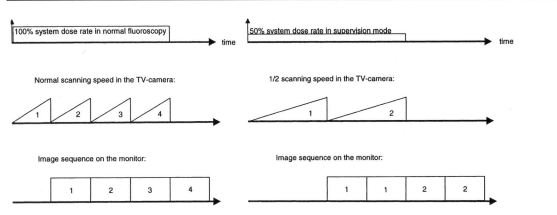

Fig. III.3.1. Principal functioning of the Supervision fluoroscopic mode

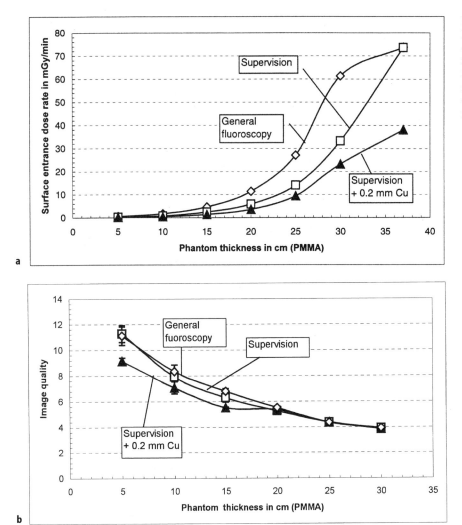

Fig. III.3.2a, b. Image-quality figure IQF$_{inv}$ and surface entrance dose dependent on phantom thickness for the normal fluoroscopy mode, the Supervision mode and the Supervision mode with additional filtration by 0.2 mm Cu. Standard deviation is indicated in Fig. III.3.2b

this fluoroscopic mode (Supervision) with an additional filtration of 0.2 mm Cu, the patients surface entrance dose can be reduced in comparison with the normal fluoroscopy mode even by about 70%.

Figure III.3.2a shows the surface entrance dose and Fig. III.3.2b the inverse image quality figure dependent on the phantom thickness obtained with the standard fluoroscopy mode, the Supervision mode and the Supervision mode in combination with an additional filter of 0.2 mm Cu. The CDRAD-phantom was always placed in the middle of the PMMA pile.

At phantom thicknesses lower than 20 cm, the image quality is somewhat reduced when using the additional filtration. This effect could be compensated if the fluoroscopic regulation curve (the so-called "anti-isowatt-curve") was corrected to a somewhat lower tube voltages (see Fig. II.6.12).

III.3.2 Image-Quality Figure in Angiography

The interest in models that can *predict* image quality *by simulation calculation* and improve the radiologist's performance for detection of lesions in medical images dependent on image parameters is understandably considerable. Cost and investigation time can be saved. One possible method is the concept of the "image-quality figure" (IQF) mentioned already in Chap. II.7.6 (Eq. II.7.29), its application with respect to angiography shall be discussed in more detail in the following:

Blood vessels can be visualised in X-ray diagnostic radiology only by injection of contrast material. *Enhancement* of attenuation against the surrounding tissue is achieved by introducing material where optimal use is made of the energy dependence of the photoelectric absorption (e.g. K-edge of iodine at 33.2 keV; see

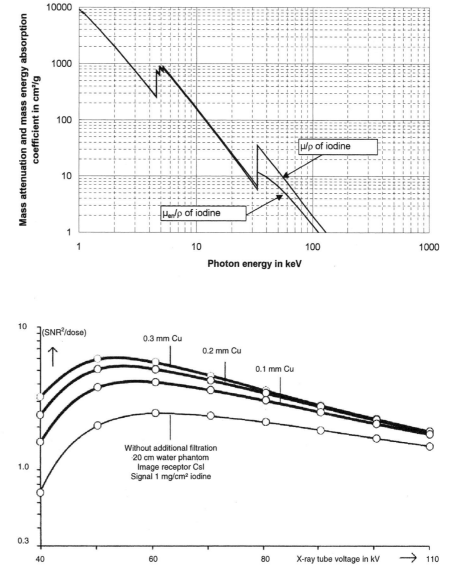

Fig. III.3.3. Mass attenuation coefficient and mass energy absorption coefficient of iodine as a function of photon energy

Fig. III.3.4. Image-quality figure for the determination of the optimal X-ray tube voltage in angiography together with additional filtration with 0.1 mm Cu, 0.2 mm Cu and 0.3 mm Cu

Fig. III.3.3). A *reduction* of attenuation against the surrounding tissue is achieved by using a gas (CO_2) because of its low mass density. It should be noted that in angiography the boundaries of the vessels are visualised, not their structure.

With regard to Fig. II.7.5, the determination of the optimal X-ray tube voltage in angiography with the help of the IQF can be explained. The detail in the phantom is supposed to represent the cross-section of a blood vessel filled with iodine contrast medium (mass density 1 mg/cm^2). Figure III.3.4 shows the IQF as a function of the X-ray tube voltage U in the case that the detail is introduced in a 20-cm-thick water phantom and an additional filtration of 0.1 mm, 0.2 mm or 0.3 mm Cu is applied to the X-ray tube assembly.

Each curve of the diagram shows a broad maximum which is caused by the K-absorption edge of iodine at a photon energy of 33.2 keV. When selecting tube voltages corresponding to the maximum values of these curves, a good compromise between image quality and radiation exposure of the patient is achieved at the system dose used for the imaging system. A more careful consideration indicates that, with increasing additional filtration, the optimal tube voltage is shifted to lower values.

Furthermore Fig. III.3.4 shows that it makes no sense to use an additional filtration of more than 0.3 mm Cu, because the IQF cannot be much improved. On the other hand the tube load is further increased. If one calculates the behaviour of the IQF for various equivalent phantom thicknesses, e.g. for 5–40 cm water, one can determine complete regulation curves for fluoroscopy and radiography (see Fig. III.3.5).

Very similar calculations can be made for an examination of the stomach. For the contrast meal, a material with high specific mass density is used (Ba, K-edge at 37.4 keV). Gagne et al. (1994) call the IQF a measure of *imaging performance per dose*. They have evaluated, for the imaging of a barium test object, that the maximum of the squared signal-to-noise ratio (SNR^2) per dose for wa-

ter-phantom thicknesses of 10–30 cm lies in the range of 59.5–63 kV. Tapiovaara et al. have used the IQF to determine improved technique factors for paediatric fluoroscopy. The optimal technique was defined as the one that minimises the absorbed dose in the patient with a constraint of constant image quality (Tapiovaara et al. 1999).

III.3.3 Image-Quality Figure in Mammography

In general X-ray diagnostics, the adaptation of the radiation quality to the medical indication is made mainly by the selection of the X-ray tube voltage and sometimes also by changing the additional filtration. In mammography the possibility of adapting radiation quality to object thickness by selection of the tube voltage alone is very limited together with the conventional molybdenum anode-molybdenum filter system (Gajewski and Reiss 1974; Jennings et al. 1981; Aichinger et al. 1994; Säbel and Aichinger 1996). The relative energy distribution of the X-ray photons (i.e. the X-ray spectrum) changes only insignificantly with an increase in the tube voltage from 25–30 kV (see Fig. III.3.6).

If, however, a *tungsten anode in combination with a molybdenum or a rhodium filter* is available, the X-ray spectrum at 30 kV is clearly shifted towards higher energies in comparison with the spectrum at 25 kV. The W/Rh-system is of advantage, especially for the imaging of thicker breasts (Klein et al. 1997).

Similar conclusions like that with regard to the Mo/Mo-system can be drawn when using an X-ray tube which contains a rhodium anode in combination with an additional filter of rhodium. The maximum flexibility in the adaptation of the radiation quality to the breast thickness or density is obtained, if one combines different anode-filter systems in one X-ray tube assembly (dual-track X-ray tube). In principle every element with absorption edges in the range of 18–25 keV is a suitable target and filter material (see Table III.3.1).

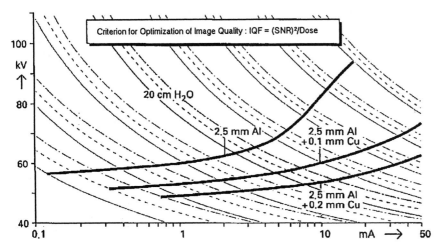

Fig. III.3.5. Characteristic curves for fluoroscopy and exposure: Calculated kilovolt-milliampere-regulation curves are shown, on which the image-quality figure IQF is always at its maximum. The crosswise running curves correspond to phantom thicknesses of 5 cm, 10 cm, 15 cm, 20 cm, 25 cm, 30 cm, 35 cm and 40 cm water. The primary X-radiation was filtered with
——— 2.5 mm Al
---- 2.5 Al + 0.1 mm Cu
--·-- 2.5 Al + 0.2 mm Cu

Fig. III.3.6. X-ray spectra of Mo/Mo-system, and W/Mo- and W/Rh-system at tube voltages of 25 kV and 30 kV, respectively

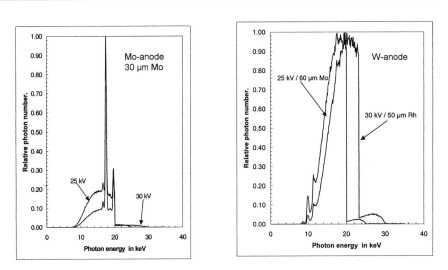

Fig. III.3.7. Mammography: Image-quality figure as a function of photon energy for various phantom thickness

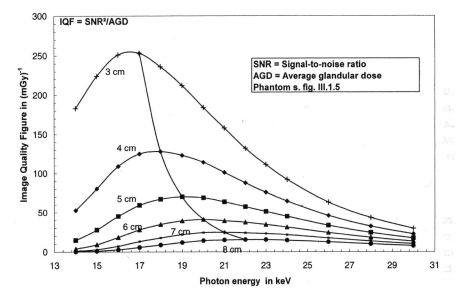

To obtain information about the optimal radiation quality for mammography, one can again use the concept of the IQF by calculating this quantity, e.g. with the assumption of mono-energetic X-radiation as a theoretical ideal. As indicated in Fig. II.7.5, one determines first the signal-to-noise ratio SNR for the display of a micro-calcification (e.g. simulated by $CaPO_4$) and then the average glandular dose AGD (see Chap. III.1.5.1.2). With these quantities one can calculate the IQF:

$$IQF = \frac{SNR^2}{AGD} \qquad (III.3.2)$$

Presently it is generally assumed that the glandular tissue is the most vulnerable in the breast with regard to the induction of cancer by ionising radiation. Therefore the average glandular dose is used as a risk-relevant dose also in Eq. III.3.2.

Table III.3.1 Elements which can be used as filter material in mammography

Chemical element	K-edge (keV)	Available as foil
Niobium	18.9	Yes
Molybdenum	20.0	Yes
Technetium	21.0	No
Ruthenium	22.1	No
Rhodium	23.2	Yes
Palladium	24.3	Yes

Figure III.3.7 shows the result of the simulation calculation dependent on the energy for various phantom thicknesses. For each object thickness, the most favourable X-ray energy, where the IQF reaches its maximum, can be read from the x-axis. For an object thickness of 3 cm, it begins at about 17 keV and reaches 22 keV for an object thickness of 8 cm. At these energies, from the radiological point of view, the most favourable compromise between

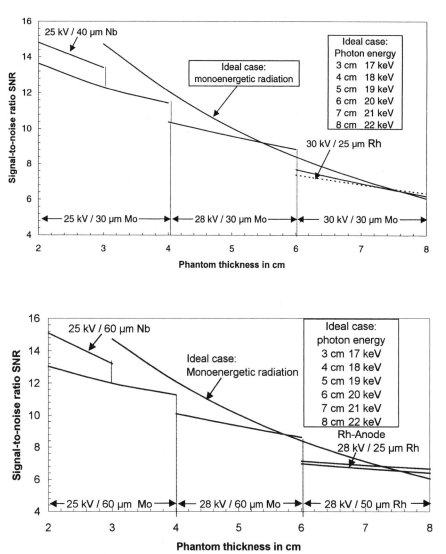

Fig. III.3.8. Signal to noise ratio dependent on the phantom thickness: Mo-anode in combination with K-edge filter

Fig. III.3.9. Signal to noise ratio dependent on the phantom thickness: W-anode in combination with K-edge filters

Table III.3.2 Anode-filter systems used in mammography and the corresponding half-value layers (HVL) of the primary radiation beam at 25 kV and 28 kV

Anode and filter materials	HVL at 25 kV (mm Al)	HVL at 28 kV (mm Al)
Mo+30 μm Mo	0.28	0.32
Mo+25 μm Mo	0.36	0.40
W+60 μm Mo	0.35	0.37
W+50 μm Rh	0.48	0.51
W+40 μm Pd	0.44	0.48
Rh+25 μm Rh	0.34	0.39

signal to noise ratio and radiation exposure of the patient is reached.

In the real case of using X-radiation from a molybdenum or tungsten tube (or a dual-track tube with two focal spots made with the anode-materials Mo/Rh or Mo/W), one must try to approximate the ideal of mono-

energetic radiation as well as possible by variation of the peak X-ray tube voltage, the filter material and the filter thickness, whereby the filter thickness in every case must be in accordance with the values shown in Table III.3.2.

Approximation of the mono-energetic case can be realised for example by simulation calculation carried out with data of measured X-ray spectra (see Part IV) and the model of a standard breast phantom (see Fig. III.1.5). Figure III.3.8 shows as an example the calculated signal to noise ratio SNR of an optimised combination of tube voltages and filters dependent on the object thickness in connection with a molybdenum anode and Fig. III.3.9 of a corresponding combination in connection with a tungsten anode.

The result is that with both anode materials (Mo and W) it is possible to obtain mammograms of high quality. For very thin breasts also the application of niobium (Nb) as filter material together with Mo- and W-anodes would be of interest (Aichinger et al. 1994; see Part IV). In clini-

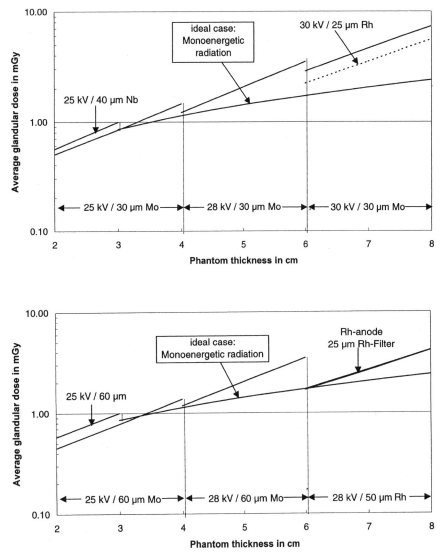

Fig. III.3.10. Average glandular dose (AGD) dependent on the phantom thickness: Mo-anode in combination with K-edge filters

Fig. III.3.11. Average glandular dose (AGD) dependent on the phantom thickness: W-anode in combination with K-edge filters

cal practice, Mo/Nb- and W/Nb-systems are not used because of their limitation to very thin breasts.

In contrast to the SNR, the resulting AGD is strongly dependent on the anode material. Figures III.3.10 and III.3.11 show the AGD dependent on the object thickness for the same anode-filter combinations as in Figs. III.3.8 and III.3.9. As a film-screen system, the Kodak MinR/SO177 or OM1/MinR system was assumed in the simulation calculation. A comparison of these results in detail reveals that – at a given system dose of about 70 µGy – the tungsten tube offers advantages with respect to the molybdenum-anode, above all within the object thickness range of 6–8 cm. Compared with the Mo/Mo-system, a reduction of the AGD to about 50% can be achieved.

References

Aichinger H, Dierker J, Säbel M, Joite-Barfuss S (1994) Image quality and dose in mammography. Electromedica 62:7–11

Brandt G-A, Boitz F, Mansfeld L, Rotte K-H (1983) Zum Stellenwert der Röntgenfilmbetrachtungsbedingungen für die Diagnosequalität bei Thoraxaufnahmen. Radiol Diagn 24:85–90

Gagne RM, Quinn PW, Jennings RJ (1994) Comparison of beam-hardening and K-edge filters for imaging barium and iodine during fluoroscopy. Med Phys 21 (1):107–121

Gajewski H, Reiss KH (1974) Physik und Technik der Weichstrahltechnik. Radiologe 10:438–446

ICRU (International Commission on Radiation Units and Measurements) (1996) Medical imaging – the assessment of image quality. Report 54. ICRU, Bethesda, MD

Jennings RJ, Eastgate RJ, Siedband MP, Ergun DL (1981) Optimal X-ray spectra for screen-film mammography. Med Phys 8:629–639

Klein R, Aichinger H, Dierker J, Jansen JTM, Joite-Barfuss S, Säbel M, Schulz-Wendtland R, Zoetelief J (1997) Determination of average glandular dose with modern mammography units for two large groups of patients. Phys Med Biol 42:641–671

Säbel M, Aichinger H (1996) Recent developments in breast imaging. Phys Med Biol 41:315–368

Tapiovaara MJ, Sandborg M, Dance DR (1999) A search for improved technique factors in paediatric fluoroscopy. Phys Med Biol 44:537–559

Thijssen MAO, Rosenbusch G, Gerlach H-J (1988) Reduktion der Strahlenexposition bei Durchleuchtung. Electromedica 56 (4):126–133

Thijssen MAO, Thijssen HOM, Merx JL, Lindeijer JM, Bijkerk KR (1989) A definition of image quality: the IQF. In: Moores BM, Wall BF, Eriskat H, Schibilla H (eds) Optimization of image quality and patient exposure in diagnostic radiology. BIR Report 20. British Institute of Radiology, London, pp 29–34

Thijssen MAO, Joosten F, Aichinger H (1997) Dose reduction during fluoroscopy with supervision: a physical and clinical evaluation. Eur Radiol (suppl 116) 7:24

Part IV   Supplement

Table IV.1 Related data in the book and on the CD-ROM

Book	CD-ROM	Reference to chapter
IV.1 X-ray spectra IV.1.1 General X-ray diagnostics: Total filtration 2.5 mm Al, target angle 10° 40 kV, 50 kV, 60 kV, 70 kV, 80 kV, 90 kV, 100 kV, 110 kV, 125 kV, 150 kV (Diagrams: 0.2 keV gradation, Tables: gradation 1 keV) X-ray spectrum at various target angles (Diagram 6°, 10° and 16°) The effect of the X-ray tube voltage wave form on the energy distribution on the photon spectra (Diagram 2P-, 6P-, 12P- and DC-Generator)	**1_X-ray spectra** General X-ray diagnostics: X-ray spectra with Be-window and filtered by 2.5 mm Al 10 kV until to 125 kV target angle 12° (Tables: gradation 0.2 keV, Data file: **QM_RS.xls**) X-ray spectra with total filtration 2.5 mm Al 40 kV, 50 kV, 60 kV, 70 kV, 80 kV, 90 kV, 100 kV, 110 kV, 125 kV, 150 kV Target angles 16°, 10°, 6° (Tables: gradation 0.2 keV, Data file: **QM_RD.xls**) Target angles 16°, 12°, 10°, 8°, 6° (Spectra with target angles 8° and 12° interpolated, all spectra with gradation 1 keV Data file: **QS_RD.xls**)	II.1
IV.1.2 Mammography: Anode-filter combinations: Mo/40 μm Nb:25 kV Mo/30 μm Mo:25 kV, 28kV Mo/30 μm Mo: 30 kV, 32 kV Mo/25 μm Rh: 28 kV, 30 kV W/80 μm Nb: 25 kV W/60 μm Mo: 25 kV, 28 kV W/50 μm Rh: 30 kV, 34 kV W/50μm Pd: 28 kV, 30 kV Rh/25 μm Rh: 25 kV, 28 kV Rh/25 μm Rh: 30 kV, 32 kV (Diagrams of spectra with target angle = 22°, emission angle 16°)	Mammography: Anode-filter combinations: Mo/40 μm Nb, Mo/30 μm Mo, Mo/25 μm Rh, Rh/25 μm Rh, W/80 μm Nb, W/60 μm Mo, W/50 μm Rh, W/50μm Pd 22 –36 kV in 1 kV steps (Tables of spectra with target angle 22°, emission angle 16°, gradation 0.1 keV, Data file: **QS_MD.xls**)	II.1
IV.2 Interaction coefficients Mass attenuation, mass energy absorption (NIST 2001) IV.2.1 Elements: Al, Si, Cu, Se, I, Gd, Pb IV.2.2 Compounds and mixtures: Air, H_2O, PMMA, CsI, Gd_2O_2S, $CaWO_4$ (Diagrams and tables of the most important materials adapted from NIST 2001)	**2_Interaction coefficients** of elemental media, compounds and mixtures (mass attenuation, mass energy absorption from NIST 2001: Data files: **MYR_NIST_Elements.xls, MYR_NIST_Compounds.xls**) Tables of the most important materials: (General diagnostics: gradation 1 keV and 0.2 keV, materials important in mammography 0.1 keV) (Data files: **MYR151.xls, MYR150_2.xls, MYR_50_1.xls**)	II.2 II.4 II.6
IV.3 Characteristics of the primary radiation beam (without scattered radiation) IV.3.1 General X-ray diagnostics: HVL, <E>, Φ/It, Ψ/It, K_a/It, relative dose, photon numbers, relative energy absorption in detector materials for X-radiation filtered by Al, Cu, H_2O (Tables IV.3.1 to VI.3.10; $CaWO_4$: 68 mg/cm², Gd_2O_2S: 68 mg/cm², 118 mg/cm² and 181 mg/cm², BaFBrI: 80 mg/cm² and CsI: 180 mg/cm²) QDE for detector materials (BaFBrI, $CaWO_4$, CsI, Gd_2O_2S) Interconnection between total filtration and half value layer (2 diagrams: AL-HVL, HVL-angle)	**3_Primary beam characteristics** (without scattered radiation) General X-ray diagnostics: HVL, <E>, Φ/It, Ψ/It, K_a/It, relative dose, photon numbers, relative energy absorption in detector materials for X-radiation filtered by Al, Cu, H_2O ($CaWO_4$: 68 mg/cm², GOS: 68 mg/cm², 118 mg/cm² and 181 mg/cm², BaFBrI: 80 mg/cm² and CsJ: 180 mg/cm², Data files: **RD_Tabs.xls**) In addition: (s. 6_Dosimetric data): Tables for water and PMMA at different additional Al and Cu filtration (Data files: **KT_water and KT_PMMA.xls**)	II.3 II.6
IV.3.2 Mammography: HVL, <E>, Φ/It, Ψ/It, K_a/It, relative dose, photon numbers, relative energy absorption in detector materials for X-radiation filtered by PMMA (Tables IV.3.11 to IV.3.18) $CaWO_4$: 34 mg/cm², Gd_2O_2S: 34 mg/ cm², BaFBrI: 40 mg/cm², a-Se: 60 mg/cm² and CsI: 80 mg/cm²) QDE for detector materials (BaFBrI, $CaWO_4$, CsI, Gd_2O_2S, Se) Relation between Al half-value layer and X-ray tube voltage (for the anode-filter combinations in IV.1.2)	Mammography: HVL, <E>, Φ/It, Ψ/It, K_a/It, relative dose, photon numbers, relative energy absorption in detector materials for X-radiation filtered by PMMA ($CaWO_4$: 34 mg/cm², Gd_2O_2S: 34 mg/ cm², BaFBrI: 40 mg/cm², a-Se: 60 mg/cm² and CsJ: 80 mg/cm², Data files: **MD_Tabs.xls**)	II.3 II.6

Book	CD-ROM	Reference to chapter
IV.4 Characteristics of the imaging radiation field	**Characteristics of the imaging radiation field**	II.5
IV.4.1 General X-ray diagnostics:	General X-ray diagnostics:	III.2
Typical values of geometrical and physical characteristics of grids used in paediatrics (Pb 8/40; Pb 15/80)	s. 4_ **Scatter**	
Characteristics of anti-scatter grids: (Selectivity, Contrast improvement factor, Bucky-factor, Improvement factor of the signal-to-noise ratio: Pb 8/40, Pb 12/40, Pb 13/75, Pb15/80 and Pb 17/70)	Backscatter factors (Data file: **Scatter \ BSF.xls**)	
Scatter fraction at image receptor with and without anti-scatter grid (Tables IV.4.2 to IV.4.8)	Scatter fraction at image receptor with and without anti-scatter grid (Data file: **Scatter \ Scafrac.xls**)	
IV.4.2 Mammography:	Mammography	II.5
Scatter fraction S in dependence on field size and on X-ray tube voltage (adapted from Barnes 1978)	—	III.2
Typical values of geometrical and physical characteristics of grids used in mammography (Pb 4/27, Pb 5/31, Pb 3/80)		
Characteristics of anti-scatter grids and air gap (Selectivity, Contrast improvement factor, Bucky-factor, Improvement factor of the signal-to-noise ratio: Pb 4/27, Pb 5/31 and Pb 3/80		
IV.5 Miscellaneous:	—	II.4
IV.5.1 Penetration and absorption of X-rays:		II.5
Typical values of attenuation ratio of material between patient and image receptor (IEC 61223-3-1 (1999)		III.2
Exposure parameters in exposure points		
IV.5.2 X-ray detectors:	X-ray detectors:	II.6
General diagnostics:	General diagnostics:	
Calculated voltage response of an image intensifier (180 mg/cm^2 CsI) (with additional filtration: 0 mm Cu, 0.1 mm Cu, 0.2 mm Cu and 0.3 mm Cu)	s. 5_**Image quality figure** file SN_CSI.xls	
Measured voltage response of an image intensifier		
Reciprocal sensitivity of an ionisation chamber		
Calculated voltage response of a Gd_2O_2S:Tb (118 mg/cm^2) (Absorber: 5 cm to 35 cm water in 5 cm steps)	file **SN_GOS.xls**	
Measured voltage response of a film-screen combination Gd_2O_2S:Tb		
Mammography	Mammography	II.3
Calculated voltage response of a Gd_2O_2S:Tb-screen (34 mg/cm^2) for Mo/Mo-, Mo/Rh-, W/Mo-, W/Rh- and Rh/Rh-anode-filter systems	s. 5_**Image quality figure** file **SN_MDGOS.xls**	II.6
IV.5.3 Image quality figure Angiography:	Image quality figure	II.7
IQF in dependence on X-ray tube voltage for various phantom thicknesses and filtration (0 mm Cu, 0.1mm Cu, 0.2 mm Cu and 0.3 mm Cu)	General diagnostics	III.3
	Data for the evaluation image quality and dose of Gd_2O_2S systems (Data files: **SN_GOS.xls**)	
	Angiography	
	IQF in dependence on X-ray tube voltage (Data file: **SN_CSI.xls**)	
Mammography	Mammography	II.7
IQF in dependence on X-ray tube voltage for a 6 cm standard breast and various anode-filter combinations	Data for the evaluation of the IQF in dependence on the anode-filter combinations (Data files: **SN_MDGOS** and **SN_M_GOS.xls**)	III.3

Book	CD-ROM	Reference to chapter
IV.6 Patient dose estimation	**6_Dosimetric data**	II.3
IV 6.1 General X-ray diagnostics:	General X-ray diagnostics:	III.1
Dose output Y_{100} (DIN 6909-7, draft 2002)	Dose output Y_{100}	
Backscatter factors (Petoussi et al. 1998)	(Data file: **KT_Water** or **KT_PMMA.xls**)	
Attenuation factor m	Backscatter factors (s. data file **Scatter \ BSF.xls**)	
(IEC 61223-3-1 1999; DIN 6809-7 Draft 2002)		
Attenuation factor $m_{patient}$ (without scatter)	Attenuation factor $m_{patient}$ (without scatter)	
(Total filtration 2.5 mm Al and 2.5 mm Al + 0.1 mm Cu)	(from tables in data file **KT_Water.xls**)	
Attenuation factor m_{grid}		
(IEC 60627 2001, DIN 6809-7 draft 2002)		
Tissue air ratios (Säbel et al. 1980)		
IV 6.2 Mammography	Mammography	
Dose output Y_{60} (table with additional 2 mm PC)	Dose output Y_{60} (Data file: **MAMM_HVL_Y60_2PC.xls**)	
DIN 6809-7 (draft 2002)		
Conversion factors g for the evaluation of the average glandular dose (Klein et al. 1997)		
Conversion factors g (CEC 1996)	Half value layers (HVL) for the most important	
Conversion factors g_{PB} (CEC 1996)	anode-filter combinations, inclusive compression plate	
HVL values for anode-filter combinations at various tube voltages	(s. data file: **QKD_MAMM.xls**)	

References:

Barnes GT and Brezowich IA (1978) The Intensity of Scattered Radiation in Mammography. Radiology 126: 243-247

CEC (Commission of the European Communities) (1996) European protocol on dosimetry in mammography, Report EUR 16263 (Luxembourg: CEC)

DIN (Deutsches Institut für Normung) (1990) Klinische Dosimetrie: Röntgendiagnostik DIN 6809, Part 3 (Berlin: Beuth)

DIN (Deutsches Institut für Normung) (2002, Draft) Klinische Dosimetrie:Verfahren zur Ermittlung der Patientendosis in der Röntgendiagnostik. DIN 6809, Part 7 (Berlin: Beuth)

Hammerstein GR, Miller DW, White DR, Masterson ME, Woodard HQ and Laughlin JS (1979) Absorbed radiation dose in mammography. Radiology 130: 485-491

Hermann KP, Geworski L, Muth M and Harder D (1985) Polyethylene-based waterequivalent phantom material for X-ray dosimetry at tube voltages from 10 kV to 100 kV. Phys. Med. Biol. 30: 1195-1200

Hubbell JH and Seltzer SM (1995) Tables of x-ray mass attenuation coefficients and mass energy-absorption coefficients 1 keV to 20 MeV for elements Z = 1 to 92 and 48 additional substances of dosimetric interest. NISTIR 5632

IEC (International Electrotechnical Commission) (1978) Characteristics of Anti-Scatter Grids used in X-ray Equipment. Publication 60627 (Geneva: IEC)

IEC (International Electrotechnical Commission) (2001) Diagnostic X-ray imaging equipment – Characteristics of general purpose and mammographic antiscatter grids. Publication 60627 (Geneva: IEC)

IEC (International Electrotechnical Commission) (1999) Evaluation and routine testing in medical imaging departments – Part 3-1: Acceptance tests – Imaging performance of X-ray equipment for radiographic and radioscopic systems. Publication 61223-3-1 (Geneva: IEC)

Klein J (1979) Zur filmmammographischen Nachweisbarkeitsgrenze von Mikroverkalkungen. Fortschr. Röntgenstr. 131: 205-210

Klein R, Aichinger H, Dierker J, Jansen J T M, Joite-Barfuß S, Säbel M, Schulz-Wendtland R und Zoetelief J (1997) Determination of average glandular dose with modern mammography units for two large groups of patients, Phys. Med. Biol. 42: 651-671

NIST (2001) Data base available under the internet adress: http://physics.nist.gov/PhysRefData/contents.html

Petoussi N, Zankl M, Drexler G, Panzer W and Regulla D (1998) Calculation of backscatter factors for diagnostic radiology using Monte Carlo methods, Phys. Med. Biol. 43: 2237-2250

Säbel, M., W. Bednar und J. Weishaar (1980) Untersuchungen zur Strahlenexposition der Leibesfrucht bei Röntgenuntersuchungen während der Schwangerschaft. 1. Mitteilung: Gewebe-Luft-Verhältnisse für Röntgenstrahlen mit Röhrenspannungen zwischen 60 kV und 120 kV, Strahlentherapie 156: 502-508

IV.1   X-ray Spectra

IV.1.1   General X-ray Diagnostics

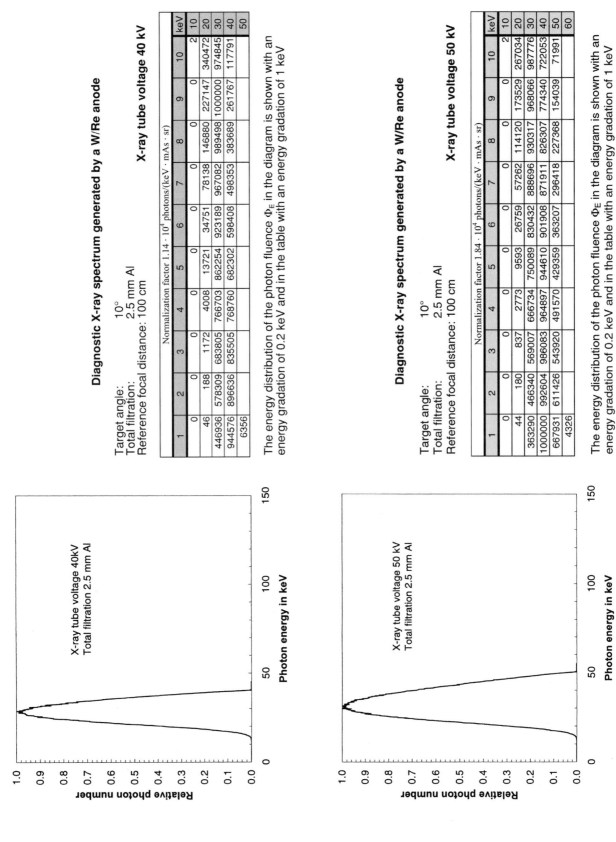

Diagnostic X-ray spectrum generated by a W/Re anode

X-ray tube voltage 40 kV

Target angle: 10°
Total filtration: 2.5 mm Al
Reference focal distance: 100 cm

Normalization factor $1.14 \cdot 10^4$ photons/(keV · mAs · sr)

1	2	3	4	5	6	7	8	9	10	keV
0	0	0	0	0	0	0	0	0	2	10
46	188	1172	4008	13721	34751	78138	146880	227147	340472	20
446936	578309	683805	766703	862254	923189	967082	989498	1000000	974845	30
944576	896636	835505	768760	682302	598408	498353	383689	261767	117791	40
6356										50

The energy distribution of the photon fluence Φ_E in the diagram is shown with an energy gradation of 0.2 keV and in the table with an energy gradation of 1 keV

Diagnostic X-ray spectrum generated by a W/Re anode

X-ray tube voltage 50 kV

Target angle: 10°
Total filtration: 2.5 mm Al
Reference focal distance: 100 cm

Normalization factor $1.84 \cdot 10^4$ photons/(keV · mAs · sr)

1	2	3	4	5	6	7	8	9	10	keV
0	0	0	0	0	0	0	0	0	2	10
44	180	837	2773	9593	26759	57262	114120	173529	267034	20
363290	466340	569007	666734	750089	830432	888696	930317	968066	987776	30
1000000	992604	986083	964897	944610	901908	871911	826307	774340	722053	40
667931	611426	543920	491570	429359	363207	296418	227368	154039	71991	50
4326										60

The energy distribution of the photon fluence Φ_E in the diagram is shown with an energy gradation of 0.2 keV and in the table with an energy gradation of 1 keV

Fig. IV.1.1

X-ray tube voltage 40kV
Total filtration 2.5 mm Al

Fig. IV.1.2

X-ray tube voltage 50 kV
Total filtration 2.5 mm Al

Diagnostic X-ray spectrum generated by a W/Re anode

X-ray tube voltage 60 kV

Target angle: 10°
Total filtration: 2.5 mm Al
Reference focal distance: 100 cm

Normalization factor $2.55 \cdot 10^4$ photons/((keV · mAs · sr))

1	2	3	4	5	6	7	8	9	10	keV
0	0	0	0	0	0	0	0	0	2	10
35	185	776	2107	7498	20384	46824	93848	143104	225251	20
302760	400193	492762	591266	679710	759412	825812	875891	926279	950378	30
973250	985141	995654	1000000	979802	977047	949813	920045	895575	868134	40
838035	803551	768453	722887	694828	649275	607518	573897	530227	493162	50
443123	409615	359176	311240	272846	230335	185584	143923	95962	45706	60
2999										70

The energy distribution of the photon fluence Φ_E in the diagram is shown with an energy gradation of 0.2 keV and in the table with an energy gradation of 1 keV

Diagnostic X-ray spectrum generated by a W/Re anode

X-ray tube voltage 70 kV

Target angle: 10°
Total filtration: 2.5 mm Al
Reference focal distance: 100 cm

Normalization factor $3.22 \cdot 10^4$ photons/((keV · mAs · sr))

1	2	3	4	5	6	7	8	9	10	keV
0	0	0	0	0	0	0	0	0	3	10
47	181	832	1808	6356	16905	38565	80002	125146	195815	20
265998	357484	450218	538697	619200	702188	769412	828222	883553	920175	30
955431	978883	994678	1000000	994380	996442	990436	968612	952442	934726	40
903555	889630	860900	829057	814643	775695	737901	706581	681073	649196	50
615128	585010	551225	528749	492453	455985	427115	390713	371200	338101	60
305792	275436	248454	216278	183154	154464	127557	95818	59779	28935	70
1886										80

The energy distribution of the photon fluence Φ_E in the diagram is shown with an energy gradation of 0.2 keV and in the table with an energy gradation of 1 keV

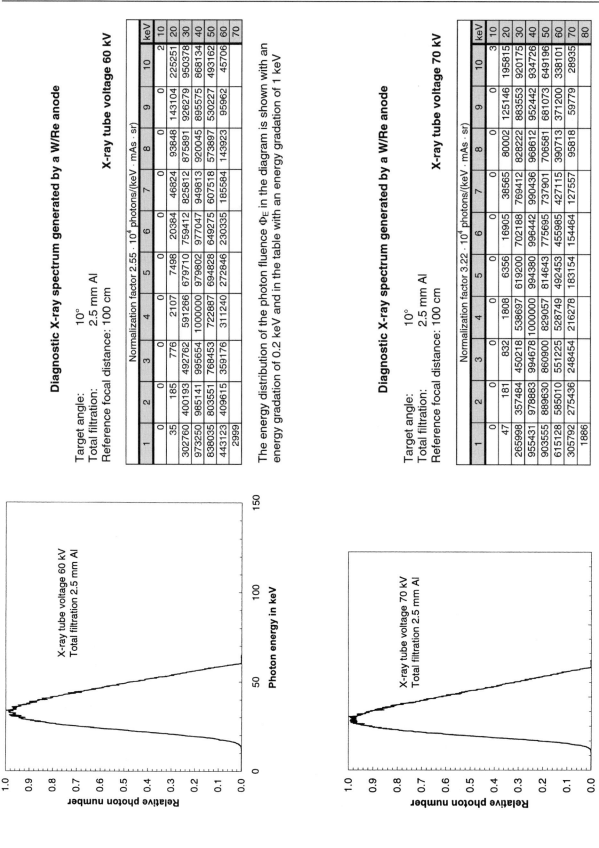

Fig. IV.1.3

Fig. IV.1.4

Fig. IV.1.5

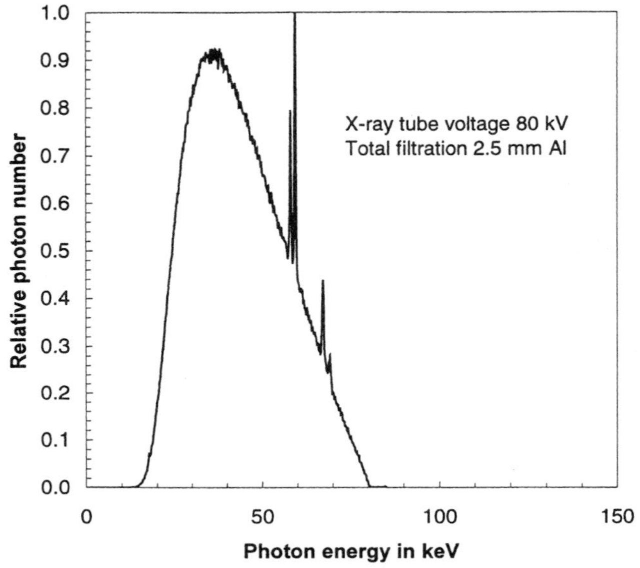

Diagnostic X-ray spectrum generated by a W/Re anode

X-ray tube voltage 80 kV

Target angle: 10°
Total filtration: 2.5 mm Al
Reference focal distance: 100 cm

Normalization factor $2.55 \cdot 10^4$ photons/(keV · mAs · sr)

1	2	3	4	5	6	7	8	9	10	keV
0	0	0	0	0	0	0	0	0	3	10
45	216	801	1796	554.5	14834	33574	71002	108858	173033	20
237816	320736	407387	495123	57233.1	654667	723462	788662	846129	888244	30
922428	949398	981689	992993	99369.1	999511	1000000	992422	990071	967930	40
950865	929203	915120	889530	86794.2	850908	816094	801329	770063	750572	50
716853	686112	665117	636488	60713.2	584454	563765	642630	591246	760723	60
462943	437385	406700	383913	36135.4	334365	348737	365391	279489	252433	70
203735	184806	163501	139874	12331.3	103922	83791	64672	43118	20499	80
2197										90

The energy distribution of the photon fluence Φ_E in the diagram is shown with an energy gradation of 0.2 keV and in the table with an energy gradation of 1 keV

Graph (Fig. IV.1.5):
X-ray tube voltage 80 kV
Total filtration 2.5 mm Al
Relative photon number (vertical axis, 0.0–1.0) vs. Photon energy in keV (horizontal axis, 0–150)

Fig. IV.1.6

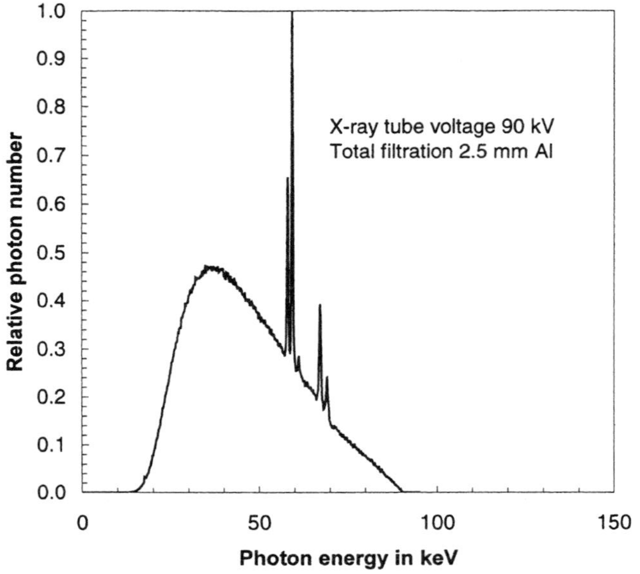

Diagnostic X-ray spectrum generated by a W/Re anode

X-ray tube voltage 90 kV

Target angle: 10°
Total filtration: 2.5 mm Al
Reference focal distance: 100 cm

Normalization factor $5.792 \cdot 10^4$ photons/(keV · mAs · sr)

1	2	3	4	5	6	7	8	9	10	keV
0	0	0	0	0	0	0	0	0	2	10
39	157	653	1383	4095	10166	22917	47778	73733	117581	20
167994	223298	288983	349337	409480	474311	527903	575483	626666	659980	30
695997	720073	738608	749503	765918	776208	773891	770329	769040	764694	40
747606	740359	726410	714506	700764	681835	663486	654415	640149	617866	50
598948	585721	568539	551554	531263	514795	492928	657931	635550	1000000	60
431037	427842	385568	372203	358748	341888	390912	460485	317590	308856	70
228940	224447	208963	197837	186796	175325	163593	152709	142631	131277	80
120508	112275	97429	86575	74108	60574	49929	38898	24603	10926	90
1215										100

The energy distribution of the photon fluence Φ_E in the diagram is shown with an energy gradation of 0.2 keV and in the table with an energy gradation of 1 keV

Graph (Fig. IV.1.6):
X-ray tube voltage 90 kV
Total filtration 2.5 mm Al
Relative photon number (vertical axis, 0.0–1.0) vs. Photon energy in keV (horizontal axis, 0–150)

Diagnostic X-ray spectrum generated by a W/Re anode

X-ray tube voltage 100 kV

Target angle: 10°
Total filtration: 2.5 mm Al
Reference focal distance: 100 cm

Normalization factor $9.168 \cdot 10^4$ photons/(keV · mAs · sr)

1	2	3	4	5	6	7	8	9	10	keV
0	0	0	0	0	0	0	0	0	0	10
2	24	123	434	942	2629	6621	14723	31296	47848	20
75917	108386	147282	187013	232358	273957	317042	358732	392792	417717	30
451957	479454	496760	513603	525633	536322	540950	552475	546783	548836	40
548653	541087	537685	529732	525172	514506	501249	492962	480670	477307	50
462693	452029	441166	426812	413765	405536	393051	386189	594842	559739	60
1000000	345936	342670	312847	302322	292874	284941	353897	427168	279839	70
272956	198974	192755	187221	176837	172509	166958	159323	152000	147098	80
140790	133951	123517	117947	113397	104866	102009	90885	86051	77766	90
70162	64487	60546	50020	44621	39511	32504	26032	19416	13513	100
5735	475									110

The energy distribution of the photon fluence Φ_E in the diagram is shown with an energy gradation of 0.2 keV and in the table with an energy gradation of 1 keV

Diagnostic X-ray spectrum generated by a W/Re anode

X-ray tube voltage 110 kV

Target angle: 10°
Total filtration: 2.5 mm Al
Reference focal distance: 100 cm

Normalization factor $1.29 \cdot 10^5$ photons/(keV · mAs · sr)

1	2	3	4	5	6	7	8	9	10	keV
0	0	0	0	0	0	0	0	0	1	10
21	93	366	685	2147	4642	10867	23423	33108	55120	20
77383	107289	136890	167549	198608	234931	264782	296734	319504	341995	30
363394	380394	391957	406058	418356	423464	429184	432910	429785	433148	40
428358	430958	422043	417677	414597	408236	401792	397551	385148	381710	50
372450	364781	353832	346368	339904	332997	327638	545283	526195	1000000	60
303355	308188	271491	262034	256154	246582	326036	421933	254350	253169	70
178802	172137	165175	159693	155124	152808	149536	141629	137067	133724	80
129175	124712	118639	116681	113069	108181	103494	98945	90326	87095	90
83987	77675	74957	68098	65506	63204	59212	53267	49534	46625	100
42332	39037	34346	30224	26765	20989	17364	12370	8571	3318	110
453										120

The energy distribution of the photon fluence Φ_E in the diagram is shown with an energy gradation of 0.2 keV and in the table with an energy gradation of 1 keV

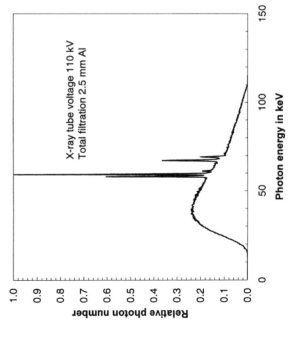

Fig. IV.1.7

X-ray tube voltage 100 kV
Total filtration 2.5 mm Al

Fig. IV.1.8

X-ray tube voltage 110 kV
Total filtration 2.5 mm Al

Fig. IV.1.9

Diagnostic X-ray spectrum generated by a W/Re anode

X-ray tube voltage 125 kV

Target angle: 10°
Total filtration: 2.5 mm Al
Reference focal distance: 100 cm

Normalization factor 1.95 · 10⁵ photons/((keV · mAs · sr))

1	2	3	4	5	6	7	8	9	10	keV
0	0	0	0	0	0	0	0	0	1	10
15	85	281	470	1456	3310	6397	15588	21931	36407	20
52676	70875	92033	116163	140333	162052	186309	210524	225812	248038	30
262029	279303	288781	300667	304896	316598	320073	321977	326573	325487	40
326559	328896	324987	322708	319902	316620	314531	311103	306467	297926	50
293210	288596	284818	273604	275718	266981	262390	506620	478327	1000000	60
256429	258424	223851	217089	213108	208990	295287	400673	227624	224533	70
145296	146003	141928	137710	135613	128824	128311	124933	124422	119324	80
113752	116460	109994	109153	107625	104977	99981	98059	89445	88261	90
86432	82867	80252	79532	76111	73282	72080	68254	65412	64476	100
60495	59330	56274	53211	51277	49340	46059	44315	43101	39410	110
35836	34740	32059	28690	28474	25276	22772	18895	16842	15205	120
12245	9908	7465	4295	1857	217					130

The energy distribution of the photon fluence Φ_E in the diagram is shown with an energy gradation of 0.2 keV and in the table with an energy gradation of 1 keV

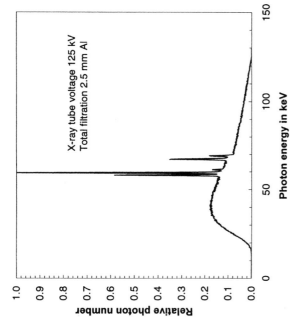

X-ray tube voltage 125 kV
Total filtration 2.5 mm Al

(Axes: Relative photon number 0.0–1.0; Photon energy in keV 0, 50, 100, 150)

Fig. IV.1.10

Diagnostic X-ray spectrum generated by a W/Re anode

X-ray tube voltage 150 kV

Target angle: 10°
Total filtration: 2.5 mm Al
Reference focal distance: 100 cm

Photon spectrum normalization factor 3.06 · 10⁵ photons/((keV·mAs·sr))

1	2	3	4	5	6	7	8	9	10	keV
0	0	0	0	0	0	0	0	0	1	10
13	65	207	397	893	2107	4099	10334	13875	23442	20
32357	45565	59864	75049	92597	107581	123769	143169	154785	167396	30
183713	191505	201046	211706	217589	225107	230978	236445	241736	241118	40
241084	246204	244152	243310	244263	242517	241621	237639	234450	236008	50
231576	228990	228108	223300	219960	215781	210100	488947	445575	1000000	60
210766	222609	187508	182949	178196	176905	278242	380686	203554	201500	70
117813	114126	114384	112603	110511	112327	110384	106389	105527	103102	80
103489	100767	96819	96966	93974	93524	95545	90348	85997	85523	90
84418	80196	80335	78145	75994	75528	72077	70726	67612	67509	100
66824	64933	67119	63047	62292	61372	60623	56801	57124	54129	110
54151	52527	49811	47617	47973	48246	44792	42515	41607	39806	120
38940	38380	37644	36719	33225	31834	32904	28272	28321	27267	130
25278	25240	22025	21077	19565	18522	17780	16434	14854	13586	140
12183	11065	9454	9407	6468	5795	5377	3147	2112	1089	150

The energy distribution of the photon fluence Φ_E in the diagram is shown with an energy gradation of 0.2 keV and in the table with an energy gradation of 1 keV

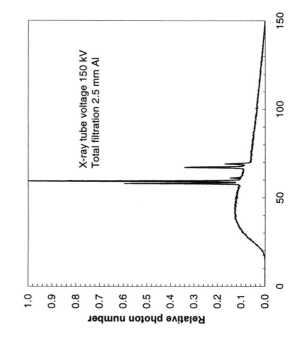

X-ray tube voltage 150 kV
Total filtration 2.5 mm Al

(Axes: Relative photon number 0.0–1.0; Photon energy in keV 0, 50, 100, 150)

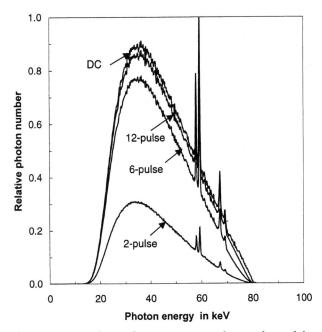

Fig. IV.1.11 Dependence of X-ray spectra on the waveform of the X-ray tube voltage (identical to Fig. II.1.6)

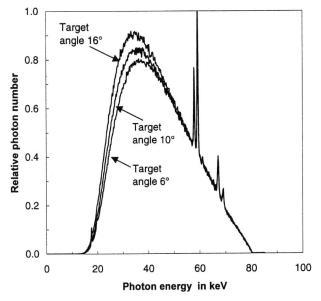

Fig. IV.1.12 Dependence of X-ray spectra on the target angle (identical to Fig. II.1.7)

IV.1.2 Mammography

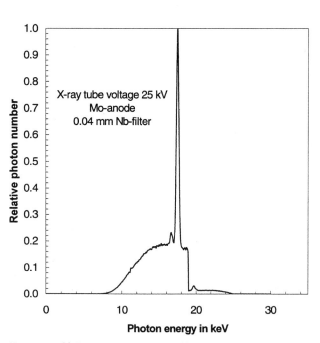

Mammographic X-rax spectrum generated by a Mo anode

Target/Emission angle:	22°/16°
Inherent filtration:	1 mm Be
Additional filtration:	0.04 mm Nb
Reference focal distance:	60 cm

X-ray tube voltage 25 kV

The energy distribution of the photon fluence Φ_E in the diagram is shown with an energy gradation of 0.1 keV, the corresponding phonton numbers in photons/(0.1 keV · mAs · sr) are given on CD-ROM

Fig. IV.1.13

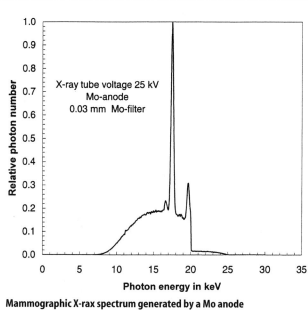

Mammographic X-rax spectrum generated by a Mo anode

Target/Emission angle:	22°/16°
Inherent filtration:	1 mm Be
Additional filtration:	0.03 mm Mo
Reference focal distance:	60 cm

X-ray tube voltage 25 kV

The energy distribution of the photon fluence Φ_E in the diagram is shown with an energy gradation of 0.1 keV, the corresponding phonton numbers in photons/(0.1 keV · mAs · sr) are given on CD-ROM

Fig. IV.1.14

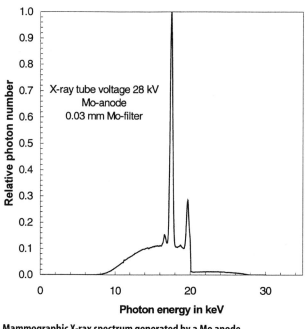

Mammographic X-rax spectrum generated by a Mo anode

Target/Emission angle:	22°/16°
Inherent filtration:	1 mm Be
Additional filtration:	0.03 mm Mo
Reference focal distance:	60 cm

X-ray tube voltage 28 kV

The energy distribution of the photon fluence Φ_E in the diagram is shown with an energy gradation of 0.1 keV, the corresponding phonton numbers in photons/(0.1 keV · mAs · sr) are given on CD-ROM

Fig. IV.1.15

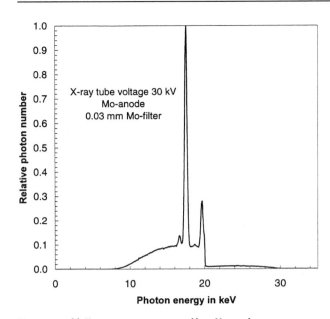

Mammographic X-rax spectrum generated by a Mo anode

Target/Emission angle: 22°/16°
Inherent filtration: 1 mm Be
Additional filtration: 0.03 mm Mo
Reference focal distance: 60 cm   **X-ray tube voltage 30 kV**

The energy distribution of the photon fluence Φ_E in the diagram is shown with an energy gradation of 0.1 keV, the corresponding phonton numbers in photons/(0.1 keV · mAs · sr) are given on CD-ROM

Fig. IV.1.16

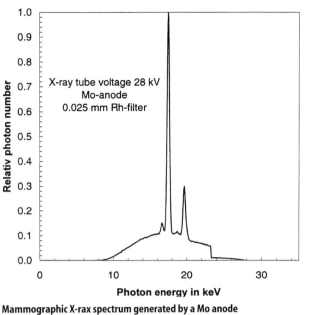

Mammographic X-rax spectrum generated by a Mo anode

Target/Emission angle: 22°/16°
Inherent filtration: 1 mm Be
Additional filtration: 0.0025 mm Rh
Reference focal distance: 60 cm   **X-ray tube voltage 28 kV**

The energy distribution of the photon fluence Φ_E in the diagram is shown with an energy gradation of 0.1 keV, the corresponding phonton numbers in photons/(0.1 keV · mAs · sr) are given on CD-ROM

Fig. IV.1.18

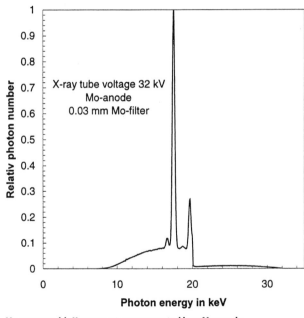

Mammographic X-rax spectrum generated by a Mo anode

Target/Emission angle: 22°/16°
Inherent filtration: 1 mm Be
Additional filtration: 0.03 mm Mo
Reference focal distance: 60 cm   **X-ray tube voltage 32 kV**

The energy distribution of the photon fluence Φ_E in the diagram is shown with an energy gradation of 0.1 keV, the corresponding phonton numbers in photons/(0.1 keV · mAs · sr) are given on CD-ROM

Fig. IV.1.17

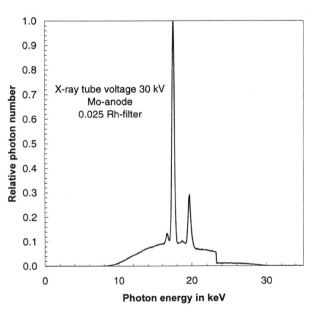

Mammographic X-rax spectrum generated by a Mo anode

Target/Emission angle: 22°/16°
Inherent filtration: 1 mm Be
Additional filtration: 0.0025 mm Rh
Reference focal distance: 60 cm   **X-ray tube voltage 30 kV**

The energy distribution of the photon fluence Φ_E in the diagram is shown with an energy gradation of 0.1 keV, the corresponding phonton numbers in photons/(0.1 keV · mAs · sr) are given on CD-ROM

Fig. IV.1.19

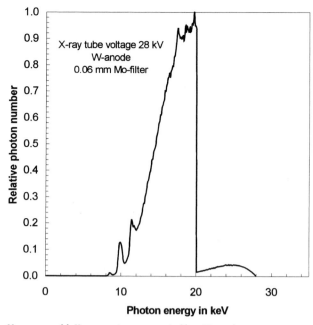

Mammographic X-rax spectrum generated by a W anode

Target/Emission angle: 22°/16°
Inherent filtration: 1 mm Be
Additional filtration: 0.08 mm Nb
Reference focal distance: 60 cm   **X-ray tube voltage 25 kV**

The energy distribution of the photon fluence Φ_E in the diagram is shown with an energy gradation of 0.1 keV, the corresponding phonton numbers in photons/(0.1 keV · mAs · sr) are given on CD-ROM

Fig. IV.1.20

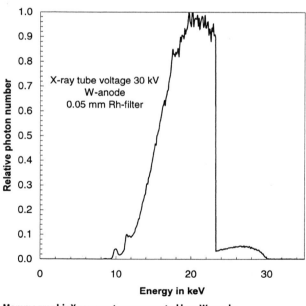

Mammographic X-rax spectrum generated by a W anode

Target/Emission angle: 22°/16°
Inherent filtration: 1 mm Be
Additional filtration: 0.06 mm Mo
Reference focal distance: 60 cm   **X-ray tube voltage 28 kV**

The energy distribution of the photon fluence Φ_E in the diagram is shown with an energy gradation of 0.1 keV, the corresponding phonton numbers in photons/(0.1 keV · mAs · sr) are given on CD-ROM

Fig. IV.1.22

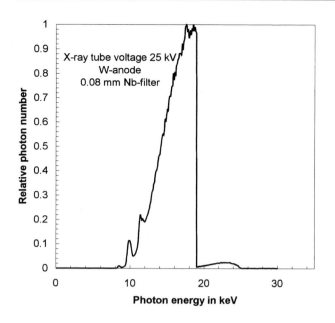

Mammographic X-rax spectrum generated by a W anode

Target/Emission angle: 22°/16°
Inherent filtration: 1 mm Be
Additional filtration: 0.06 mm Mo
Reference focal distance: 60 cm   **X-ray tube voltage 25 kV**

The energy distribution of the photon fluence Φ_E in the diagram is shown with an energy gradation of 0.1 keV, the corresponding phonton numbers in photons/(0.1 keV · mAs · sr) are given on CD-ROM

Fig. IV.1.21

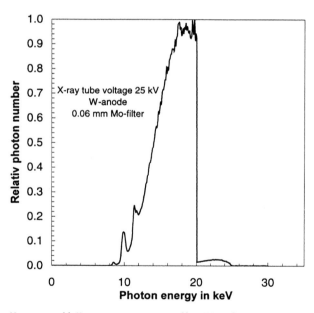

Mammographic X-rax spectrum generated by a W anode

Target/Emission angle: 22°/16°
Inherent filtration: 1 mm Be
Additional filtration: 0.05 mm Rh
Reference focal distance: 60 cm   **X-ray tube voltage 30 kV**

The energy distribution of the photon fluence Φ_E in the diagram is shown with an energy gradation of 0.1 keV, the corresponding phonton numbers in photons/(0.1 keV · mAs · sr) are given on CD-ROM

Fig. IV.1.23

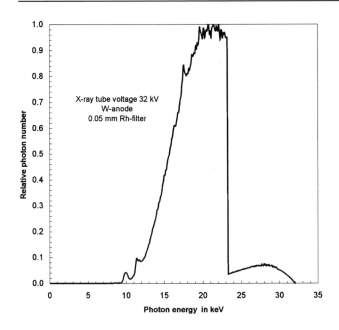

Mammographic X-rax spectrum generated by a W anode

Target/Emission angle: 22°/16°
Inherent filtration: 1 mm Be
Additional filtration: 0.05 mm Rh
Reference focal distance: 60 cm   **X-ray tube voltage 32 kV**

The energy distribution of the photon fluence Φ_E in the diagram is shown with an energy gradation of 0.1 keV, the corresponding phonton numbers in photons/(0.1 keV · mAs · sr) are given on CD-ROM

Fig. IV.1.24

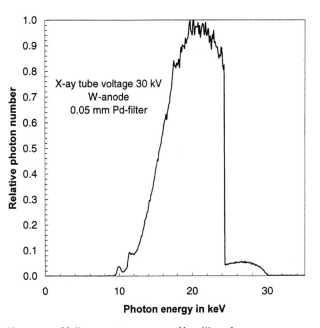

Mammographic X-rax spectrum generated by a W anode

Target/Emission angle: 22°/16°
Inherent filtration: 1 mm Be
Additional filtration: 0.05 mm Pd
Reference focal distance: 60 cm   **X-ray tube voltage 30 kV**

The energy distribution of the photon fluence Φ_E in the diagram is shown with an energy gradation of 0.1 keV, the corresponding phonton numbers in photons/(0.1 keV · mAs · sr) are given on CD-ROM

Fig. IV.1.26

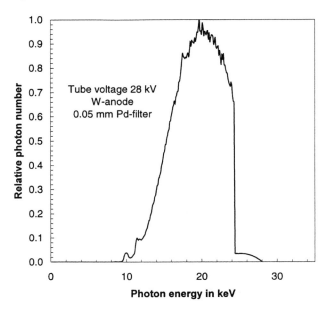

Mammographic X-rax spectrum generated by a W anode

Target/Emission angle: 22°/16°
Inherent filtration: 1 mm Be
Additional filtration: 0.05 mm Pd
Reference focal distance: 60 cm   **X-ray tube voltage 26 kV**

The energy distribution of the photon fluence Φ_E in the diagram is shown with an energy gradation of 0.1 keV, the corresponding phonton numbers in photons/(0.1 keV · mAs · sr) are given on CD-ROM

Fig. IV.1.25

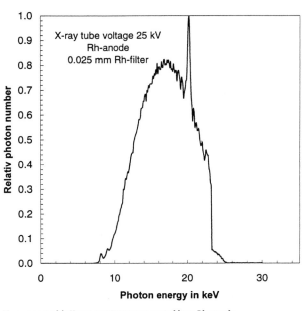

Mammographic X-rax spectrum generated by a Rh anode

Emission angle: 16°
Inherent filtration: 1 mm Be
Additional filtration: 0.025 mm Rh
Reference focal distance: 60 cm   **X-ray tube voltage 25 kV**

The energy distribution of the photon fluence Φ_E in the diagram is shown with an energy gradation of 0.1 keV, the corresponding phonton numbers in photons/(0.1 keV · mAs · sr) are given on CD-ROM

Fig. IV.1.27

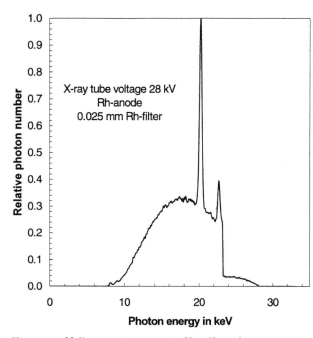

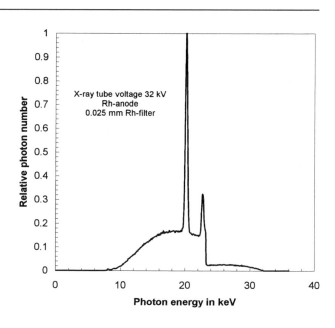

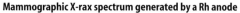

Mammographic X-rax spectrum generated by a Rh anode

Emission angle: 16°
Inherent filtration: 1 mm Be
Additional filtration: 0.025 mm Rh
Reference focal distance: 60 cm   **X-ray tube voltage 28 kV**

The energy distribution of the photon fluence Φ_E in the diagram is shown with an energy gradation of 0.1 keV, the corresponding phonton numbers in photons/(0.1 keV · mAs · sr) are given on CD-ROM

Fig. IV.1.28

Mammographic X-rax spectrum generated by a Rh anode

Emission angle: 16°
Inherent filtration: 1 mm Be
Additional filtration: 0.025 mm Rh
Reference focal distance: 60 cm   **X-ray tube voltage 32 kV**

The energy distribution of the photon fluence Φ_E in the diagram is shown with an energy gradation of 0.1 keV, the corresponding phonton numbers in photons/(0.1 keV · mAs · sr) are given on CD-ROM

Fig. IV.1.30

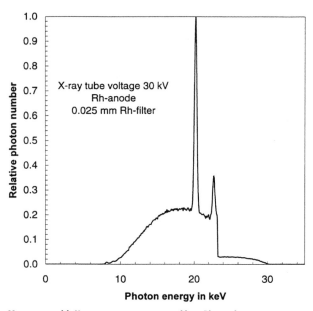

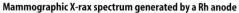

Mammographic X-rax spectrum generated by a Rh anode

Emission angle: 16°
Inherent filtration: 1 mm Be
Additional filtration: 0.025 mm Nb
Reference focal distance: 60 cm   **X-ray tube voltage 30 kV**

The energy distribution of the photon fluence Φ_E in the diagram is shown with an energy gradation of 0.1 keV, the corresponding phonton numbers in photons/(0.1 keV · mAs · sr) are given on CD-ROM

Fig. IV.1.29

IV.2 Interaction Coefficients

IV.2.1 Elements

Mass attenuation and mass energy absorption coefficient (NIST) of Aluminum

Atomic number 13; mass density 2.699 g/cm³

Energy keV	μ/ρ cm²/g	μ_en/ρ cm²/g	Energy keV	μ/ρ cm²/g	μ_en/ρ cm²/g	Energy keV	μ/ρ cm²/g	μ_en/ρ cm²/g
1	1.185E+03	1.183E+03	10	2.623E+01	2.543E+01	150	1.378E-01	2.827E-02
1.50	4.022E+02	4.001E+02	15	7.955E+00	7.487E+00	200	1.223E-01	2.745E-02
1.56	3.621E+02	3.600E+02	20	3.441E+00	3.094E+00	300	1.042E-01	2.816E-02
1.56	3.957E+03	3.829E+03	30	1.128E+00	8.778E-01	400	9.276E-02	2.862E-02
2	2.263E+03	2.204E+03	40	5.685E-01	3.601E-01	500	8.445E-02	2.868E-02
3	7.880E+02	7.732E+02	50	3.681E-01	1.840E-01	600	7.802E-02	2.851E-02
4	3.605E+02	3.545E+02	60	2.778E-01	1.099E-01	800	6.841E-02	2.778E-02
5	1.934E+02	1.902E+02	80	2.018E-01	5.511E-02	1000	6.146E-02	2.686E-02
6	1.153E+02	1.133E+02	100	1.704E-01	3.794E-02			
8	5.033E+01	4.918E+01						

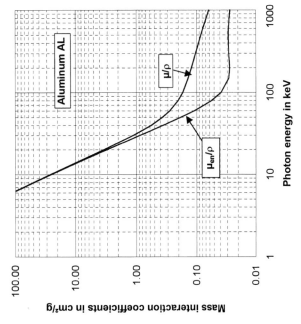

Fig. IV.2.1

Mass attenuation and mass energy absorption coefficient (NIST) of Silicon

Atomic number 14; mass density 2.33 g/cm³

Energy keV	μ/ρ cm²/g	μ_en/ρ cm²/g	Energy keV	μ/ρ cm²/g	μ_en/ρ cm²/g	Energy keV	μ/ρ cm²/g	μ_en/ρ cm²/g
1	1.570E+03	1.567E+03	8	6.468E+01	6.313E+01	100	1.835E-01	4.513E-02
1.5	5.355E+02	5.331E+02	10	3.389E+01	3.289E+01	150	1.448E-01	3.086E-02
1.8389	3.092E+02	3.070E+02	15	1.034E+01	9.794E+00	200	1.275E-01	2.905E-02
1.8389	3.192E+03	3.059E+03	20	4.464E+00	4.076E+00	300	1.082E-01	2.932E-02
2	2.777E+03	2.669E+03	30	1.436E+00	1.164E+00	400	9.614E-02	2.968E-02
3	9.784E+02	9.516E+02	40	7.012E-01	4.782E-01	500	8.748E-02	2.971E-02
4	4.529E+02	4.427E+02	50	4.385E-01	2.430E-01	600	8.077E-02	2.951E-02
5	2.450E+02	2.400E+02	60	3.207E-01	1.434E-01	800	7.082E-02	2.875E-02
6	1.470E+02	1.439E+02	80	2.228E-01	6.896E-02	1000	6.361E-02	2.778E-02

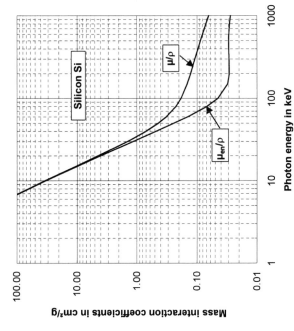

Fig. IV.2.2

Mass attenuation and mass energy absorption coefficient (NIST) of Copper

Atomic number 29; mass density 8.96 g/cm³

Energy keV	μ/ρ cm²/g	μ_en/ρ cm²/g	Energy keV	μ/ρ cm²/g	μ_en/ρ cm²/g	Energy keV	μ/ρ cm²/g	μ_en/ρ cm²/g
1	1.057E+04	1.049E+04	10	2.159E+02	1.484E+02	150	2.217E-01	1.027E-01
1.0961	8.242E+03	8.186E+03	15	7.405E+01	5.788E+01	200	1.559E-01	5.781E-02
1.0961	9.347E+03	9.282E+03	20	3.379E+01	2.788E+01	300	1.119E-01	3.617E-02
1.5	4.418E+03	4.393E+03	30	1.092E+01	9.349E+00	400	9.413E-02	3.121E-02
2	2.154E+03	2.142E+03	40	4.862E+00	4.163E+00	500	8.362E-02	2.933E-02
3	7.488E+02	7.430E+02	50	2.613E+00	2.192E+00	600	7.625E-02	2.826E-02
4	3.473E+02	3.432E+02	60	1.593E+00	1.290E+00	800	6.605E-02	2.681E-02
5	1.899E+02	1.866E+02	80	7.630E-01	5.581E-01	1000	5.901E-02	2.562E-02
6	1.156E+02	1.128E+02	100	4.584E-01	2.949E-01			
8	5.255E+01	5.054E+01						
8.9789	3.829E+01	3.652E+01						
8.9789	2.784E+02	1.824E+02						

Mass attenuation and mass energy absorption coefficient (NIST) of Selenium

Atomic number 34; mass density 4.5 g/cm³

Energy keV	μ/ρ cm²/g	μ_en/ρ cm²/g	Energy keV	μ/ρ cm²/g	μ_en/ρ cm²/g	Energy keV	μ/ρ cm²/g	μ_en/ρ cm²/g
1	2.317E+03	2.312E+03	6	1.773E+02	1.737E+02	80	1.090E+00	8.332E-01
1.4358	9.814E+02	9.760E+02	8	8.116E+01	7.865E+01	100	6.278E-01	4.426E-01
1.4358	4.347E+03	4.287E+03	10	4.414E+01	4.221E+01	150	2.703E-01	1.483E-01
1.4762	3.907E+03	3.855E+03	12.6578	2.318E+01	2.173E+01	200	1.742E-01	7.695E-02
1.4762	5.186E+03	5.112E+03	12.6578	1.589E+02	8.599E+01	300	1.144E-01	4.113E-02
1.5	5.336E+03	5.260E+03	15	1.033E+02	6.270E+01	400	9.299E-02	3.261E-02
1.6539	4.342E+03	4.284E+03	20	4.818E+01	3.352E+01	500	8.129E-02	2.941E-02
1.6539	4.915E+03	4.849E+03	30	1.596E+01	1.240E+01	600	7.350E-02	2.775E-02
2	3.098E+03	3.062E+03	40	7.184E+00	5.796E+00	800	6.314E-02	2.581E-02
3	1.116E+03	1.104E+03	50	3.864E+00	3.143E+00	1000	5.619E-02	2.446E-02
4	5.252E+02	5.187E+02	60	2.341E+00	1.886E+00			
5	2.896E+02	2.851E+02						

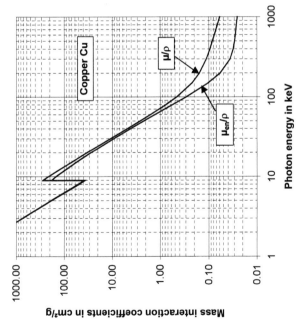

Fig. IV.2.3

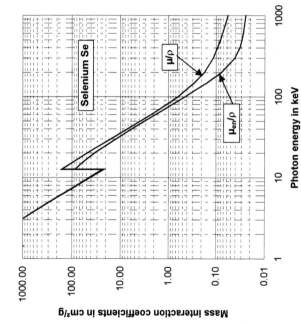

Fig. IV.2.4

Mass attenuation and mass energy absorption coefficient (NIST) of iodine

Atomic number 53; mass density 4.93 g/cm³

Energy keV	μ/ρ cm²/g	μ_{en}/ρ cm²/g
1	9.096E+03	9.078E+03
1.0721	7.863E+03	7.847E+03
1.0721	8.198E+03	8.181E+03
1.5	3.919E+03	3.908E+03
2	1.997E+03	1.988E+03
3	7.420E+02	7.351E+02
4	3.607E+02	3.548E+02
4.5571	2.592E+02	2.537E+02
4.5571	7.550E+02	7.121E+02
4.8521	6.636E+02	6.270E+02
4.8521	8.943E+02	8.375E+02
5	8.430E+02	7.903E+02
5.1881	7.665E+02	7.198E+02
5.1881	8.837E+02	8.283E+02
6	6.173E+02	5.822E+02
8	2.922E+02	2.777E+02
10	1.626E+02	1.548E+02
15	5.512E+01	5.208E+01
20	2.543E+01	2.363E+01
30	8.561E+00	7.622E+00
33.1694	6.553E+00	5.744E+00
33.1694	3.582E+01	1.188E+01
40	2.210E+01	9.616E+00
50	1.232E+01	6.573E+00
60	7.579E+00	4.518E+00
80	3.510E+00	2.331E+00
100	1.942E+00	1.342E+00
150	6.978E-01	4.742E-01
200	3.663E-01	2.295E-01
300	1.771E-01	9.257E-02
400	9.701E-02	5.650E-02
500	8.313E-02	4.267E-02
600	6.749E-02	3.598E-02
800	6.749E-02	2.962E-02
1000	5.841E-02	2.646E-02

Fig. IV.2.5

Mass attenuation and mass energy absorption coefficient (NIST) of Gadolinium

Atomic number 64; mass density 7.895 g/cm³

Energy keV	μ/ρ cm²/g	μ_{en}/ρ cm²/g
1	2.291E+03	2.281E+03
1.1852	1.668E+03	1.658E+03
1.1852	1.931E+03	1.919E+03
1.2172	3.961E+03	3.934E+03
1.2172	4.764E+03	4.731E+03
1.5	5.041E+03	5.008E+03
1.544	4.701E+03	4.670E+03
1.544	5.432E+03	5.395E+03
1.6883	4.421E+03	4.391E+03
1.6883	4.694E+03	4.662E+03
1.8808	3.691E+03	3.666E+03
1.8808	3.854E+03	3.829E+03
2	3.360E+03	3.337E+03
3	1.292E+03	1.280E+03
4	6.380E+02	6.296E+02
5	3.653E+02	3.584E+02
6	2.305E+02	2.246E+02
7.2428	1.429E+02	1.379E+02
7.2428	3.844E+02	3.452E+02
7.9303	3.049E+02	2.755E+02
7.9303	4.149E+02	3.665E+02
8	4.068E+02	3.595E+02
8.3756	3.631E+02	3.223E+02
8.3756	4.190E+02	3.702E+02
10	2.693E+02	2.416E+02
15	9.335E+01	8.538E+01
20	4.363E+01	3.994E+01
30	1.484E+01	1.333E+01
40	6.920E+00	6.033E+00
50	3.859E+00	3.242E+00
50.2391	3.812E+00	3.199E+00
50.2391	1.864E+01	5.585E+00
60	1.175E+01	4.722E+00
80	5.573E+00	2.937E+00
100	3.109E+00	1.849E+00
150	1.100E+00	7.197E-01
200	5.534E-01	3.584E-01
300	2.410E-01	1.409E-01
400	1.517E-01	8.039E-02
500	1.139E-01	5.650E-02
600	9.371E-02	4.483E-02
800	7.252E-02	3.399E-02
1000	6.120E-02	2.893E-02

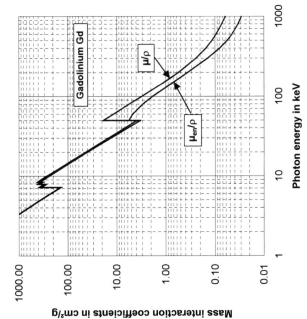

Fig. IV.2.6

Mass attenuation and mass energy absorption coefficient (NIST) of lead

Atomic number 82; mass density 11.35 g/cm³

Energy keV	μ/ρ cm²/g	μen/ρ cm²/g	Energy keV	μ/ρ cm²/g	μen/ρ cm²/g	Energy keV	μ/ρ cm²/g	μen/ρ cm²/g
1	5.210E+03	5.197E+03	5	7.304E+02	7.124E+02	60	5.021E+00	4.149E+00
1.5	2.356E+03	2.344E+03	6	4.672E+02	4.546E+02	80	2.419E+00	1.916E+00
2	1.285E+03	1.274E+03	8	2.287E+02	2.207E+02	88.0045	1.910E+00	1.482E+00
2.484	8.006E+02	7.895E+02	10	1.306E+02	1.247E+02	88.0045	7.683E+00	2.160E+00
2.484	1.397E+03	1.366E+03	13.0352	6.701E+01	6.270E+01	100	5.549E+00	1.976E+00
2.5856	1.944E+03	1.895E+03	13.0352	1.621E+02	1.291E+02	150	2.014E+00	1.056E+00
2.5856	2.458E+03	2.390E+03	15	1.116E+02	9.100E+01	200	9.985E-01	5.870E-01
3	1.965E+03	1.913E+03	15.2	1.078E+02	8.807E+01	300	4.031E-01	2.455E-01
3.0664	1.857E+03	1.808E+03	15.2	1.485E+02	1.131E+02	400	2.323E-01	1.370E-01
3.0664	2.146E+03	2.090E+03	15.8608	1.344E+02	1.032E+02	500	1.614E-01	9.128E-02
3.5542	1.496E+03	1.459E+03	15.8608	1.548E+02	1.180E+02	600	1.248E-01	6.819E-02
3.5542	1.585E+03	1.546E+03	20	8.636E+01	6.899E+01	800	8.870E-02	4.644E-02
3.8507	1.311E+03	1.279E+03	30	3.032E+01	2.536E+01	1000	7.102E-02	3.654E-02
3.8507	1.368E+03	1.335E+03	40	1.436E+01	1.211E+01	1500	5.222E-02	2.640E-02
4	1.251E+03	1.221E+03	50	8.041E+00	6.740E+00			

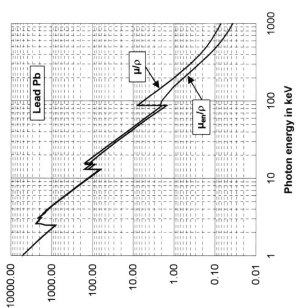

Fig. IV.2.7

Lead Pb. μ/ρ and μen/ρ. Mass interaction coefficients in cm²/g vs Photon energy in keV.

IV.2.2 Compounds and Mixtures

Mass attenuation and mass energy absorption coefficient (NIST) of air

Mass density (sea level and 20° C) 0.001205 g/cm³

Energy keV	μ/ρ cm²/g	μen/ρ cm²/g	Energy keV	μ/ρ cm²/g	μen/ρ cm²/g	Energy keV	μ/ρ cm²/g	μen/ρ cm²/g
1	3.606E+03	3.599E+03	8	9.921E+00	9.446E+00	100	1.541E-01	2.325E-02
1.5	1.191E+03	1.188E+03	10	5.120E+00	4.742E+00	150	1.356E-01	2.496E-02
2	5.279E+02	5.262E+02	15	1.614E+00	1.334E+00	200	1.233E-01	2.672E-02
3	1.625E+02	1.614E+02	20	7.779E-01	5.389E-01	300	1.067E-01	2.872E-02
3.2029	1.340E+02	1.330E+02	30	3.538E-01	1.537E-01	400	9.549E-02	2.949E-02
3.2029	1.485E+02	1.460E+02	40	2.485E-01	6.833E-02	500	8.712E-02	2.966E-02
4	7.788E+01	7.636E+01	50	2.080E-01	4.098E-02	600	8.055E-02	2.953E-02
5	4.027E+01	3.931E+01	60	1.875E-01	3.041E-02	800	7.074E-02	2.882E-02
6	2.341E+01	2.270E+01	80	1.662E-01	2.407E-02	1000	6.358E-02	2.789E-02

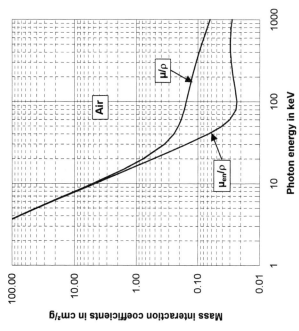

Fig. IV.2.8

Air. μ/ρ and μen/ρ. Mass interaction coefficients in cm²/g vs Photon energy in keV.

Mass attenuation and mass energy absorption coefficient (NIST) of water

Mass density 1.00 g/cm³

Energy keV	μ/ρ cm²/g	μen/ρ cm²/g	Energy keV	μ/ρ cm²/g	μen/ρ cm²/g	Energy keV	μ/ρ cm²/g	μen/ρ cm²/g
1	4.078E+03	4.065E+03	15	1.673E+00	1.374E+00	150	1.505E-01	2.764E-02
1.5	1.376E+03	1.372E+03	20	8.096E-01	5.503E-01	200	1.370E-01	2.967E-02
2	6.173E+02	6.152E+02	30	3.756E-01	1.557E-01	300	1.186E-01	3.192E-02
3	1.929E+02	1.917E+02	40	2.683E-01	6.947E-02	400	1.061E-01	3.279E-02
4	8.278E+01	8.191E+01	50	2.269E-01	4.223E-02	500	9.687E-02	3.299E-02
5	4.258E+01	4.188E+01	60	2.059E-01	3.190E-02	600	8.956E-02	3.284E-02
6	2.464E+01	2.405E+01	80	1.837E-01	2.597E-02	800	7.865E-02	3.206E-02
8	1.037E+01	9.915E+00	100	1.707E-01	2.546E-02	1000	7.072E-02	3.103E-02
10	5.329E+00	4.944E+00						

Mass attenuation and mass energy absorption coefficient (NIST) of PMMA

Mass density 1.19 g/cm³

Energy keV	μ/ρ cm²/g	μen/ρ cm²/g	Energy keV	μ/ρ cm²/g	μen/ρ cm²/g	Energy keV	μ/ρ cm²/g	μen/ρ cm²/g
1	2.794E+03	2.788E+03	15	1.101E+00	8.324E-01	150	1.456E-01	2.657E-02
1.5	9.153E+02	9.131E+02	20	5.714E-01	3.328E-01	200	1.328E-01	2.872E-02
2	4.037E+02	4.024E+02	30	3.032E-01	9.645E-02	300	1.152E-01	3.099E-02
3	1.236E+02	1.228E+02	40	2.350E-01	4.599E-02	400	1.031E-01	3.185E-02
4	5.247E+01	5.181E+01	50	2.074E-01	3.067E-02	500	9.410E-02	3.206E-02
5	2.681E+01	2.627E+01	60	1.924E-01	2.530E-02	600	8.701E-02	3.191E-02
6	1.545E+01	1.498E+01	80	1.751E-01	2.302E-02	800	7.641E-02	3.116E-02
8	6.494E+00	6.114E+00	100	1.641E-01	2.368E-02	1000	6.870E-02	3.015E-02
10	3.357E+00	3.026E+00						

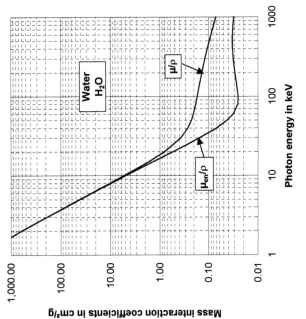

Fig. IV.2.9

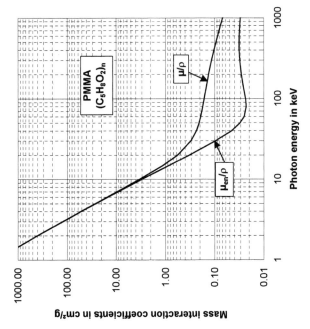

Fig. IV.2.10

Mass attenuation and mass energy absorption coefficient (NIST) of CsI

Mass density 4.51 g/cm³

Energy keV	μ/ρ cm²/g	μen/ρ cm²/g	Energy keV	μ/ρ cm²/g	μen/ρ cm²/g	Energy keV	μ/ρ cm²/g	μen/ρ cm²/g
1	9.234E+03	9.213E+03	5.0119	5.268E+02	4.985E+02	33.1694	6.923E+00	6.088E+00
1.065	8.098E+03	8.080E+03	5.0119	7.511E+02	7.037E+02	33.1694	2.122E+01	9.086E+00
1.065	8.339E+03	8.320E+03	5.1881	6.881E+02	6.457E+02	35.9846	1.719E+01	7.990E+00
1.0721	8.224E+03	8.205E+03	5.1881	7.453E+02	6.987E+02	35.9846	3.027E+01	1.059E+01
1.0721	8.387E+03	8.368E+03	5.3594	6.875E+02	6.454E+02	40	2.297E+01	9.395E+00
1.2171	6.413E+03	6.398E+03	5.3594	7.923E+02	7.397E+02	50	1.287E+01	6.596E+00
1.2171	6.569E+03	6.553E+03	5.7143	6.761E+02	6.331E+02	60	7.921E+00	4.586E+00
1.5	4.132E+03	4.120E+03	5.7143	7.268E+02	6.795E+02	80	3.677E+00	2.399E+00
2	2.114E+03	2.104E+03	6	6.448E+02	6.043E+02	100	2.035E+00	1.391E+00
3	7.880E+02	7.809E+02	8	3.071E+02	2.906E+02	150	7.290E-01	4.951E-01
4	3.836E+02	3.776E+02	10	1.711E+02	1.624E+02	200	3.805E-01	2.401E-01
4.5571	2.752E+02	2.696E+02	15	5.815E+01	5.486E+01	300	1.818E-01	9.634E-02
4.5571	5.174E+02	4.936E+02	20	2.686E+01	2.496E+01	400	1.237E-01	5.828E-02
4.8521	4.510E+02	4.303E+02	30	9.045E+00	8.071E+00	500	9.809E-02	4.366E-02
4.8521	5.637E+02	5.332E+02				600	8.373E-02	3.657E-02
5	5.296E+02	5.012E+02				800	6.769E-02	2.987E-02
						1000	5.848E-02	2.657E-02

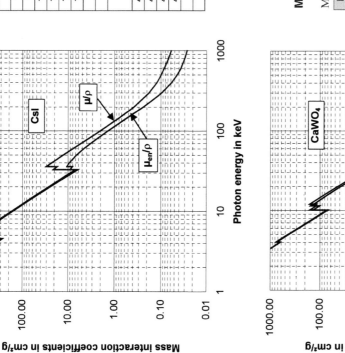

Fig. IV.2.11

Mass attenuation and mass energy absorption coefficient (NIST) of CaWO₄

Mass density 6.062 g/cm³

Energy keV	μ/ρ cm²/g	μen/ρ cm²/g	Energy keV	μ/ρ cm²/g	μen/ρ cm²/g	Energy keV	μ/ρ cm²/g	μen/ρ cm²/g
2	2.770E+03	2.725E+03	8	1.355E+02	1.297E+02	60	2.505E+00	2.030E+00
2.281	1.990E+03	1.959E+03	10	7.621E+01	7.218E+01	69.525	1.737E+00	1.356E+00
2.281	2.278E+03	2.243E+03	10.204	7.228E+01	6.837E+01	69.525	7.278E+00	2.098E+00
2.575	1.692E+03	1.667E+03	10.204	1.626E+02	1.382E+02	80	5.075E+00	1.871E+00
2.575	1.791E+03	1.764E+03	11.541	1.174E+02	1.011E+02	100	2.904E+00	1.361E+00
2.82	1.445E+03	1.423E+03	11.541	1.572E+02	1.295E+02	150	1.063E+00	6.112E-01
2.82	1.503E+03	1.480E+03	12.098	1.402E+02	1.163E+02	200	5.475E-01	3.247E-01
3	1.300E+03	1.281E+03	12.098	1.605E+02	1.322E+02	300	2.461E-01	1.368E-01
4	6.483E+02	6.377E+02	15	9.325E+01	7.908E+01	400	1.578E-01	8.106E-02
4.037	6.337E+02	6.232E+02	20	4.398E+01	3.821E+01	500	1.197E-01	5.831E-02
4.037	7.596E+02	7.307E+02	30	1.517E+01	1.326E+01	600	9.907E-02	4.694E-02
5	4.479E+02	4.315E+02	40	7.125E+00	6.131E+00	800	7.717E-02	3.615E-02
6	2.825E+02	2.721E+02	50	3.988E+00	3.343E+00	1000	6.531E-02	3.100E-02

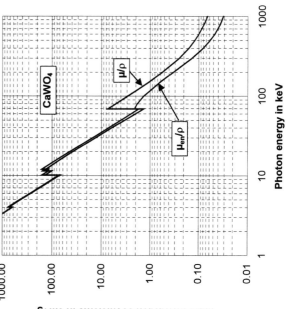

Fig. IV.2.12

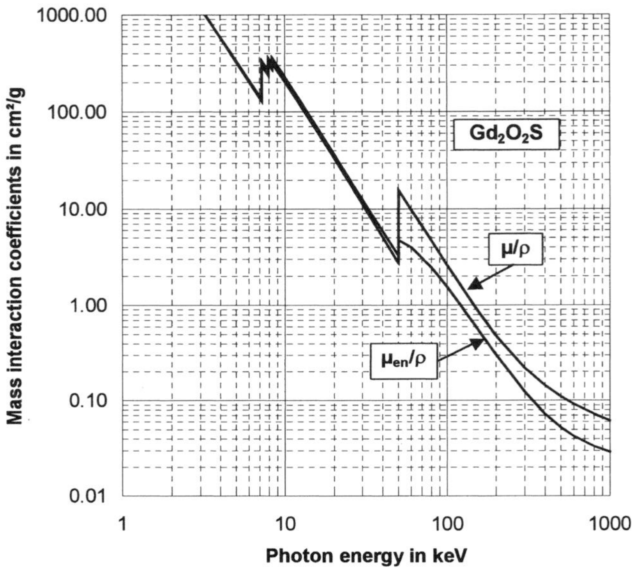

Mass attenuation and mass energy absorption coefficient (NIST) of GD$_2$O$_2$S:Tb

Mass density 7.34 g/cm^3

Energy keV	μ/ρ cm²/g	μ_{er}/ρ cm²/g	Energy keV	μ/ρ cm²/g	μ_{er}/ρ cm²/g	Energy keV	μ/ρ cm²/g	μ_{er}/ρ cm²/g
1	2.497E+03	2.487E+03	4	5.916E+02	5.822E+02	50	3.274E+00	2.732E+00
1.1852	1.765E+03	1.756E+03	5	3.371E+02	3.302E+02	50.2391	3.234E+00	2.695E+00
1.1852	1.984E+03	1.973E+03	6	2.118E+02	2.062E+02	50.2391	1.555E+01	4.677E+00
1.2172	3.644E+03	3.621E+03	7.2428	1.306E+02	1.261E+02	60	9.815E+00	3.946E+00
1.2172	4.311E+03	4.282E+03	7.2428	3.312E+02	2.983E+02	80	4.666E+00	2.453E+00
1.5	4.390E+03	4.361E+03	7.9303	2.625E+02	2.378E+02	100	2.613E+00	1.545E+00
1.544	4.092E+03	4.065E+03	7.9303	3.539E+02	3.134E+02	150	9.381E-01	6.043E-01
1.544	4.699E+03	4.668E+03	8	3.470E+02	3.074E+02	200	4.812E-01	3.035E-01
1.6883	3.819E+03	3.793E+03	8.3756	3.096E+02	2.754E+02	300	2.185E-01	1.225E-01
1.6883	4.046E+03	4.019E+03	8.3756	3.560E+02	3.152E+02	400	1.423E-01	7.211E-02
1.8808	3.175E+03	3.154E+03	10	2.285E+02	2.053E+02	500	1.095E-01	5.218E-02
1.8808	3.311E+03	3.289E+03	15	7.902E+01	7.232E+01	600	9.153E-02	4.240E-02
2	2.883E+03	2.863E+03	20	3.689E+01	3.376E+01	800	7.225E-02	3.322E-02
2.472	1.752E+03	1.739E+03	30	1.254E+01	1.125E+01	1000	6.163E-02	2.883E-02
2.472	1.909E+03	1.885E+03	40	5.854E+00	5.084E+00			
3	1.205E+03	1.189E+03						

Fig. IV.2.13

IV.3 Characteristics of the Primary Radiation Beam

IV.3.1 General X-ray Diagnostics

Table IV.3.1

Characteristic X-ray beam quantities without scattered radiation and relative energy absorption in typical image detectors at variable absorber or phantom materials for diagnostic X-ray tube assemblies with W/Re anode at 10° target angle and an inherent (quality equivalent) filtration of 2.5 mm Al

Reference focal distance: 100 cm   X-ray tube voltage: 40 kV

Absorber material: Al

Thick-ness mm	HVL mm Al	\<E\> keV	Φ/It 1/As/cm²	Ψ/It J/As/cm²	Ka/It mGy/As	Relative dose	Φ/Ka photons /cm²/µGy	68 CaWO4	68 GOS	118 GOS	181 GOS	80 SPF	180 CsI
0.0	1.44	28.4	1.71E+10	7.81E-05	1.61E+01	1.00E+00	1.07E+06	.623	.570	.746	.861	.576	.832
.5	1.57	29.0	1.39E+10	6.47E-05	1.23E+01	7.69E-01	1.13E+06	.607	.554	.733	.852	.561	.824
1.0	1.69	29.5	1.15E+10	5.42E-05	9.72E+00	6.06E-01	1.18E+06	.594	.540	.721	.844	.549	.819
1.5	1.79	29.9	9.55E+09	4.58E-05	7.80E+00	4.85E-01	1.23E+06	.582	.527	.710	.836	.538	.814
2.0	1.89	30.3	8.03E+09	3.90E-05	6.34E+00	3.95E-01	1.27E+06	.572	.517	.701	.829	.529	.809
2.5	1.98	30.7	6.79E+09	3.34E-05	5.21E+00	3.24E-01	1.30E+06	.562	.507	.692	.823	.521	.806
3.0	2.06	31.0	5.78E+09	2.87E-05	4.32E+00	2.69E-01	1.34E+06	.554	.498	.684	.817	.514	.803
3.5	2.13	31.3	4.95E+09	2.48E-05	3.62E+00	2.25E-01	1.37E+06	.546	.491	.677	.812	.508	.800
4.0	2.20	31.5	4.25E+09	2.15E-05	3.05E+00	1.90E-01	1.40E+06	.539	.483	.670	.806	.502	.798
4.5	2.26	31.8	3.67E+09	1.87E-05	2.58E+00	1.61E-01	1.42E+06	.533	.477	.664	.802	.497	.797
5.0	2.32	32.0	3.18E+09	1.63E-05	2.20E+00	1.37E-01	1.45E+06	.527	.471	.658	.797	.492	.795
6.0	2.43	32.4	2.41E+09	1.25E-05	1.61E+00	1.01E-01	1.49E+06	.516	.460	.647	.788	.484	.793
7.0	2.53	32.8	1.84E+09	9.66E-06	1.20E+00	7.49E-02	1.53E+06	.506	.450	.638	.781	.478	.791
8.0	2.62	33.1	1.42E+09	7.52E-06	9.07E-01	5.65E-02	1.56E+06	.498	.442	.629	.774	.472	.790
9.0	2.70	33.4	1.10E+09	5.90E-06	6.91E-01	4.30E-02	1.60E+06	.491	.435	.621	.767	.467	.790
10.0	2.77	33.6	8.63E+08	4.65E-06	5.31E-01	3.31E-02	1.62E+06	.484	.428	.614	.761	.463	.790
12.0	2.90	34.1	5.36E+08	2.93E-06	3.20E-01	1.99E-02	1.67E+06	.472	.417	.602	.751	.457	.791
14.0	3.00	34.4	3.39E+08	1.87E-06	1.97E-01	1.23E-02	1.72E+06	.463	.407	.592	.742	.452	.792
16.0	3.10	34.8	2.17E+08	1.21E-06	1.24E-01	7.71E-03	1.75E+06	.454	.399	.583	.734	.448	.794
18.0	3.18	35.1	1.41E+08	7.91E-07	7.89E-02	4.91E-03	1.78E+06	.447	.392	.575	.727	.446	.796
20.0	3.25	35.3	9.21E+07	5.21E-07	5.08E-02	3.16E-03	1.81E+06	.441	.386	.568	.720	.444	.798
22.0	3.32	35.6	6.08E+07	3.46E-07	3.31E-02	2.06E-03	1.84E+06	.435	.381	.562	.715	.442	.800
25.0	3.40	35.9	3.30E+07	1.90E-07	1.76E-02	1.10E-03	1.87E+06	.428	.374	.554	.707	.441	.803
30.0	3.52	36.3	1.23E+07	7.12E-08	6.41E-03	3.99E-04	1.91E+06	.419	.365	.543	.697	.441	.808
35.0	3.62	36.6	4.68E+06	2.75E-08	2.40E-03	1.50E-04	1.95E+06	.411	.357	.534	.688	.441	.812
40.0	3.70	36.9	1.83E+06	1.08E-08	9.23E-04	5.75E-05	1.98E+06	.405	.351	.527	.681	.442	.815
45.0	3.76	37.1	7.25E+05	4.31E-09	3.62E-04	2.25E-05	2.00E+06	.399	.346	.521	.675	.443	.818
50.0	3.82	37.3	2.92E+05	1.74E-09	1.44E-04	8.97E-06	2.02E+06	.395	.342	.516	.670	.445	.820
55.0	3.87	37.5	1.19E+05	7.13E-10	5.81E-05	3.62E-06	2.04E+06	.391	.339	.511	.665	.447	.821
60.0	3.91	37.7	4.88E+04	2.94E-10	2.37E-05	1.48E-06	2.06E+06	.388	.336	.507	.661	.448	.823

Absorber material: Cu

Thick-ness mm	HVL mm Al	\<E\> keV	Φ/It 1/As/cm²	Ψ/It J/As/cm²	Ka/It mGy/As	Relative dose	Φ/Ka photons /cm²/µGy	68 CaWO4	68 GOS	118 GOS	181 GOS	80 SPF	180 CsI
0.00	1.44	28.4	1.71E+10	7.81E-05	1.61E+01	1.00E+00	1.07E+06	.623	.570	.746	.861	.576	.832
.02	1.61	29.2	1.31E+10	6.15E-05	1.15E+01	7.15E-01	1.15E+06	.602	.549	.728	.849	.557	.822
.04	1.76	29.8	1.03E+10	4.93E-05	8.52E+00	5.30E-01	1.21E+06	.585	.531	.713	.838	.541	.815
.06	1.89	30.4	8.23E+09	4.00E-05	6.49E+00	4.04E-01	1.27E+06	.571	.516	.700	.829	.528	.809
.08	2.01	30.8	6.66E+09	3.29E-05	5.05E+00	3.14E-01	1.32E+06	.558	.503	.688	.820	.518	.805
.10	2.11	31.2	5.44E+09	2.72E-05	3.99E+00	2.49E-01	1.36E+06	.547	.491	.677	.812	.509	.801
.20	2.52	32.8	2.22E+09	1.16E-05	1.45E+00	9.03E-02	1.53E+06	.506	.450	.637	.780	.478	.792
.30	2.80	33.8	1.01E+09	5.46E-06	6.16E-01	3.84E-02	1.64E+06	.480	.424	.610	.758	.461	.790
.40	3.01	34.5	4.92E+08	2.72E-06	2.86E-01	1.78E-02	1.72E+06	.462	.406	.590	.741	.452	.793
.50	3.17	35.1	2.50E+08	1.41E-06	1.41E-01	8.75E-03	1.78E+06	.448	.393	.575	.727	.446	.796
.60	3.30	35.5	1.32E+08	7.49E-07	7.19E-02	4.48E-03	1.83E+06	.437	.382	.563	.716	.443	.800
.70	3.41	35.9	7.09E+07	4.07E-07	3.79E-02	2.36E-03	1.87E+06	.428	.374	.553	.707	.441	.804
.80	3.50	36.2	3.89E+07	2.25E-07	2.04E-02	1.27E-03	1.90E+06	.421	.367	.545	.699	.441	.807
.90	3.57	36.5	2.17E+07	1.27E-07	1.12E-02	6.98E-04	1.93E+06	.414	.361	.538	.692	.441	.810
1.00	3.64	36.7	1.22E+07	7.19E-08	6.25E-03	3.89E-04	1.96E+06	.409	.356	.532	.686	.441	.813
1.20	3.74	37.1	4.01E+06	2.38E-08	2.01E-03	1.25E-04	2.00E+06	.401	.348	.522	.677	.443	.817
1.40	3.83	37.4	1.36E+06	8.14E-09	6.71E-04	4.18E-05	2.03E+06	.394	.342	.515	.669	.445	.820
1.60	3.89	37.6	4.73E+05	2.85E-09	2.30E-04	1.43E-05	2.05E+06	.389	.337	.509	.663	.448	.822
1.80	3.95	37.8	1.67E+05	1.01E-09	8.07E-05	5.02E-06	2.07E+06	.384	.333	.503	.657	.450	.824
2.00	4.00	38.0	6.02E+04	3.66E-10	2.88E-05	1.79E-06	2.09E+06	.381	.329	.499	.653	.452	.825
2.50	4.09	38.3	4.93E+03	3.02E-11	2.32E-06	1.44E-07	2.12E+06	.374	.323	.491	.645	.456	.826
3.00	4.15	38.6	4.26E+02	2.63E-12	1.98E-07	1.24E-08	2.15E+06	.369	.318	.485	.638	.459	.826
3.50	4.20	38.7	3.84E+01	2.38E-13	1.77E-08	1.10E-09	2.17E+06	.365	.315	.481	.634	.460	.825
4.00	4.24	38.9	3.56E+00	2.21E-14	1.63E-09	1.02E-10	2.18E+06	.363	.312	.477	.630	.460	.825
4.50	4.26	39.0	3.37E-01	2.11E-15	1.54E-10	9.59E-12	2.19E+06	.360	.310	.475	.627	.460	.824
5.00	4.29	39.1	3.25E-02	2.04E-16	1.48E-11	9.21E-13	2.20E+06	.359	.308	.473	.625	.460	.823
6.00	4.32	39.2	3.15E-04	1.98E-18	1.43E-13	8.88E-15	2.21E+06	.356	.306	.470	.622	.460	.822
7.00	4.35	39.3	3.17E-06	2.00E-20	1.43E-15	8.90E-17	2.22E+06	.354	.305	.467	.620	.459	.822
8.00	4.36	39.3	3.27E-08	2.06E-22	1.47E-17	9.16E-19	2.23E+06	.353	.303	.466	.618	.458	.821

Reference focal distance: 100 cm   X-ray tube voltage: 40 kV

Absorber material: Water

Thickness mm	HVL mm Al	$\langle E \rangle$ keV	Φ/It 1/As/cm²	Ψ/It J/As/cm²	Ka/It mGy/As	Relative dose	Φ/Ka photons /cm²/µGy	68 CaWO4	68 GOS	118 GOS	181 GOS	80 SPF	180 CsI

Relative energy absorption in the image detector (coverage mg/cm²)

No added filter

Thickness	HVL	$\langle E \rangle$	Φ/It	Ψ/It	Ka/It	Rel. dose	Φ/Ka	68 CaWO4	68 GOS	118 GOS	181 GOS	80 SPF	180 CsI
0.0	1.44	28.4	1.71E+10	7.81E-05	1.61E+01	1.00E+00	1.07E+06	.623	.570	.746	.861	.576	.832
2.5	1.84	30.1	5.83E+09	2.81E-05	4.67E+00	2.91E-01	1.25E+06	.577	.522	.705	.833	.534	.811
5.0	2.13	31.3	2.19E+09	1.10E-05	1.60E+00	9.99E-02	1.37E+06	.546	.490	.677	.811	.508	.800
7.5	2.35	32.1	8.75E+08	4.50E-06	6.00E-01	3.74E-02	1.46E+06	.524	.467	.655	.794	.490	.794
10.0	2.53	32.8	3.63E+08	1.90E-06	2.37E-01	1.47E-02	1.53E+06	.506	.450	.637	.780	.478	.791
12.5	2.68	33.3	1.54E+08	8.23E-07	9.70E-02	6.04E-03	1.59E+06	.492	.436	.623	.768	.468	.790
15.0	2.81	33.8	6.70E+07	3.62E-07	4.08E-02	2.54E-03	1.64E+06	.480	.424	.610	.758	.461	.790
17.5	2.92	34.1	2.95E+07	1.62E-07	1.76E-02	1.09E-03	1.68E+06	.470	.415	.600	.749	.456	.791
20.0	3.01	34.5	1.32E+07	7.29E-08	7.68E-03	4.78E-04	1.72E+06	.462	.406	.591	.741	.452	.792
22.5	3.10	34.8	5.96E+06	3.32E-08	3.41E-03	2.12E-04	1.75E+06	.455	.399	.583	.734	.449	.794
25.0	3.17	35.0	2.72E+06	1.53E-08	1.53E-03	9.51E-05	1.78E+06	.448	.393	.576	.727	.446	.796
27.5	3.24	35.3	1.25E+06	7.06E-09	6.92E-04	4.31E-05	1.81E+06	.442	.388	.570	.722	.445	.798
30.0	3.30	35.5	5.78E+05	3.28E-09	3.16E-04	1.97E-05	1.83E+06	.437	.383	.564	.716	.443	.800
32.5	3.35	35.7	2.69E+05	1.54E-09	1.45E-04	9.05E-06	1.85E+06	.433	.378	.559	.712	.442	.802
35.0	3.40	35.9	1.26E+05	7.22E-10	6.73E-05	4.19E-06	1.87E+06	.428	.374	.554	.707	.441	.803

Added filter 0.5 mm Al

Thickness	HVL	$\langle E \rangle$	Φ/It	Ψ/It	Ka/It	Rel. dose	Φ/Ka	68 CaWO4	68 GOS	118 GOS	181 GOS	80 SPF	180 CsI
0.0	1.57	29.0	1.39E+10	6.47E-05	1.23E+01	1.00E+00	1.13E+06	.607	.554	.733	.852	.561	.824
2.5	1.93	30.5	4.91E+09	2.40E-05	3.82E+00	3.09E-01	1.29E+06	.567	.512	.696	.826	.525	.808
5.0	2.20	31.5	1.89E+09	9.53E-06	1.35E+00	1.09E-01	1.40E+06	.539	.483	.670	.806	.502	.798
7.5	2.41	32.3	7.62E+08	3.94E-06	5.14E-01	4.16E-02	1.48E+06	.518	.462	.649	.790	.486	.793
10.0	2.58	32.9	3.18E+08	1.68E-06	2.05E-01	1.66E-02	1.55E+06	.502	.446	.633	.777	.475	.791
12.5	2.72	33.4	1.36E+08	7.30E-07	8.49E-02	6.87E-03	1.61E+06	.488	.432	.619	.765	.466	.790
15.0	2.84	33.9	5.94E+07	3.23E-07	3.60E-02	2.91E-03	1.65E+06	.477	.421	.607	.755	.460	.790
17.5	2.94	34.2	2.63E+07	1.44E-07	1.55E-02	1.26E-03	1.69E+06	.468	.412	.597	.747	.455	.791
20.0	3.04	34.6	1.18E+07	6.54E-08	6.83E-03	5.53E-04	1.73E+06	.460	.404	.588	.739	.451	.793
22.5	3.12	34.9	5.35E+06	2.98E-08	3.04E-03	2.46E-04	1.76E+06	.453	.397	.581	.732	.448	.794
25.0	3.19	35.1	2.44E+06	1.37E-08	1.37E-03	1.11E-04	1.79E+06	.446	.391	.574	.726	.446	.796
27.5	3.25	35.3	1.12E+06	6.36E-09	6.20E-04	5.02E-05	1.81E+06	.441	.386	.568	.720	.444	.798
30.0	3.31	35.5	5.21E+05	2.97E-09	2.84E-04	2.30E-05	1.83E+06	.436	.381	.562	.715	.443	.800
32.5	3.37	35.7	2.43E+05	1.39E-09	1.31E-04	1.06E-05	1.85E+06	.431	.377	.557	.710	.442	.802
35.0	3.42	35.9	1.14E+05	6.54E-10	6.07E-05	4.91E-06	1.87E+06	.427	.373	.553	.706	.441	.804

Added filter 0.1 mm Cu

Thickness	HVL	$\langle E \rangle$	Φ/It	Ψ/It	Ka/It	Rel. dose	Φ/Ka	68 CaWO4	68 GOS	118 GOS	181 GOS	80 SPF	180 CsI
0.0	2.11	31.2	5.44E+09	2.72E-05	3.99E+00	1.00E+00	1.36E+06	.547	.491	.677	.812	.509	.801
2.5	2.34	32.1	2.17E+09	1.11E-05	1.49E+00	3.72E-01	1.46E+06	.524	.468	.655	.795	.491	.795
5.0	2.53	32.8	8.96E+08	4.70E-06	5.86E-01	1.47E-01	1.53E+06	.506	.450	.637	.780	.478	.791
7.5	2.68	33.3	3.81E+08	2.03E-06	2.40E-01	6.00E-02	1.59E+06	.492	.436	.623	.768	.468	.790
10.0	2.81	33.8	1.65E+08	8.95E-07	1.01E-01	2.53E-02	1.64E+06	.480	.424	.610	.758	.461	.790
12.5	2.92	34.1	7.30E+07	3.99E-07	4.34E-02	1.09E-02	1.68E+06	.470	.415	.600	.749	.456	.791
15.0	3.01	34.5	3.26E+07	1.80E-07	1.90E-02	4.75E-03	1.72E+06	.462	.406	.591	.741	.452	.792
17.5	3.10	34.8	1.47E+07	8.21E-08	8.41E-03	2.11E-03	1.75E+06	.454	.399	.583	.734	.449	.794
20.0	3.17	35.0	6.72E+06	3.77E-08	3.77E-03	9.45E-04	1.78E+06	.448	.393	.576	.727	.446	.796
22.5	3.24	35.3	3.09E+06	1.74E-08	1.71E-03	4.28E-04	1.81E+06	.442	.387	.569	.722	.444	.798
25.0	3.30	35.5	1.43E+06	8.12E-09	7.81E-04	1.95E-04	1.83E+06	.437	.382	.564	.716	.443	.800
27.5	3.35	35.7	6.64E+05	3.80E-09	3.59E-04	8.99E-05	1.85E+06	.433	.378	.559	.712	.442	.802
30.0	3.40	35.9	3.11E+05	1.78E-09	1.66E-04	4.16E-05	1.87E+06	.428	.374	.554	.707	.441	.803
32.5	3.45	36.0	1.46E+05	8.42E-10	7.74E-05	1.94E-05	1.89E+06	.425	.370	.550	.703	.441	.805
35.0	3.49	36.2	6.89E+04	3.99E-10	3.62E-05	9.07E-06	1.90E+06	.421	.367	.546	.699	.441	.807

Added filter 0.2 mm Cu

Thickness	HVL	$\langle E \rangle$	Φ/It	Ψ/It	Ka/It	Rel. dose	Φ/Ka	68 CaWO4	68 GOS	118 GOS	181 GOS	80 SPF	180 CsI
0.0	2.52	32.8	2.22E+09	1.16E-05	1.45E+00	1.00E+00	1.53E+06	.506	.450	.637	.780	.478	.792
2.5	2.68	33.3	9.42E+08	5.02E-06	5.93E-01	4.09E-01	1.59E+06	.492	.436	.623	.768	.468	.790
5.0	2.81	33.8	4.09E+08	2.21E-06	2.49E-01	1.72E-01	1.64E+06	.480	.424	.610	.758	.461	.790
7.5	2.92	34.1	1.80E+08	9.86E-07	1.07E-01	7.39E-02	1.68E+06	.470	.415	.600	.749	.456	.791
10.0	3.01	34.5	8.06E+07	4.45E-07	4.69E-02	3.23E-02	1.72E+06	.462	.406	.591	.741	.452	.792
12.5	3.10	34.8	3.64E+07	2.03E-07	2.08E-02	1.43E-02	1.75E+06	.454	.399	.583	.734	.449	.794
15.0	3.17	35.0	1.66E+07	9.32E-08	9.32E-03	6.43E-03	1.78E+06	.448	.393	.576	.727	.446	.796
17.5	3.24	35.3	7.63E+06	4.31E-08	4.22E-03	2.91E-03	1.81E+06	.442	.387	.569	.722	.444	.798
20.0	3.30	35.5	3.53E+06	2.01E-08	1.93E-03	1.33E-03	1.83E+06	.437	.382	.564	.716	.443	.800
22.5	3.35	35.7	1.64E+06	9.39E-09	8.87E-04	6.12E-04	1.85E+06	.432	.378	.559	.711	.442	.802
25.0	3.40	35.9	7.68E+05	4.41E-09	4.11E-04	2.83E-04	1.87E+06	.428	.374	.554	.707	.441	.803
27.5	3.45	36.0	3.61E+05	2.08E-09	1.91E-04	1.32E-04	1.89E+06	.424	.370	.550	.703	.441	.805
30.0	3.49	36.2	1.70E+05	9.87E-10	8.95E-05	6.18E-05	1.90E+06	.421	.367	.546	.699	.441	.807
32.5	3.53	36.3	8.07E+04	4.69E-10	4.21E-05	2.90E-05	1.92E+06	.418	.364	.542	.696	.441	.808
35.0	3.57	36.4	3.84E+04	2.24E-10	1.99E-05	1.37E-05	1.93E+06	.415	.361	.539	.693	.441	.810

Table IV.3.2

Characteristic X-ray beam quantities without scattered radiation and relative energy absorption in typical image detectors at variable absorber or phantom materials for diagnostic X-ray tube assemblies with W/Re anode at 10° target angle and an inherent (quality equivalent) filtration of 2.5 mm Al

Reference focal distance: 100 cm   X-ray tube voltage: 50 kV

Absorber material: Al

Thickness mm	HVL mm Al	<E> keV	Φ/It 1/As/cm²	Ψ/It J/As/cm²	Ka/It mGy/As	Relative dose	Φ/Ka photons /cm²/µGy	68 CaWO4	68 GOS	118 GOS	181 GOS	80 SPF	180 CsI
0.0	1.83	32.7	3.85E+10	2.02E-04	2.90E+01	1.00E+00	1.33E+06	.498	.447	.615	.744	.507	.810
.5	2.01	33.4	3.29E+10	1.76E-04	2.34E+01	8.07E-01	1.41E+06	.483	.432	.600	.731	.496	.806
1.0	2.17	34.0	2.84E+10	1.55E-04	1.93E+01	6.65E-01	1.47E+06	.469	.418	.587	.720	.487	.802
1.5	2.32	34.5	2.48E+10	1.37E-04	1.62E+01	5.57E-01	1.53E+06	.457	.406	.575	.710	.479	.799
2.0	2.45	35.0	2.17E+10	1.22E-04	1.37E+01	4.72E-01	1.59E+06	.447	.396	.564	.701	.472	.796
2.5	2.58	35.4	1.92E+10	1.09E-04	1.17E+01	4.04E-01	1.64E+06	.437	.386	.554	.692	.467	.794
3.0	2.69	35.8	1.70E+10	9.74E-05	1.01E+01	3.49E-01	1.68E+06	.429	.378	.545	.684	.462	.793
3.5	2.80	36.1	1.51E+10	8.76E-05	8.79E+00	3.03E-01	1.72E+06	.421	.370	.537	.677	.457	.791
4.0	2.90	36.5	1.35E+10	7.90E-05	7.69E+00	2.65E-01	1.76E+06	.414	.363	.529	.670	.454	.790
4.5	3.00	36.8	1.21E+10	7.15E-05	6.76E+00	2.33E-01	1.79E+06	.407	.357	.522	.663	.450	.789
5.0	3.09	37.1	1.09E+10	6.48E-05	5.97E+00	2.06E-01	1.83E+06	.401	.351	.515	.657	.447	.788
6.0	3.25	37.6	8.88E+09	5.35E-05	4.71E+00	1.62E-01	1.89E+06	.390	.341	.503	.646	.442	.787
7.0	3.40	38.1	7.30E+09	4.45E-05	3.76E+00	1.30E-01	1.94E+06	.381	.331	.493	.636	.438	.786
8.0	3.53	38.5	6.04E+09	3.72E-05	3.03E+00	1.05E-01	1.99E+06	.372	.323	.483	.626	.434	.785
9.0	3.65	38.9	5.02E+09	3.13E-05	2.47E+00	8.52E-02	2.03E+06	.365	.316	.475	.618	.431	.785
10.0	3.77	39.2	4.20E+09	2.64E-05	2.03E+00	6.99E-02	2.07E+06	.358	.310	.467	.610	.429	.784
12.0	3.97	39.8	2.98E+09	1.90E-05	1.39E+00	4.79E-02	2.14E+06	.346	.299	.454	.596	.425	.783
14.0	4.14	40.4	2.14E+09	1.38E-05	9.72E-01	3.35E-02	2.20E+06	.337	.290	.442	.585	.421	.782
16.0	4.29	40.8	1.55E+09	1.02E-05	6.91E-01	2.38E-02	2.25E+06	.328	.282	.432	.574	.419	.781
18.0	4.42	41.2	1.14E+09	7.53E-06	4.97E-01	1.71E-02	2.29E+06	.321	.275	.424	.565	.416	.780
20.0	4.54	41.6	8.43E+08	5.62E-06	3.61E-01	1.24E-02	2.33E+06	.314	.269	.416	.557	.414	.779
22.0	4.65	42.0	6.27E+08	4.22E-06	2.65E-01	9.13E-03	2.37E+06	.309	.264	.409	.549	.412	.777
25.0	4.80	42.4	4.07E+08	2.77E-06	1.69E-01	5.81E-03	2.42E+06	.301	.257	.400	.539	.410	.775
30.0	5.00	43.0	2.03E+08	1.40E-06	8.19E-02	2.82E-03	2.48E+06	.291	.248	.388	.525	.406	.772
35.0	5.16	43.6	1.04E+08	7.23E-07	4.09E-02	1.41E-03	2.53E+06	.283	.240	.378	.514	.402	.768
40.0	5.30	44.0	5.38E+07	3.79E-07	2.09E-02	7.20E-04	2.57E+06	.277	.234	.369	.505	.399	.764
45.0	5.42	44.4	2.83E+07	2.02E-07	1.09E-02	3.74E-04	2.61E+06	.271	.229	.362	.497	.396	.761
50.0	5.52	44.7	1.51E+07	1.08E-07	5.72E-03	1.97E-04	2.64E+06	.266	.225	.356	.490	.393	.758
55.0	5.61	45.0	8.14E+06	5.87E-08	3.05E-03	1.05E-04	2.67E+06	.262	.221	.351	.484	.390	.755
60.0	5.69	45.3	4.42E+06	3.21E-08	1.64E-03	5.66E-05	2.69E+06	.258	.218	.347	.478	.388	.752

Absorber material: Cu

Thickness mm	HVL mm Al	<E> keV	Φ/It 1/As/cm²	Ψ/It J/As/cm²	Ka/It mGy/As	Relative dose	Φ/Ka photons /cm²/µGy	68 CaWO4	68 GOS	118 GOS	181 GOS	80 SPF	180 CsI
0.00	1.83	32.7	3.85E+10	2.02E-04	2.90E+01	1.00E+00	1.33E+06	.498	.447	.615	.744	.507	.810
.02	2.07	33.6	3.16E+10	1.70E-04	2.21E+01	7.63E-01	1.43E+06	.477	.426	.595	.727	.492	.804
.04	2.28	34.4	2.65E+10	1.46E-04	1.75E+01	6.02E-01	1.52E+06	.460	.408	.577	.712	.481	.800
.06	2.47	35.0	2.25E+10	1.26E-04	1.41E+01	4.86E-01	1.59E+06	.445	.394	.562	.699	.471	.796
.08	2.64	35.6	1.93E+10	1.10E-04	1.16E+01	4.00E-01	1.66E+06	.432	.381	.548	.687	.464	.793
.10	2.80	36.2	1.67E+10	9.68E-05	9.71E+00	3.35E-01	1.72E+06	.420	.370	.536	.676	.457	.791
.20	3.41	38.1	8.90E+09	5.44E-05	4.57E+00	1.58E-01	1.95E+06	.379	.330	.491	.634	.437	.786
.30	3.85	39.5	5.20E+09	3.29E-05	2.48E+00	8.54E-02	2.10E+06	.353	.305	.461	.604	.427	.784
.40	4.18	40.5	3.21E+09	2.08E-05	1.45E+00	5.01E-02	2.21E+06	.334	.287	.439	.581	.421	.782
.50	4.44	41.3	2.06E+09	1.36E-05	8.96E-01	3.09E-02	2.30E+06	.320	.274	.422	.563	.416	.779
.60	4.65	42.0	1.35E+09	9.11E-06	5.72E-01	1.97E-02	2.37E+06	.309	.264	.409	.549	.412	.777
.70	4.83	42.5	9.10E+08	6.20E-06	3.75E-01	1.29E-02	2.43E+06	.300	.256	.398	.537	.409	.774
.80	4.98	43.0	6.21E+08	4.28E-06	2.51E-01	8.65E-03	2.47E+06	.292	.249	.389	.527	.406	.772
.90	5.11	43.4	4.29E+08	2.98E-06	1.71E-01	5.89E-03	2.51E+06	.286	.243	.381	.518	.403	.769
1.00	5.22	43.8	3.00E+08	2.10E-06	1.18E-01	4.06E-03	2.55E+06	.280	.238	.374	.510	.401	.766
1.20	5.41	44.4	1.50E+08	1.07E-06	5.77E-02	1.99E-03	2.61E+06	.271	.230	.363	.497	.396	.761
1.40	5.56	44.8	7.75E+07	5.57E-07	2.92E-02	1.01E-03	2.65E+06	.265	.224	.354	.487	.392	.757
1.60	5.68	45.3	4.08E+07	2.96E-07	1.52E-02	5.23E-04	2.69E+06	.259	.219	.347	.479	.388	.753
1.80	5.78	45.6	2.19E+07	1.60E-07	8.03E-03	2.77E-04	2.72E+06	.254	.214	.341	.472	.385	.749
2.00	5.87	45.9	1.19E+07	8.72E-08	4.32E-03	1.49E-04	2.75E+06	.250	.211	.336	.466	.382	.746
2.50	6.05	46.5	2.70E+06	2.01E-08	9.66E-04	3.33E-05	2.80E+06	.243	.204	.326	.454	.376	.739
3.00	6.17	46.9	6.47E+05	4.86E-09	2.28E-04	7.87E-06	2.83E+06	.237	.199	.319	.445	.371	.734
3.50	6.27	47.3	1.60E+05	1.22E-09	5.61E-05	1.93E-06	2.86E+06	.233	.195	.314	.439	.367	.730
4.00	6.34	47.6	4.09E+04	3.12E-10	1.42E-05	4.90E-07	2.88E+06	.230	.193	.310	.434	.364	.726
4.50	6.41	47.8	1.07E+04	8.17E-11	3.68E-06	1.27E-07	2.90E+06	.227	.190	.306	.429	.362	.724
5.00	6.46	48.0	2.83E+03	2.17E-11	9.72E-07	3.35E-08	2.91E+06	.225	.188	.304	.426	.360	.721
6.00	6.54	48.3	2.07E+02	1.60E-12	7.06E-08	2.43E-09	2.93E+06	.222	.185	.299	.420	.357	.718
7.00	6.60	48.5	1.58E+01	1.23E-13	5.36E-09	1.85E-10	2.95E+06	.219	.183	.296	.416	.354	.715
8.00	6.64	48.7	1.24E+00	9.69E-15	4.20E-10	1.45E-11	2.96E+06	.217	.182	.294	.413	.353	.713

Reference focal distance: 100 cm   X-ray tube voltage: 50 kV

Absorber material: Thick- ness mm	HVL mm Al	\<E> keV	Water Φ/It 1/As/cm²	Ψ/It J/As/cm²	Ka/It mGy/As	Relative dose	Φ/Ka photons /cm²/µGy	Relative energy absorption in the image detector (coverage mg/cm²) 68 CaWO4	68 GOS	118 GOS	181 GOS	80 SPF	180 CsI
No added filter													
0.0	1.83	32.7	3.85E+10	2.02E-04	2.90E+01	1.00E+00	1.33E+06	.498	.447	.615	.744	.507	.810
2.5	2.39	34.8	1.55E+10	8.61E-05	9.90E+00	3.41E-01	1.56E+06	.451	.400	.568	.705	.475	.797
5.0	2.81	36.2	6.75E+09	3.91E-05	3.91E+00	1.35E-01	1.72E+06	.420	.369	.536	.676	.457	.791
7.5	3.14	37.3	3.09E+09	1.84E-05	1.67E+00	5.75E-02	1.85E+06	.397	.347	.511	.653	.445	.788
10.0	3.42	38.1	1.46E+09	8.90E-06	7.47E-01	2.58E-02	1.95E+06	.379	.330	.491	.634	.437	.786
12.5	3.65	38.9	7.03E+08	4.37E-06	3.46E-01	1.19E-02	2.03E+06	.365	.316	.475	.618	.431	.784
15.0	3.85	39.5	3.45E+08	2.18E-06	1.64E-01	5.66E-03	2.10E+06	.353	.305	.461	.604	.427	.783
17.5	4.02	40.0	1.71E+08	1.10E-06	7.93E-02	2.73E-03	2.16E+06	.343	.295	.449	.592	.423	.782
20.0	4.17	40.5	8.62E+07	5.59E-07	3.89E-02	1.34E-03	2.21E+06	.334	.287	.439	.581	.420	.781
22.5	4.31	40.9	4.37E+07	2.87E-07	1.94E-02	6.67E-04	2.26E+06	.326	.280	.430	.572	.418	.780
25.0	4.44	41.3	2.23E+07	1.48E-07	9.72E-03	3.35E-04	2.30E+06	.320	.274	.422	.563	.416	.779
27.5	4.55	41.7	1.15E+07	7.67E-08	4.92E-03	1.70E-04	2.34E+06	.314	.269	.415	.556	.414	.778
30.0	4.65	42.0	5.95E+06	4.00E-08	2.51E-03	8.66E-05	2.37E+06	.308	.264	.409	.549	.412	.777
32.5	4.74	42.3	3.10E+06	2.10E-08	1.29E-03	4.45E-05	2.40E+06	.304	.259	.403	.542	.410	.775
35.0	4.83	42.5	1.62E+06	1.10E-08	6.67E-04	2.30E-05	2.43E+06	.299	.255	.397	.536	.409	.774
Added filter 0.5 mm Al													
0.0	2.01	33.4	3.29E+10	1.76E-04	2.34E+01	1.00E+00	1.41E+06	.483	.432	.600	.731	.496	.806
2.5	2.52	35.2	1.36E+10	7.68E-05	8.44E+00	3.60E-01	1.61E+06	.441	.390	.558	.696	.469	.795
5.0	2.91	36.5	6.03E+09	3.53E-05	3.42E+00	1.46E-01	1.76E+06	.413	.362	.528	.669	.453	.790
7.5	3.22	37.5	2.79E+09	1.68E-05	1.48E+00	6.34E-02	1.88E+06	.391	.342	.505	.647	.443	.787
10.0	3.48	38.4	1.32E+09	8.14E-06	6.71E-01	2.87E-02	1.97E+06	.375	.326	.486	.629	.435	.785
12.5	3.71	39.0	6.42E+08	4.02E-06	3.13E-01	1.34E-02	2.05E+06	.361	.313	.471	.614	.430	.784
15.0	3.90	39.6	3.16E+08	2.01E-06	1.49E-01	6.38E-03	2.12E+06	.350	.302	.458	.600	.426	.783
17.5	4.07	40.2	1.58E+08	1.02E-06	7.25E-02	3.10E-03	2.17E+06	.340	.293	.446	.589	.423	.782
20.0	4.21	40.6	7.95E+07	5.18E-07	3.57E-02	1.53E-03	2.23E+06	.332	.285	.436	.578	.420	.781
22.5	4.35	41.0	4.04E+07	2.66E-07	1.78E-02	7.61E-04	2.27E+06	.324	.278	.428	.569	.417	.780
25.0	4.47	41.4	2.07E+07	1.37E-07	8.97E-03	3.83E-04	2.31E+06	.318	.272	.420	.561	.415	.779
27.5	4.58	41.7	1.07E+07	7.14E-08	4.55E-03	1.94E-04	2.35E+06	.312	.267	.413	.554	.413	.778
30.0	4.68	42.1	5.53E+06	3.73E-08	2.33E-03	9.94E-05	2.38E+06	.307	.262	.407	.547	.411	.776
32.5	4.77	42.3	2.88E+06	1.95E-08	1.20E-03	5.11E-05	2.41E+06	.302	.258	.401	.541	.410	.775
35.0	4.85	42.6	1.51E+06	1.03E-08	6.20E-04	2.65E-05	2.43E+06	.298	.254	.396	.535	.408	.774
Added filter 0.1 mm Cu													
0.0	2.80	36.2	1.67E+10	9.68E-05	9.71E+00	1.00E+00	1.72E+06	.420	.370	.536	.676	.457	.791
2.5	3.14	37.3	7.63E+09	4.56E-05	4.13E+00	4.26E-01	1.85E+06	.397	.347	.511	.653	.446	.788
5.0	3.41	38.1	3.60E+09	2.20E-05	1.85E+00	1.90E-01	1.95E+06	.379	.330	.491	.634	.437	.786
7.5	3.65	38.9	1.74E+09	1.08E-05	8.55E-01	8.81E-02	2.03E+06	.365	.316	.475	.618	.431	.785
10.0	3.85	39.5	8.52E+08	5.39E-06	4.06E-01	4.18E-02	2.10E+06	.353	.305	.461	.604	.427	.783
12.5	4.02	40.0	4.24E+08	2.72E-06	1.96E-01	2.02E-02	2.16E+06	.343	.295	.449	.592	.423	.782
15.0	4.17	40.5	2.13E+08	1.38E-06	9.62E-02	9.91E-03	2.21E+06	.334	.287	.439	.581	.420	.781
17.5	4.31	40.9	1.08E+08	7.08E-07	4.78E-02	4.93E-03	2.26E+06	.326	.280	.430	.572	.418	.780
20.0	4.44	41.3	5.52E+07	3.65E-07	2.40E-02	2.47E-03	2.30E+06	.320	.274	.422	.563	.416	.779
22.5	4.55	41.7	2.84E+07	1.90E-07	1.22E-02	1.25E-03	2.34E+06	.314	.269	.415	.556	.414	.778
25.0	4.65	42.0	1.47E+07	9.89E-08	6.21E-03	6.39E-04	2.37E+06	.309	.264	.409	.549	.412	.777
27.5	4.74	42.3	7.65E+06	5.18E-08	3.19E-03	3.28E-04	2.40E+06	.304	.259	.403	.542	.410	.775
30.0	4.83	42.5	4.00E+06	2.72E-08	1.65E-03	1.70E-04	2.43E+06	.299	.255	.398	.536	.409	.774
32.5	4.91	42.8	2.10E+06	1.44E-08	8.56E-04	8.81E-05	2.45E+06	.295	.252	.393	.531	.407	.773
35.0	4.98	43.0	1.10E+06	7.61E-09	4.46E-04	4.60E-05	2.48E+06	.292	.248	.388	.526	.406	.771
Added filter 0.2 mm Cu													
0.0	3.41	38.1	8.90E+09	5.44E-05	4.57E+00	1.00E+00	1.95E+06	.379	.330	.491	.634	.437	.786
2.5	3.64	38.9	4.29E+09	2.67E-05	2.11E+00	4.62E-01	2.03E+06	.365	.316	.475	.618	.432	.785
5.0	3.85	39.5	2.11E+09	1.33E-05	1.00E+00	2.19E-01	2.10E+06	.353	.305	.461	.604	.427	.784
7.5	4.02	40.0	1.05E+09	6.71E-06	4.85E-01	1.06E-01	2.16E+06	.343	.295	.449	.592	.423	.783
10.0	4.17	40.5	5.26E+08	3.41E-06	2.38E-01	5.20E-02	2.21E+06	.334	.287	.439	.581	.421	.782
12.5	4.31	40.9	2.67E+08	1.75E-06	1.18E-01	2.59E-02	2.26E+06	.326	.280	.430	.572	.418	.780
15.0	4.44	41.3	1.36E+08	9.03E-07	5.93E-02	1.30E-02	2.30E+06	.320	.274	.422	.563	.416	.779
17.5	4.55	41.7	7.02E+07	4.68E-07	3.01E-02	6.58E-03	2.34E+06	.314	.269	.415	.556	.414	.778
20.0	4.65	42.0	3.63E+07	2.44E-07	1.53E-02	3.36E-03	2.37E+06	.309	.264	.409	.549	.412	.777
22.5	4.74	42.3	1.89E+07	1.28E-07	7.88E-03	1.72E-03	2.40E+06	.304	.259	.403	.542	.410	.776
25.0	4.83	42.5	9.88E+06	6.73E-08	4.07E-03	8.91E-04	2.43E+06	.299	.255	.398	.537	.409	.774
27.5	4.91	42.8	5.18E+06	3.55E-08	2.11E-03	4.63E-04	2.45E+06	.295	.252	.393	.531	.407	.773
30.0	4.98	43.0	2.73E+06	1.88E-08	1.10E-03	2.41E-04	2.47E+06	.292	.248	.388	.526	.406	.771
32.5	5.05	43.2	1.44E+06	9.98E-09	5.78E-04	1.26E-04	2.50E+06	.289	.245	.384	.522	.404	.770
35.0	5.11	43.4	7.64E+05	5.32E-09	3.04E-04	6.64E-05	2.52E+06	.285	.243	.380	.517	.403	.769

Table IV.3.3

Characteristic X-ray beam quantities without scattered radiation and relative energy absorption in typical image detectors at variable absorber or phantom materials for diagnostic X-ray tube assemblies with W/Re anode at 10° target angle and an inherent (quality equivalent) filtration of 2.5 mm Al

Reference focal distance: 100 cm   X-ray tube voltage: 60 kV

Absorber material: Al

Thick-ness mm	HVL mm Al	<E> keV	Φ/It 1/As/cm²	Ψ/It J/As/cm²	Ka/It mGy/As	Relative dose	Φ/Ka photons /cm²/µGy	68 CaWO4	68 GOS	118 GOS	181 GOS	80 SPF	180 CsI
0.0	2.17	36.5	6.65E+10	3.89E-04	4.31E+01	1.00E+00	1.54E+06	.411	.381	.541	.673	.456	.777
.5	2.40	37.2	5.84E+10	3.49E-04	3.58E+01	8.30E-01	1.63E+06	.396	.368	.527	.661	.446	.772
1.0	2.60	37.9	5.18E+10	3.15E-04	3.03E+01	7.03E-01	1.71E+06	.384	.356	.515	.651	.438	.768
1.5	2.78	38.5	4.63E+10	2.85E-04	2.60E+01	6.03E-01	1.78E+06	.373	.346	.505	.642	.431	.764
2.0	2.95	39.0	4.16E+10	2.60E-04	2.25E+01	5.23E-01	1.84E+06	.363	.338	.495	.633	.425	.761
2.5	3.11	39.5	3.75E+10	2.38E-04	1.97E+01	4.58E-01	1.90E+06	.354	.330	.487	.626	.420	.759
3.0	3.25	40.0	3.40E+10	2.18E-04	1.74E+01	4.04E-01	1.95E+06	.346	.323	.480	.619	.416	.756
3.5	3.39	40.4	3.09E+10	2.00E-04	1.55E+01	3.59E-01	2.00E+06	.339	.317	.473	.612	.412	.754
4.0	3.52	40.8	2.82E+10	1.84E-04	1.38E+01	3.20E-01	2.04E+06	.332	.312	.467	.606	.408	.752
4.5	3.64	41.2	2.58E+10	1.70E-04	1.24E+01	2.87E-01	2.08E+06	.326	.307	.461	.601	.405	.750
5.0	3.75	41.5	2.37E+10	1.57E-04	1.11E+01	2.59E-01	2.12E+06	.320	.303	.456	.596	.402	.749
6.0	3.96	42.1	2.00E+10	1.35E-04	9.13E+00	2.12E-01	2.19E+06	.310	.295	.446	.587	.396	.745
7.0	4.15	42.7	1.71E+10	1.17E-04	7.57E+00	1.76E-01	2.25E+06	.301	.288	.438	.579	.392	.743
8.0	4.33	43.2	1.46E+10	1.01E-04	6.33E+00	1.47E-01	2.31E+06	.293	.282	.432	.572	.388	.740
9.0	4.49	43.7	1.26E+10	8.81E-05	5.34E+00	1.24E-01	2.36E+06	.286	.277	.425	.566	.384	.737
10.0	4.63	44.1	1.09E+10	7.70E-05	4.53E+00	1.05E-01	2.40E+06	.280	.273	.420	.561	.381	.735
12.0	4.90	44.9	8.25E+09	5.93E-05	3.32E+00	7.71E-02	2.48E+06	.269	.266	.411	.552	.375	.730
14.0	5.13	45.5	6.32E+09	4.61E-05	2.48E+00	5.75E-02	2.55E+06	.260	.260	.404	.544	.371	.726
16.0	5.33	46.1	4.89E+09	3.61E-05	1.87E+00	4.35E-02	2.61E+06	.252	.256	.399	.538	.366	.721
18.0	5.51	46.7	3.81E+09	2.85E-05	1.43E+00	3.33E-02	2.66E+06	.245	.253	.395	.534	.362	.717
20.0	5.67	47.1	2.99E+09	2.26E-05	1.11E+00	2.57E-02	2.70E+06	.239	.250	.391	.530	.359	.713
22.0	5.81	47.6	2.36E+09	1.80E-05	8.62E-01	2.00E-02	2.74E+06	.234	.248	.388	.527	.355	.710
25.0	6.01	48.2	1.67E+09	1.29E-05	5.99E-01	1.39E-02	2.79E+06	.227	.245	.385	.523	.351	.704
30.0	6.28	49.0	9.61E+08	7.54E-06	3.35E-01	7.78E-03	2.87E+06	.217	.242	.381	.519	.344	.696
35.0	6.51	49.7	5.63E+08	4.48E-06	1.93E-01	4.47E-03	2.92E+06	.210	.241	.379	.517	.338	.689
40.0	6.70	50.3	3.35E+08	2.70E-06	1.13E-01	2.62E-03	2.97E+06	.203	.240	.378	.516	.332	.682
45.0	6.86	50.8	2.02E+08	1.64E-06	6.70E-02	1.56E-03	3.01E+06	.198	.240	.378	.516	.327	.676
50.0	7.00	51.3	1.23E+08	1.01E-06	4.03E-02	9.36E-04	3.04E+06	.193	.240	.378	.517	.323	.671
55.0	7.13	51.7	7.53E+07	6.24E-07	2.45E-02	5.70E-04	3.07E+06	.189	.241	.379	.517	.319	.666
60.0	7.24	52.1	4.66E+07	3.89E-07	1.51E-02	3.50E-04	3.09E+06	.186	.241	.380	.518	.316	.661

Absorber material: Cu

Thick-ness mm	HVL mm Al	<E> keV	Ψ/It 1/As/cm±	|/It J/As/cm±	Ka/It mGy/As	Relative dose	Ψ/Ka photons /cm±/èGy	68 CaWO4	68 GOS	118 GOS	181 GOS	80 SPF	180 CsI
0.00	2.17	36.5	6.65E+10	3.89E-04	4.31E+01	1.00E+00	1.54E+06	.411	.381	.541	.673	.456	.777
.02	2.47	37.5	5.68E+10	3.41E-04	3.42E+01	7.93E-01	1.66E+06	.391	.363	.522	.657	.443	.770
.04	2.74	38.4	4.93E+10	3.03E-04	2.79E+01	6.48E-01	1.76E+06	.374	.348	.506	.643	.433	.765
.06	2.97	39.2	4.33E+10	2.72E-04	2.34E+01	5.42E-01	1.85E+06	.360	.336	.493	.631	.424	.761
.08	3.19	39.9	3.84E+10	2.45E-04	1.99E+01	4.61E-01	1.93E+06	.348	.325	.482	.621	.417	.757
.10	3.39	40.5	3.43E+10	2.22E-04	1.71E+01	3.97E-01	2.00E+06	.337	.317	.472	.611	.411	.754
.20	4.18	42.8	2.10E+10	1.44E-04	9.28E+00	2.15E-01	2.26E+06	.299	.287	.437	.577	.391	.742
.30	4.75	44.5	1.40E+10	9.95E-05	5.72E+00	1.33E-01	2.44E+06	.274	.270	.416	.556	.378	.732
.40	5.19	45.8	9.73E+09	7.13E-05	3.79E+00	8.79E-02	2.57E+06	.257	.259	.403	.542	.369	.724
.50	5.54	46.8	6.99E+09	5.24E-05	2.62E+00	6.08E-02	2.67E+06	.243	.252	.394	.533	.361	.716
.60	5.82	47.6	5.14E+09	3.92E-05	1.87E+00	4.35E-02	2.75E+06	.233	.248	.388	.527	.355	.709
.70	6.06	48.3	3.85E+09	2.98E-05	1.37E+00	3.18E-02	2.81E+06	.225	.245	.384	.522	.349	.702
.80	6.27	49.0	2.92E+09	2.29E-05	1.02E+00	2.37E-02	2.86E+06	.218	.243	.381	.520	.344	.696
.90	6.44	49.5	2.24E+09	1.78E-05	7.71E-01	1.79E-02	2.91E+06	.212	.241	.380	.518	.339	.691
1.00	6.60	50.0	1.74E+09	1.39E-05	5.89E-01	1.37E-02	2.94E+06	.207	.241	.379	.517	.335	.685
1.20	6.85	50.8	1.06E+09	8.67E-06	3.54E-01	8.22E-03	3.01E+06	.198	.240	.378	.516	.328	.676
1.40	7.06	51.5	6.69E+08	5.52E-06	2.19E-01	5.08E-03	3.05E+06	.191	.240	.379	.517	.321	.668
1.60	7.24	52.1	4.28E+08	3.57E-06	1.38E-01	3.21E-03	3.09E+06	.186	.241	.380	.518	.316	.661
1.80	7.38	52.6	2.78E+08	2.34E-06	8.91E-02	2.07E-03	3.12E+06	.182	.242	.381	.520	.312	.655
2.00	7.51	53.0	1.83E+08	1.55E-06	5.81E-02	1.35E-03	3.15E+06	.178	.243	.382	.522	.308	.650
2.50	7.76	53.9	6.70E+07	5.78E-07	2.09E-02	4.86E-04	3.20E+06	.170	.245	.385	.525	.300	.639
3.00	7.94	54.6	2.56E+07	2.24E-07	7.92E-03	1.84E-04	3.24E+06	.165	.246	.387	.528	.293	.631
3.50	8.08	55.1	1.01E+07	8.94E-08	3.11E-03	7.21E-05	3.26E+06	.161	.247	.388	.529	.289	.624
4.00	8.20	55.5	4.11E+06	3.66E-08	1.25E-03	2.91E-05	3.28E+06	.158	.247	.389	.530	.285	.619
4.50	8.29	55.9	1.70E+06	1.52E-08	5.15E-04	1.20E-05	3.30E+06	.155	.247	.389	.530	.282	.614
5.00	8.37	56.2	7.14E+05	6.43E-09	2.16E-04	5.01E-06	3.31E+06	.153	.247	.388	.530	.279	.610
6.00	8.50	56.7	1.31E+05	1.19E-09	3.93E-05	9.13E-07	3.33E+06	.149	.246	.387	.528	.275	.604
7.00	8.59	57.1	2.50E+04	2.28E-10	7.46E-06	1.73E-07	3.34E+06	.147	.245	.386	.527	.271	.599
8.00	8.66	57.4	4.90E+03	4.50E-11	1.46E-06	3.39E-08	3.35E+06	.145	.244	.385	.525	.269	.595

Reference focal distance: 100 cm   X-ray tube voltage: 60 kV

Absorber material:			Water				Φ/Ka	Relative energy absorption in the					
Thick-								image detector (coverage mg/cm²)					
ness	HVL	<E>	Φ/It	Ψ/It	Ka/It	Relative	photons	68	68	118	181	80	180
mm	mm Al	keV	1/As/cm²	J/As/cm²	mGy/As	dose	/cm²/μGy	CaWO4	GOS	GOS	GOS	SPF	CsI

No added filter

0.0	2.17	36.5	6.65E+10	3.89E-04	4.31E+01	1.00E+00	1.54E+06	.411	.381	.541	.673	.456	.777
2.5	2.87	38.8	2.93E+10	1.82E-04	1.62E+01	3.75E-01	1.82E+06	.367	.341	.499	.637	.428	.762
5.0	3.41	40.5	1.39E+10	9.00E-05	6.92E+00	1.61E-01	2.01E+06	.337	.316	.471	.611	.411	.753
7.5	3.84	41.8	6.83E+09	4.58E-05	3.18E+00	7.37E-02	2.15E+06	.315	.299	.451	.592	.399	.747
10.0	4.19	42.9	3.46E+09	2.37E-05	1.52E+00	3.54E-02	2.27E+06	.298	.286	.436	.577	.390	.741
12.5	4.50	43.8	1.78E+09	1.25E-05	7.54E-01	1.75E-02	2.36E+06	.284	.277	.425	.565	.383	.736
15.0	4.77	44.6	9.33E+08	6.66E-06	3.81E-01	8.85E-03	2.45E+06	.273	.269	.416	.556	.377	.731
17.5	5.00	45.3	4.94E+08	3.58E-06	1.96E-01	4.56E-03	2.52E+06	.264	.263	.408	.548	.372	.727
20.0	5.21	45.9	2.64E+08	1.94E-06	1.03E-01	2.38E-03	2.58E+06	.255	.259	.402	.542	.368	.722
22.5	5.40	46.4	1.42E+08	1.06E-06	5.42E-02	1.26E-03	2.63E+06	.248	.255	.398	.537	.364	.718
25.0	5.57	46.9	7.73E+07	5.81E-07	2.89E-02	6.70E-04	2.68E+06	.242	.252	.394	.533	.360	.714
27.5	5.72	47.4	4.22E+07	3.20E-07	1.55E-02	3.60E-04	2.72E+06	.236	.250	.390	.529	.356	.710
30.0	5.86	47.8	2.31E+07	1.77E-07	8.40E-03	1.95E-04	2.76E+06	.231	.248	.388	.527	.353	.707
32.5	5.99	48.2	1.28E+07	9.84E-08	4.57E-03	1.06E-04	2.79E+06	.227	.246	.386	.524	.350	.703
35.0	6.11	48.5	7.06E+06	5.49E-08	2.50E-03	5.81E-05	2.82E+06	.223	.245	.384	.523	.347	.700

Added filter 0.5 mm Al

0.0	2.40	37.2	5.84E+10	3.49E-04	3.58E+01	1.00E+00	1.63E+06	.396	.368	.527	.661	.446	.772
2.5	3.04	39.3	2.64E+10	1.67E-04	1.41E+01	3.94E-01	1.88E+06	.357	.333	.490	.629	.422	.760
5.0	3.53	40.9	1.27E+10	8.30E-05	6.18E+00	1.73E-01	2.05E+06	.330	.311	.465	.605	.407	.751
7.5	3.94	42.1	6.29E+09	4.24E-05	2.88E+00	8.04E-02	2.19E+06	.310	.295	.447	.587	.396	.745
10.0	4.28	43.1	3.20E+09	2.21E-05	1.39E+00	3.89E-02	2.30E+06	.294	.283	.433	.573	.388	.740
12.5	4.58	44.0	1.66E+09	1.17E-05	6.94E-01	1.94E-02	2.39E+06	.281	.275	.422	.562	.382	.735
15.0	4.83	44.8	8.71E+08	6.24E-06	3.53E-01	9.87E-03	2.47E+06	.270	.268	.413	.554	.376	.730
17.5	5.06	45.4	4.62E+08	3.36E-06	1.83E-01	5.10E-03	2.53E+06	.261	.262	.407	.546	.371	.726
20.0	5.26	46.0	2.48E+08	1.83E-06	9.56E-02	2.67E-03	2.59E+06	.253	.258	.401	.541	.367	.721
22.5	5.44	46.5	1.34E+08	9.98E-07	5.06E-02	1.42E-03	2.64E+06	.246	.254	.396	.536	.363	.717
25.0	5.61	47.0	7.27E+07	5.48E-07	2.71E-02	7.56E-04	2.69E+06	.240	.251	.393	.532	.359	.713
27.5	5.76	47.5	3.98E+07	3.02E-07	1.46E-02	4.07E-04	2.73E+06	.235	.249	.390	.529	.355	.710
30.0	5.90	47.9	2.18E+07	1.68E-07	7.90E-03	2.21E-04	2.76E+06	.230	.247	.387	.526	.352	.706
32.5	6.02	48.3	1.21E+07	9.32E-08	4.31E-03	1.20E-04	2.80E+06	.226	.246	.385	.524	.349	.702
35.0	6.14	48.6	6.68E+06	5.20E-08	2.36E-03	6.60E-05	2.83E+06	.222	.244	.384	.522	.346	.699

Added filter 0.1 mm Cu

0.0	3.39	40.5	3.43E+10	2.22E-04	1.71E+01	1.00E+00	2.00E+06	.337	.317	.472	.611	.411	.754
2.5	3.83	41.8	1.69E+10	1.13E-04	7.85E+00	4.58E-01	2.15E+06	.315	.299	.452	.592	.399	.747
5.0	4.19	42.9	8.52E+09	5.85E-05	3.76E+00	2.20E-01	2.27E+06	.298	.286	.437	.577	.391	.741
7.5	4.50	43.8	4.39E+09	3.08E-05	1.86E+00	1.09E-01	2.36E+06	.285	.277	.425	.565	.384	.736
10.0	4.76	44.5	2.30E+09	1.64E-05	9.41E-01	5.49E-02	2.44E+06	.273	.269	.416	.556	.378	.732
12.5	5.00	45.2	1.22E+09	8.82E-06	4.84E-01	2.83E-02	2.51E+06	.264	.264	.408	.548	.373	.727
15.0	5.21	45.8	6.51E+08	4.78E-06	2.53E-01	1.48E-02	2.57E+06	.256	.259	.402	.542	.368	.723
17.5	5.39	46.4	3.51E+08	2.61E-06	1.33E-01	7.79E-03	2.63E+06	.248	.255	.398	.537	.364	.719
20.0	5.56	46.9	1.90E+08	1.43E-06	7.11E-02	4.15E-03	2.67E+06	.242	.252	.394	.533	.360	.715
22.5	5.71	47.3	1.04E+08	7.87E-07	3.82E-02	2.23E-03	2.72E+06	.237	.250	.390	.529	.357	.711
25.0	5.85	47.8	5.69E+07	4.36E-07	2.07E-02	1.21E-03	2.75E+06	.232	.248	.388	.527	.353	.707
27.5	5.98	48.1	3.14E+07	2.42E-07	1.13E-02	6.57E-04	2.79E+06	.227	.246	.386	.524	.350	.704
30.0	6.10	48.5	1.74E+07	1.35E-07	6.16E-03	3.60E-04	2.82E+06	.223	.245	.384	.523	.347	.700
32.5	6.21	48.8	9.64E+06	7.54E-08	3.39E-03	1.98E-04	2.85E+06	.219	.244	.383	.521	.345	.697
35.0	6.31	49.1	5.37E+06	4.23E-08	1.87E-03	1.09E-04	2.87E+06	.216	.243	.382	.520	.342	.694

Added filter 0.2 mm Cu

0.0	4.18	42.8	2.10E+10	1.44E-04	9.28E+00	1.00E+00	2.26E+06	.299	.287	.437	.577	.391	.742
2.5	4.49	43.7	1.08E+10	7.59E-05	4.59E+00	4.94E-01	2.36E+06	.285	.277	.425	.565	.384	.737
5.0	4.76	44.5	5.67E+09	4.04E-05	2.32E+00	2.50E-01	2.44E+06	.274	.270	.416	.556	.378	.732
7.5	4.99	45.2	3.00E+09	2.17E-05	1.19E+00	1.29E-01	2.51E+06	.264	.264	.408	.548	.373	.727
10.0	5.20	45.8	1.60E+09	1.18E-05	6.23E-01	6.71E-02	2.57E+06	.256	.259	.402	.542	.368	.723
12.5	5.39	46.4	8.63E+08	6.41E-06	3.29E-01	3.54E-02	2.63E+06	.249	.255	.398	.537	.364	.719
15.0	5.55	46.9	4.68E+08	3.52E-06	1.75E-01	1.89E-02	2.67E+06	.242	.252	.394	.533	.360	.715
17.5	5.71	47.3	2.56E+08	1.94E-06	9.41E-02	1.01E-02	2.71E+06	.237	.250	.390	.529	.357	.711
20.0	5.85	47.7	1.40E+08	1.07E-06	5.09E-02	5.48E-03	2.75E+06	.232	.248	.388	.527	.354	.707
22.5	5.98	48.1	7.72E+07	5.95E-07	2.77E-02	2.98E-03	2.79E+06	.227	.246	.386	.524	.350	.704
25.0	6.09	48.5	4.27E+07	3.32E-07	1.52E-02	1.63E-03	2.82E+06	.223	.245	.384	.522	.348	.701
27.5	6.20	48.8	2.37E+07	1.85E-07	8.33E-03	8.97E-04	2.85E+06	.219	.244	.383	.521	.345	.697
30.0	6.30	49.1	1.32E+07	1.04E-07	4.60E-03	4.95E-04	2.87E+06	.216	.243	.382	.520	.342	.694
32.5	6.40	49.4	7.38E+06	5.84E-08	2.55E-03	2.74E-04	2.90E+06	.213	.242	.381	.519	.340	.691
35.0	6.49	49.7	4.13E+06	3.29E-08	1.42E-03	1.53E-04	2.92E+06	.210	.241	.380	.518	.337	.688

Table IV.3.4

Characteristic X-ray beam quantities without scattered radiation and relative energy absorption in typical image detectors at variable absorber or phantom materials for diagnostic X-ray tube assemblies with W/Re anode at 10° target angle and an inherent (quality equivalent) filtration of 2.5 mm Al

Reference focal distance: 100 cm   X-ray tube voltage: 70 kV

Absorber material: Al

Thick-ness mm	HVL mm Al	\<E\> keV	Φ/It 1/As/cm²	Ψ/It J/As/cm²	Ka/It mGy/As	Relative dose	Φ/Ka photons /cm²/µGy	68 CaWO4	68 GOS	118 GOS	181 GOS	80 SPF	180 CsI
0.0	2.50	39.9	9.92E+10	6.35E-04	5.76E+01	1.00E+00	1.72E+06	.348	.342	.496	.631	.412	.736
.5	2.76	40.8	8.87E+10	5.79E-04	4.88E+01	8.47E-01	1.82E+06	.335	.331	.485	.620	.403	.730
1.0	2.99	41.5	8.00E+10	5.32E-04	4.20E+01	7.30E-01	1.90E+06	.323	.322	.475	.611	.396	.726
1.5	3.21	42.1	7.25E+10	4.90E-04	3.67E+01	6.37E-01	1.98E+06	.313	.314	.466	.603	.389	.721
2.0	3.41	42.7	6.61E+10	4.53E-04	3.24E+01	5.62E-01	2.04E+06	.304	.307	.458	.596	.384	.718
2.5	3.59	43.3	6.05E+10	4.20E-04	2.88E+01	5.00E-01	2.10E+06	.296	.301	.451	.590	.379	.714
3.0	3.76	43.8	5.56E+10	3.90E-04	2.57E+01	4.47E-01	2.16E+06	.289	.295	.445	.584	.374	.711
3.5	3.92	44.3	5.12E+10	3.63E-04	2.32E+01	4.03E-01	2.21E+06	.282	.291	.440	.579	.370	.708
4.0	4.07	44.7	4.73E+10	3.39E-04	2.10E+01	3.64E-01	2.26E+06	.276	.286	.435	.574	.367	.705
4.5	4.21	45.1	4.38E+10	3.17E-04	1.91E+01	3.31E-01	2.30E+06	.270	.282	.430	.569	.363	.702
5.0	4.35	45.5	4.07E+10	2.96E-04	1.74E+01	3.02E-01	2.34E+06	.265	.279	.426	.565	.360	.700
6.0	4.60	46.2	3.52E+10	2.60E-04	1.46E+01	2.53E-01	2.41E+06	.256	.273	.419	.558	.355	.695
7.0	4.82	46.8	3.07E+10	2.30E-04	1.24E+01	2.15E-01	2.48E+06	.248	.268	.413	.552	.350	.690
8.0	5.03	47.4	2.68E+10	2.04E-04	1.06E+01	1.84E-01	2.53E+06	.241	.263	.408	.547	.345	.686
9.0	5.22	48.0	2.36E+10	1.81E-04	9.13E+00	1.59E-01	2.59E+06	.234	.260	.403	.542	.341	.682
10.0	5.39	48.5	2.08E+10	1.62E-04	7.91E+00	1.37E-01	2.63E+06	.228	.257	.399	.538	.337	.678
12.0	5.71	49.3	1.64E+10	1.30E-04	6.04E+00	1.05E-01	2.71E+06	.218	.252	.393	.531	.331	.671
14.0	5.98	50.1	1.30E+10	1.05E-04	4.69E+00	8.14E-02	2.78E+06	.210	.248	.388	.526	.325	.664
16.0	6.22	50.8	1.04E+10	8.51E-05	3.68E+00	6.39E-02	2.84E+06	.203	.245	.384	.522	.320	.658
18.0	6.44	51.5	8.44E+09	6.96E-05	2.92E+00	5.07E-02	2.89E+06	.197	.242	.381	.519	.315	.652
20.0	6.63	52.0	6.86E+09	5.71E-05	2.34E+00	4.06E-02	2.93E+06	.191	.240	.378	.516	.311	.647
22.0	6.81	52.6	5.60E+09	4.71E-05	1.88E+00	3.27E-02	2.97E+06	.186	.239	.376	.514	.306	.641
25.0	7.04	53.3	4.16E+09	3.55E-05	1.38E+00	2.39E-02	3.02E+06	.180	.237	.374	.511	.301	.634
30.0	7.38	54.3	2.59E+09	2.25E-05	8.38E-01	1.46E-02	3.09E+06	.171	.235	.371	.508	.293	.623
35.0	7.65	55.2	1.64E+09	1.45E-05	5.21E-01	9.06E-03	3.14E+06	.164	.234	.370	.506	.286	.613
40.0	7.88	55.9	1.05E+09	9.42E-06	3.30E-01	5.73E-03	3.18E+06	.158	.233	.368	.505	.280	.605
45.0	8.08	56.6	6.82E+08	6.18E-06	2.12E-01	3.68E-03	3.22E+06	.153	.232	.367	.504	.274	.597
50.0	8.25	57.2	4.46E+08	4.09E-06	1.38E-01	2.39E-03	3.25E+06	.149	.232	.367	.503	.270	.590
55.0	8.41	57.7	2.95E+08	2.73E-06	9.01E-02	1.57E-03	3.27E+06	.145	.231	.366	.502	.265	.584
60.0	8.54	58.2	1.96E+08	1.83E-06	5.95E-02	1.03E-03	3.29E+06	.142	.231	.365	.501	.262	.579

Absorber material: Cu

Thick-ness mm	HVL mm Al	\<E\> keV	Φ/It 1/As/cm²	Ψ/It J/As/cm²	Ka/It mGy/As	Relative dose	Φ/Ka photons /cm²/µGy	68 CaWO4	68 GOS	118 GOS	181 GOS	80 SPF	180 CsI
0.00	2.50	39.9	9.92E+10	6.35E-04	5.76E+01	1.00E+00	1.72E+06	.348	.342	.496	.631	.412	.736
.02	2.85	41.1	8.69E+10	5.72E-04	4.69E+01	8.15E-01	1.85E+06	.330	.327	.480	.616	.400	.728
.04	3.16	42.0	7.71E+10	5.20E-04	3.93E+01	6.83E-01	1.96E+06	.315	.315	.467	.604	.390	.722
.06	3.44	42.9	6.92E+10	4.76E-04	3.36E+01	5.84E-01	2.06E+06	.302	.305	.456	.594	.382	.716
.08	3.70	43.7	6.26E+10	4.38E-04	2.92E+01	5.08E-01	2.14E+06	.290	.297	.447	.585	.375	.711
.10	3.93	44.3	5.70E+10	4.05E-04	2.57E+01	4.47E-01	2.21E+06	.281	.290	.439	.578	.370	.707
.20	4.87	47.0	3.81E+10	2.87E-04	1.53E+01	2.65E-01	2.49E+06	.246	.267	.411	.550	.348	.689
.30	5.55	48.9	2.73E+10	2.14E-04	1.02E+01	1.77E-01	2.67E+06	.223	.254	.396	.535	.334	.674
.40	6.07	50.4	2.04E+10	1.65E-04	7.27E+00	1.26E-01	2.80E+06	.207	.247	.386	.525	.323	.662
.50	6.49	51.6	1.56E+10	1.29E-04	5.39E+00	9.37E-02	2.90E+06	.195	.242	.380	.518	.314	.650
.60	6.83	52.6	1.23E+10	1.03E-04	4.12E+00	7.15E-02	2.98E+06	.185	.239	.376	.514	.306	.640
.70	7.12	53.5	9.75E+09	8.36E-05	3.21E+00	5.58E-02	3.04E+06	.178	.237	.373	.511	.299	.631
.80	7.36	54.3	7.85E+09	6.83E-05	2.54E+00	4.42E-02	3.09E+06	.171	.235	.372	.509	.293	.623
.90	7.57	54.9	6.39E+09	5.62E-05	2.04E+00	3.55E-02	3.13E+06	.166	.234	.370	.507	.288	.616
1.00	7.76	55.5	5.24E+09	4.66E-05	1.66E+00	2.88E-02	3.16E+06	.161	.233	.369	.506	.283	.609
1.20	8.07	56.6	3.59E+09	3.26E-05	1.12E+00	1.94E-02	3.22E+06	.153	.232	.367	.504	.275	.597
1.40	8.33	57.4	2.52E+09	2.32E-05	7.73E-01	1.34E-02	3.26E+06	.147	.231	.366	.503	.268	.588
1.60	8.54	58.2	1.79E+09	1.67E-05	5.45E-01	9.47E-03	3.29E+06	.142	.231	.365	.501	.262	.579
1.80	8.71	58.8	1.30E+09	1.22E-05	3.91E-01	6.79E-03	3.31E+06	.138	.230	.364	.500	.257	.571
2.00	8.87	59.4	9.45E+08	9.00E-06	2.84E-01	4.92E-03	3.33E+06	.135	.229	.363	.499	.253	.565
2.50	9.17	60.6	4.47E+08	4.33E-06	1.33E-01	2.30E-03	3.37E+06	.128	.227	.360	.495	.244	.551
3.00	9.40	61.5	2.20E+08	2.16E-06	6.47E-02	1.12E-03	3.39E+06	.123	.225	.357	.492	.237	.541
3.50	9.58	62.2	1.11E+08	1.11E-06	3.26E-02	5.67E-04	3.41E+06	.119	.223	.354	.488	.232	.532
4.00	9.72	62.8	5.76E+07	5.80E-07	1.68E-02	2.92E-04	3.42E+06	.116	.221	.351	.484	.228	.525
4.50	9.84	63.3	3.04E+07	3.08E-07	8.86E-03	1.54E-04	3.43E+06	.113	.219	.348	.481	.224	.519
5.00	9.94	63.8	1.62E+07	1.66E-07	4.73E-03	8.22E-05	3.43E+06	.111	.217	.346	.478	.221	.514
6.00	10.10	64.5	4.82E+06	4.98E-08	1.40E-03	2.43E-05	3.44E+06	.108	.214	.342	.474	.216	.506
7.00	10.22	65.1	1.48E+06	1.54E-08	4.29E-04	7.45E-06	3.45E+06	.105	.212	.339	.470	.212	.499
8.00	10.32	65.5	4.66E+05	4.90E-09	1.35E-04	2.35E-06	3.45E+06	.103	.210	.336	.466	.210	.494

Reference focal distance: 100 cm   X-ray tube voltage: 70 kV

Absorber material: Water

Thickness mm	HVL mm Al	$<E>$ keV	Φ/It 1/As/cm²	Ψ/It J/As/cm²	Ka/It mGy/As	Relative dose	Φ/Ka photons /cm²/µGy	68 CaWO4	68 GOS	118 GOS	181 GOS	80 SPF	180 CsI
								Relative energy absorption in the image detector (coverage mg/cm²)					

No added filter

0.0	2.50	39.9	9.92E+10	6.35E-04	5.76E+01	1.00E+00	1.72E+06	.348	.342	.496	.631	.412	.736
2.5	3.32	42.5	4.65E+10	3.17E-04	2.30E+01	4.00E-01	2.02E+06	.307	.309	.461	.599	.386	.719
5.0	3.95	44.4	2.31E+10	1.65E-04	1.04E+01	1.81E-01	2.22E+06	.280	.289	.438	.577	.369	.706
7.5	4.47	45.9	1.19E+10	8.78E-05	5.02E+00	8.72E-02	2.38E+06	.260	.276	.422	.561	.356	.696
10.0	4.90	47.2	6.31E+09	4.77E-05	2.52E+00	4.38E-02	2.50E+06	.244	.266	.411	.550	.347	.687
12.5	5.27	48.2	3.39E+09	2.62E-05	1.30E+00	2.26E-02	2.60E+06	.231	.259	.402	.541	.339	.679
15.0	5.59	49.2	1.85E+09	1.45E-05	6.88E-01	1.19E-02	2.68E+06	.221	.253	.395	.534	.332	.671
17.5	5.88	50.0	1.02E+09	8.14E-06	3.69E-01	6.40E-03	2.76E+06	.212	.249	.390	.528	.325	.664
20.0	6.13	50.7	5.64E+08	4.59E-06	2.00E-01	3.48E-03	2.82E+06	.204	.246	.385	.524	.320	.658
22.5	6.36	51.4	3.16E+08	2.60E-06	1.10E-01	1.91E-03	2.87E+06	.197	.243	.382	.520	.315	.651
25.0	6.57	52.0	1.78E+08	1.48E-06	6.09E-02	1.06E-03	2.92E+06	.192	.241	.379	.517	.310	.646
27.5	6.76	52.6	1.01E+08	8.47E-07	3.40E-02	5.90E-04	2.96E+06	.186	.240	.377	.515	.306	.640
30.0	6.93	53.1	5.72E+07	4.87E-07	1.91E-02	3.31E-04	3.00E+06	.182	.238	.375	.513	.302	.635
32.5	7.09	53.6	3.27E+07	2.80E-07	1.08E-02	1.87E-04	3.03E+06	.177	.237	.374	.511	.298	.630
35.0	7.24	54.0	1.87E+07	1.62E-07	6.12E-03	1.06E-04	3.06E+06	.174	.236	.372	.510	.294	.625

Added filter 0.5 mm Al

0.0	2.76	40.8	8.87E+10	5.79E-04	4.88E+01	1.00E+00	1.82E+06	.335	.331	.485	.620	.403	.730
2.5	3.51	43.1	4.25E+10	2.93E-04	2.04E+01	4.18E-01	2.08E+06	.299	.303	.454	.592	.380	.715
5.0	4.10	44.9	2.14E+10	1.54E-04	9.43E+00	1.93E-01	2.27E+06	.274	.285	.433	.572	.365	.703
7.5	4.59	46.3	1.11E+10	8.24E-05	4.61E+00	9.45E-02	2.41E+06	.255	.273	.419	.558	.354	.693
10.0	5.00	47.5	5.90E+09	4.49E-05	2.33E+00	4.79E-02	2.53E+06	.240	.264	.408	.547	.344	.685
12.5	5.36	48.5	3.19E+09	2.47E-05	1.21E+00	2.49E-02	2.62E+06	.228	.257	.400	.539	.337	.677
15.0	5.67	49.4	1.74E+09	1.38E-05	6.43E-01	1.32E-02	2.71E+06	.218	.252	.393	.532	.330	.669
17.5	5.95	50.2	9.60E+08	7.72E-06	3.46E-01	7.10E-03	2.77E+06	.210	.248	.388	.527	.324	.663
20.0	6.20	50.9	5.34E+08	4.36E-06	1.89E-01	3.87E-03	2.83E+06	.202	.245	.385	.523	.319	.656
22.5	6.42	51.6	2.99E+08	2.47E-06	1.04E-01	2.13E-03	2.88E+06	.196	.243	.381	.519	.314	.650
25.0	6.62	52.2	1.69E+08	1.41E-06	5.76E-02	1.18E-03	2.93E+06	.190	.241	.379	.516	.309	.644
27.5	6.81	52.7	9.57E+07	8.08E-07	3.22E-02	6.60E-04	2.97E+06	.185	.239	.377	.514	.305	.639
30.0	6.97	53.2	5.45E+07	4.64E-07	1.81E-02	3.71E-04	3.01E+06	.181	.238	.375	.512	.301	.633
32.5	7.13	53.7	3.11E+07	2.68E-07	1.02E-02	2.10E-04	3.04E+06	.176	.237	.373	.511	.297	.628
35.0	7.27	54.1	1.79E+07	1.55E-07	5.82E-03	1.19E-04	3.07E+06	.173	.236	.372	.509	.294	.624

Added filter 0.1 mm Cu

0.0	3.93	44.3	5.70E+10	4.05E-04	2.57E+01	1.00E+00	2.21E+06	.281	.290	.439	.578	.370	.707
2.5	4.45	45.9	2.94E+10	2.16E-04	1.24E+01	4.81E-01	2.37E+06	.260	.276	.423	.562	.357	.697
5.0	4.88	47.1	1.55E+10	1.17E-04	6.21E+00	2.41E-01	2.50E+06	.245	.266	.411	.550	.347	.688
7.5	5.26	48.2	8.32E+09	6.42E-05	3.21E+00	1.25E-01	2.60E+06	.232	.259	.402	.541	.339	.680
10.0	5.58	49.1	4.53E+09	3.56E-05	1.69E+00	6.57E-02	2.68E+06	.221	.254	.395	.534	.332	.672
12.5	5.86	49.9	2.49E+09	1.99E-05	9.05E-01	3.52E-02	2.75E+06	.213	.249	.390	.528	.326	.665
15.0	6.12	50.6	1.38E+09	1.12E-05	4.92E-01	1.91E-02	2.81E+06	.205	.246	.386	.524	.321	.659
17.5	6.35	51.3	7.73E+08	6.36E-06	2.70E-01	1.05E-02	2.87E+06	.198	.244	.382	.520	.315	.652
20.0	6.55	51.9	4.35E+08	3.62E-06	1.49E-01	5.80E-03	2.92E+06	.192	.241	.380	.517	.311	.646
22.5	6.74	52.5	2.46E+08	2.07E-06	8.32E-02	3.24E-03	2.96E+06	.187	.240	.377	.515	.306	.641
25.0	6.92	53.0	1.40E+08	1.19E-06	4.67E-02	1.82E-03	2.99E+06	.182	.238	.375	.513	.302	.636
27.5	7.07	53.5	7.98E+07	6.84E-07	2.64E-02	1.03E-03	3.03E+06	.178	.237	.374	.511	.299	.631
30.0	7.22	54.0	4.57E+07	3.95E-07	1.50E-02	5.82E-04	3.06E+06	.174	.236	.373	.510	.295	.626
32.5	7.36	54.4	2.63E+07	2.29E-07	8.53E-03	3.32E-04	3.08E+06	.171	.235	.372	.508	.292	.621
35.0	7.48	54.8	1.52E+07	1.33E-07	4.88E-03	1.90E-04	3.11E+06	.167	.235	.371	.507	.289	.617

Added filter 0.2 mm Cu

0.0	4.87	47.0	3.81E+10	2.87E-04	1.53E+01	1.00E+00	2.49E+06	.246	.267	.411	.550	.348	.689
2.5	5.24	48.1	2.04E+10	1.57E-04	7.88E+00	5.16E-01	2.59E+06	.233	.259	.402	.541	.340	.681
5.0	5.56	49.0	1.11E+10	8.73E-05	4.15E+00	2.72E-01	2.68E+06	.222	.254	.396	.534	.333	.673
7.5	5.85	49.8	6.11E+09	4.88E-05	2.22E+00	1.46E-01	2.75E+06	.213	.250	.390	.529	.327	.666
10.0	6.10	50.6	3.39E+09	2.75E-05	1.21E+00	7.89E-02	2.81E+06	.206	.246	.386	.524	.321	.660
12.5	6.33	51.2	1.89E+09	1.56E-05	6.61E-01	4.33E-02	2.86E+06	.199	.244	.382	.520	.316	.653
15.0	6.54	51.9	1.07E+09	8.85E-06	3.66E-01	2.39E-02	2.91E+06	.193	.242	.380	.517	.312	.647
17.5	6.73	52.4	6.02E+08	5.06E-06	2.04E-01	1.33E-02	2.95E+06	.188	.240	.377	.515	.307	.642
20.0	6.90	52.9	3.42E+08	2.90E-06	1.14E-01	7.49E-03	2.99E+06	.183	.238	.376	.513	.303	.637
22.5	7.06	53.4	1.95E+08	1.67E-06	6.46E-02	4.23E-03	3.02E+06	.179	.237	.374	.511	.299	.632
25.0	7.20	53.9	1.12E+08	9.65E-07	3.66E-02	2.40E-03	3.05E+06	.175	.236	.373	.510	.296	.627
27.5	7.34	54.3	6.42E+07	5.59E-07	2.09E-02	1.36E-03	3.08E+06	.171	.235	.372	.509	.292	.622
30.0	7.46	54.7	3.70E+07	3.24E-07	1.19E-02	7.80E-04	3.10E+06	.168	.235	.371	.508	.289	.618
32.5	7.58	55.1	2.14E+07	1.89E-07	6.84E-03	4.48E-04	3.13E+06	.165	.234	.370	.507	.286	.614
35.0	7.69	55.4	1.24E+07	1.10E-07	3.94E-03	2.58E-04	3.15E+06	.162	.234	.369	.506	.283	.610

Table IV.3.5

Characteristic X-ray beam quantities without scattered radiation and relative energy absorption in typical image detectors at variable absorber or phantom materials for diagnostic X-ray tube assemblies with W/Re anode at 10° target angle and an inherent (quality equivalent) filtration of 2.5 mm Al

Reference focal distance: 100 cm   X-ray tube voltage: 80 kV

Absorber material: Al

Thick-ness mm	HVL mm Al	<E> keV	Φ/It 1/As/cm²	Ψ/It J/As/cm²	Ka/It mGy/As	Relative dose	Φ/Ka photons /cm²/µGy	68 CaWO4	68 GOS	118 GOS	181 GOS	80 SPF	180 CsI
0.0	2.86	43.5	1.38E+11	9.59E-04	7.30E+01	1.00E+00	1.89E+06	.299	.311	.459	.593	.372	.690
.5	3.15	44.3	1.25E+11	8.88E-04	6.28E+01	8.61E-01	1.99E+06	.287	.302	.449	.584	.364	.684
1.0	3.42	45.1	1.14E+11	8.24E-04	5.50E+01	7.53E-01	2.07E+06	.277	.294	.441	.576	.357	.679
1.5	3.67	45.8	1.05E+11	7.68E-04	4.87E+01	6.67E-01	2.15E+06	.268	.288	.434	.569	.350	.674
2.0	3.89	46.5	9.65E+10	7.18E-04	4.35E+01	5.96E-01	2.22E+06	.261	.282	.427	.563	.345	.670
2.5	4.10	47.0	8.92E+10	6.73E-04	3.91E+01	5.36E-01	2.28E+06	.253	.277	.421	.558	.340	.665
3.0	4.30	47.6	8.28E+10	6.31E-04	3.55E+01	4.86E-01	2.34E+06	.247	.272	.416	.553	.336	.662
3.5	4.48	48.1	7.70E+10	5.93E-04	3.23E+01	4.42E-01	2.39E+06	.241	.268	.411	.548	.332	.658
4.0	4.65	48.6	7.18E+10	5.59E-04	2.95E+01	4.05E-01	2.43E+06	.236	.265	.407	.544	.329	.655
4.5	4.81	49.0	6.71E+10	5.27E-04	2.71E+01	3.72E-01	2.48E+06	.231	.262	.403	.540	.325	.651
5.0	4.97	49.4	6.28E+10	4.97E-04	2.50E+01	3.42E-01	2.52E+06	.227	.259	.400	.537	.322	.648
6.0	5.25	50.2	5.53E+10	4.45E-04	2.14E+01	2.93E-01	2.59E+06	.219	.254	.394	.530	.316	.642
7.0	5.51	50.9	4.90E+10	3.99E-04	1.85E+01	2.53E-01	2.65E+06	.212	.250	.389	.525	.311	.637
8.0	5.74	51.5	4.35E+10	3.59E-04	1.61E+01	2.20E-01	2.71E+06	.206	.246	.384	.521	.307	.632
9.0	5.96	52.1	3.89E+10	3.24E-04	1.41E+01	1.93E-01	2.76E+06	.200	.243	.381	.517	.303	.627
10.0	6.16	52.7	3.48E+10	2.94E-04	1.24E+01	1.70E-01	2.80E+06	.195	.241	.377	.513	.299	.622
12.0	6.51	53.6	2.81E+10	2.42E-04	9.78E+00	1.34E-01	2.88E+06	.187	.237	.372	.507	.292	.614
14.0	6.82	54.5	2.30E+10	2.01E-04	7.81E+00	1.07E-01	2.94E+06	.180	.233	.368	.503	.286	.606
16.0	7.08	55.3	1.89E+10	1.67E-04	6.31E+00	8.65E-02	2.99E+06	.174	.231	.364	.499	.281	.599
18.0	7.32	56.0	1.56E+10	1.40E-04	5.14E+00	7.05E-02	3.04E+06	.169	.229	.361	.495	.276	.592
20.0	7.54	56.6	1.30E+10	1.18E-04	4.22E+00	5.79E-02	3.07E+06	.164	.227	.359	.493	.271	.586
22.0	7.73	57.2	1.08E+10	9.94E-05	3.49E+00	4.78E-02	3.11E+06	.160	.225	.357	.490	.267	.580
25.0	7.99	58.0	8.34E+09	7.75E-05	2.65E+00	3.63E-02	3.15E+06	.155	.223	.354	.487	.261	.572
30.0	8.35	59.1	5.46E+09	5.17E-05	1.70E+00	2.34E-02	3.20E+06	.148	.221	.351	.483	.253	.560
35.0	8.65	60.1	3.63E+09	3.50E-05	1.12E+00	1.53E-02	3.25E+06	.143	.219	.348	.480	.246	.549
40.0	8.90	61.0	2.45E+09	2.39E-05	7.46E-01	1.02E-02	3.28E+06	.139	.217	.345	.477	.240	.540
45.0	9.11	61.7	1.66E+09	1.64E-05	5.04E-01	6.90E-03	3.30E+06	.135	.216	.343	.474	.235	.532
50.0	9.30	62.4	1.14E+09	1.14E-05	3.43E-01	4.70E-03	3.32E+06	.133	.214	.341	.472	.230	.524
55.0	9.46	63.0	7.86E+08	7.94E-06	2.36E-01	3.23E-03	3.34E+06	.130	.213	.339	.469	.226	.518
60.0	9.61	63.6	5.45E+08	5.56E-06	1.63E-01	2.23E-03	3.35E+06	.128	.211	.337	.467	.223	.512

Absorber material: Cu

Thick-ness mm	HVL mm Al	<E> keV	Φ/It 1/As/cm²	Ψ/It J/As/cm²	Ka/It mGy/As	Relative dose	Φ/Ka photons /cm²/µGy	68 CaWO4	68 GOS	118 GOS	181 GOS	80 SPF	180 CsI
0.00	2.86	43.5	1.38E+11	9.59E-04	7.30E+01	1.00E+00	1.89E+06	.299	.311	.459	.593	.372	.690
.02	3.26	44.7	1.23E+11	8.81E-04	6.08E+01	8.33E-01	2.02E+06	.283	.299	.446	.581	.360	.682
.04	3.62	45.7	1.11E+11	8.14E-04	5.20E+01	7.13E-01	2.14E+06	.270	.289	.435	.570	.351	.675
.06	3.94	46.7	1.01E+11	7.57E-04	4.54E+01	6.22E-01	2.23E+06	.258	.280	.425	.561	.344	.668
.08	4.23	47.5	9.30E+10	7.07E-04	4.01E+01	5.50E-01	2.32E+06	.249	.274	.417	.554	.337	.662
.10	4.50	48.2	8.59E+10	6.63E-04	3.59E+01	4.92E-01	2.39E+06	.240	.268	.411	.547	.331	.657
.20	5.57	51.1	6.11E+10	5.01E-04	2.29E+01	3.14E-01	2.67E+06	.210	.249	.388	.524	.310	.635
.30	6.34	53.2	4.63E+10	3.94E-04	1.63E+01	2.23E-01	2.84E+06	.190	.239	.374	.510	.295	.617
.40	6.92	54.8	3.63E+10	3.19E-04	1.23E+01	1.68E-01	2.96E+06	.177	.232	.366	.501	.284	.603
.50	7.38	56.2	2.91E+10	2.62E-04	9.57E+00	1.31E-01	3.05E+06	.167	.228	.361	.495	.274	.590
.60	7.76	57.3	2.38E+10	2.19E-04	7.66E+00	1.05E-01	3.11E+06	.160	.225	.357	.490	.266	.579
.70	8.07	58.2	1.97E+10	1.84E-04	6.24E+00	8.55E-02	3.16E+06	.154	.223	.354	.487	.260	.569
.80	8.33	59.1	1.65E+10	1.56E-04	5.16E+00	7.07E-02	3.20E+06	.149	.221	.351	.484	.254	.560
.90	8.56	59.8	1.39E+10	1.34E-04	4.31E+00	5.91E-02	3.23E+06	.145	.219	.349	.481	.248	.552
1.00	8.76	60.5	1.19E+10	1.15E-04	3.64E+00	4.99E-02	3.26E+06	.141	.218	.347	.479	.244	.545
1.20	9.10	61.7	8.72E+09	8.62E-05	2.64E+00	3.62E-02	3.30E+06	.136	.216	.343	.475	.236	.532
1.40	9.37	62.7	6.53E+09	6.56E-05	1.96E+00	2.69E-02	3.33E+06	.132	.214	.340	.471	.229	.522
1.60	9.59	63.5	4.96E+09	5.05E-05	1.48E+00	2.03E-02	3.35E+06	.128	.212	.337	.467	.223	.512
1.80	9.78	64.3	3.81E+09	3.93E-05	1.13E+00	1.55E-02	3.36E+06	.126	.210	.335	.464	.218	.504
2.00	9.95	64.9	2.96E+09	3.08E-05	8.77E-01	1.20E-02	3.37E+06	.124	.208	.332	.461	.214	.497
2.50	10.28	66.3	1.62E+09	1.72E-05	4.78E-01	6.55E-03	3.39E+06	.121	.204	.326	.453	.205	.482
3.00	10.53	67.5	9.19E+08	9.93E-06	2.70E-01	3.71E-03	3.40E+06	.119	.200	.320	.446	.198	.470
3.50	10.73	68.4	5.35E+08	5.86E-06	1.57E-01	2.16E-03	3.40E+06	.118	.196	.315	.440	.193	.460
4.00	10.90	69.2	3.18E+08	3.53E-06	9.36E-02	1.28E-03	3.40E+06	.117	.193	.311	.435	.188	.452
4.50	11.04	69.9	1.92E+08	2.16E-06	5.67E-02	7.76E-04	3.40E+06	.117	.191	.307	.430	.184	.444
5.00	11.16	70.5	1.18E+08	1.33E-06	3.48E-02	4.77E-04	3.39E+06	.117	.188	.304	.425	.181	.438
6.00	11.35	71.6	4.59E+07	5.26E-07	1.35E-02	1.86E-04	3.39E+06	.118	.185	.298	.418	.176	.428
7.00	11.50	72.4	1.85E+07	2.14E-07	5.47E-03	7.49E-05	3.38E+06	.119	.181	.293	.412	.171	.420
8.00	11.63	73.1	7.64E+06	8.94E-08	2.27E-03	3.11E-05	3.37E+06	.120	.179	.289	.408	.168	.413

Reference focal distance: 100 cm   X-ray tube voltage: 80 kV

Absorber material:		Water					Φ/Ka	Relative energy absorption in the image detector (coverage mg/cm²)					
Thick-													
ness	HVL	<E>	Φ/It	Ψ/It	Ka/It	Relative	photons	68	68	118	181	80	180
mm	mm Al	keV	1/As/cm²	J/As/cm²	mGy/As	dose	/cm²/μGy	CaWO4	GOS	GOS	GOS	SPF	CsI

No added filter

0.0	2.86	43.5	1.38E+11	9.59E-04	7.30E+01	1.00E+00	1.89E+06	.299	.311	.459	.593	.372	.690
2.5	3.80	46.3	6.76E+10	5.02E-04	3.08E+01	4.23E-01	2.19E+06	.263	.284	.429	.565	.346	.670
5.0	4.53	48.4	3.50E+10	2.72E-04	1.46E+01	2.00E-01	2.40E+06	.239	.267	.410	.546	.330	.655
7.5	5.13	50.0	1.87E+10	1.50E-04	7.33E+00	1.00E-01	2.56E+06	.221	.256	.396	.532	.317	.642
10.0	5.63	51.4	1.02E+10	8.42E-05	3.82E+00	5.23E-02	2.68E+06	.207	.248	.386	.522	.307	.631
12.5	6.05	52.6	5.67E+09	4.78E-05	2.04E+00	2.80E-02	2.78E+06	.196	.242	.379	.514	.299	.621
15.0	6.43	53.6	3.18E+09	2.74E-05	1.11E+00	1.53E-02	2.86E+06	.187	.237	.373	.508	.291	.612
17.5	6.75	54.6	1.80E+09	1.58E-05	6.17E-01	8.46E-03	2.92E+06	.180	.234	.368	.503	.285	.604
20.0	7.04	55.4	1.03E+09	9.14E-06	3.46E-01	4.74E-03	2.98E+06	.174	.231	.364	.498	.279	.596
22.5	7.31	56.2	5.92E+08	5.33E-06	1.96E-01	2.68E-03	3.03E+06	.168	.228	.361	.495	.274	.588
25.0	7.54	56.9	3.42E+08	3.12E-06	1.12E-01	1.53E-03	3.07E+06	.163	.226	.358	.492	.269	.581
27.5	7.75	57.5	1.99E+08	1.83E-06	6.40E-02	8.78E-04	3.11E+06	.159	.225	.356	.489	.264	.575
30.0	7.95	58.1	1.16E+08	1.08E-06	3.70E-02	5.07E-04	3.14E+06	.155	.223	.354	.487	.260	.569
32.5	8.12	58.7	6.79E+07	6.38E-07	2.15E-02	2.94E-04	3.16E+06	.152	.222	.352	.484	.256	.563
35.0	8.29	59.2	3.99E+07	3.78E-07	1.25E-02	1.72E-04	3.19E+06	.149	.220	.350	.482	.253	.558

Added filter   0.5 mm Al

0.0	3.15	44.3	1.25E+11	8.88E-04	6.28E+01	1.00E+00	1.99E+06	.287	.302	.449	.584	.364	.684
2.5	4.02	46.9	6.25E+10	4.70E-04	2.77E+01	4.40E-01	2.26E+06	.255	.278	.423	.559	.341	.666
5.0	4.71	48.8	3.27E+10	2.56E-04	1.34E+01	2.13E-01	2.45E+06	.233	.264	.405	.542	.326	.652
7.5	5.27	50.4	1.76E+10	1.42E-04	6.79E+00	1.08E-01	2.59E+06	.217	.253	.393	.529	.315	.640
10.0	5.75	51.7	9.65E+09	8.00E-05	3.57E+00	5.68E-02	2.71E+06	.204	.246	.384	.520	.305	.629
12.5	6.16	52.9	5.37E+09	4.55E-05	1.92E+00	3.05E-02	2.80E+06	.194	.241	.377	.512	.297	.619
15.0	6.52	53.9	3.02E+09	2.61E-05	1.05E+00	1.67E-02	2.88E+06	.185	.236	.371	.506	.290	.610
17.5	6.83	54.8	1.72E+09	1.51E-05	5.84E-01	9.29E-03	2.94E+06	.178	.233	.367	.502	.283	.602
20.0	7.11	55.6	9.82E+08	8.74E-06	3.28E-01	5.22E-03	2.99E+06	.172	.230	.363	.497	.278	.594
22.5	7.37	56.3	5.65E+08	5.10E-06	1.86E-01	2.96E-03	3.04E+06	.167	.228	.360	.494	.273	.587
25.0	7.59	57.0	3.27E+08	2.99E-06	1.06E-01	1.69E-03	3.08E+06	.162	.226	.358	.491	.268	.580
27.5	7.80	57.7	1.90E+08	1.76E-06	6.11E-02	9.72E-04	3.11E+06	.158	.224	.355	.488	.263	.574
30.0	7.99	58.2	1.11E+08	1.04E-06	3.53E-02	5.62E-04	3.14E+06	.155	.223	.353	.486	.259	.568
32.5	8.17	58.8	6.51E+07	6.13E-07	2.05E-02	3.27E-04	3.17E+06	.151	.221	.351	.484	.255	.562
35.0	8.33	59.3	3.83E+07	3.64E-07	1.20E-02	1.91E-04	3.19E+06	.149	.220	.350	.482	.252	.557

Added filter   0.1 mm Cu

0.0	4.50	48.2	8.59E+10	6.63E-04	3.59E+01	1.00E+00	2.39E+06	.240	.268	.411	.547	.331	.657
2.5	5.10	49.9	4.59E+10	3.66E-04	1.80E+01	5.01E-01	2.55E+06	.222	.256	.397	.533	.319	.644
5.0	5.60	51.3	2.50E+10	2.05E-04	9.36E+00	2.61E-01	2.67E+06	.208	.248	.387	.523	.309	.633
7.5	6.02	52.4	1.39E+10	1.16E-04	5.00E+00	1.39E-01	2.77E+06	.197	.242	.379	.515	.300	.623
10.0	6.40	53.5	7.77E+09	6.66E-05	2.72E+00	7.59E-02	2.85E+06	.188	.238	.373	.509	.293	.614
12.5	6.72	54.4	4.40E+09	3.83E-05	1.51E+00	4.20E-02	2.92E+06	.181	.234	.369	.503	.286	.605
15.0	7.01	55.3	2.51E+09	2.22E-05	8.43E-01	2.35E-02	2.97E+06	.174	.231	.365	.499	.280	.597
17.5	7.27	56.0	1.44E+09	1.29E-05	4.77E-01	1.33E-02	3.02E+06	.169	.229	.361	.495	.275	.590
20.0	7.51	56.7	8.32E+08	7.56E-06	2.72E-01	7.57E-03	3.06E+06	.164	.227	.359	.492	.270	.583
22.5	7.72	57.4	4.83E+08	4.44E-06	1.56E-01	4.34E-03	3.10E+06	.160	.225	.356	.490	.265	.577
25.0	7.91	58.0	2.82E+08	2.62E-06	8.99E-02	2.50E-03	3.13E+06	.156	.223	.354	.487	.261	.571
27.5	8.09	58.5	1.65E+08	1.55E-06	5.22E-02	1.45E-03	3.16E+06	.153	.222	.352	.485	.257	.565
30.0	8.26	59.0	9.68E+07	9.15E-07	3.04E-02	8.47E-04	3.18E+06	.150	.221	.351	.483	.254	.559
32.5	8.41	59.5	5.70E+07	5.44E-07	1.78E-02	4.95E-04	3.21E+06	.147	.220	.349	.481	.250	.554
35.0	8.55	60.0	3.37E+07	3.24E-07	1.04E-02	2.91E-04	3.23E+06	.145	.219	.348	.479	.247	.549

Added filter   0.2 mm Cu

0.0	5.57	51.1	6.11E+10	5.01E-04	2.29E+01	1.00E+00	2.67E+06	.210	.249	.388	.524	.310	.635
2.5	6.00	52.3	3.38E+10	2.84E-04	1.22E+01	5.34E-01	2.76E+06	.199	.243	.380	.516	.301	.625
5.0	6.37	53.3	1.90E+10	1.62E-04	6.66E+00	2.90E-01	2.85E+06	.189	.238	.374	.509	.294	.616
7.5	6.69	54.3	1.07E+10	9.32E-05	3.68E+00	1.60E-01	2.91E+06	.182	.234	.369	.504	.287	.607
10.0	6.98	55.1	6.11E+09	5.40E-05	2.06E+00	8.97E-02	2.97E+06	.175	.232	.365	.500	.281	.599
12.5	7.24	55.9	3.51E+09	3.14E-05	1.16E+00	5.07E-02	3.02E+06	.170	.229	.362	.496	.276	.592
15.0	7.47	56.6	2.02E+09	1.83E-05	6.61E-01	2.88E-02	3.06E+06	.165	.227	.359	.493	.271	.585
17.5	7.69	57.2	1.17E+09	1.08E-05	3.79E-01	1.65E-02	3.10E+06	.161	.225	.357	.490	.266	.578
20.0	7.88	57.8	6.84E+08	6.34E-06	2.19E-01	9.53E-03	3.13E+06	.157	.224	.355	.488	.262	.572
22.5	8.06	58.4	4.00E+08	3.74E-06	1.27E-01	5.52E-03	3.16E+06	.153	.222	.353	.486	.258	.566
25.0	8.22	58.9	2.35E+08	2.21E-06	7.38E-02	3.22E-03	3.18E+06	.150	.221	.351	.484	.255	.561
27.5	8.38	59.4	1.38E+08	1.31E-06	4.31E-02	1.88E-03	3.20E+06	.148	.220	.350	.482	.251	.556
30.0	8.52	59.9	8.15E+07	7.82E-07	2.53E-02	1.10E-03	3.22E+06	.145	.219	.348	.480	.248	.551
32.5	8.65	60.3	4.83E+07	4.66E-07	1.49E-02	6.50E-04	3.24E+06	.143	.218	.347	.478	.245	.546
35.0	8.77	60.7	2.86E+07	2.78E-07	8.80E-03	3.83E-04	3.25E+06	.141	.217	.345	.477	.242	.542

Table IV.3.6

Characteristic X-ray beam quantities without scattered radiation and relative energy absorption in typical image detectors at variable absorber or phantom materials for diagnostic X-ray tube assemblies with W/Re anode at 10° target angle and an inherent (quality equivalent) filtration of 2.5 mm Al

Reference focal distance: 100 cm   X-ray tube voltage:　90 kV

Absorber material: Al

Thick-ness mm	HVL mm Al	\<E\> keV	Φ/It 1/As/cm²	Ψ/It J/As/cm²	Ka/It mGy/As	Relative dose	Φ/Ka photons /cm²/µGy	68 CaWO4	68 GOS	118 GOS	181 GOS	80 SPF	180 CsI
0.0	3.23	46.7	1.82E+11	1.36E-03	8.98E+01	1.00E+00	2.03E+06	.265	.287	.430	.562	.339	.649
.5	3.56	47.6	1.67E+11	1.27E-03	7.84E+01	8.73E-01	2.13E+06	.255	.279	.421	.554	.331	.642
1.0	3.85	48.4	1.54E+11	1.19E-03	6.94E+01	7.73E-01	2.21E+06	.246	.273	.413	.547	.325	.636
1.5	4.12	49.1	1.42E+11	1.12E-03	6.22E+01	6.92E-01	2.29E+06	.238	.267	.407	.540	.319	.631
2.0	4.37	49.8	1.32E+11	1.05E-03	5.61E+01	6.25E-01	2.35E+06	.232	.262	.401	.535	.314	.627
2.5	4.60	50.4	1.23E+11	9.94E-04	5.10E+01	5.68E-01	2.41E+06	.225	.258	.396	.530	.309	.622
3.0	4.82	51.0	1.15E+11	9.39E-04	4.66E+01	5.20E-01	2.47E+06	.220	.254	.392	.525	.305	.618
3.5	5.02	51.5	1.08E+11	8.89E-04	4.28E+01	4.77E-01	2.52E+06	.215	.251	.388	.521	.302	.614
4.0	5.21	52.0	1.01E+11	8.42E-04	3.95E+01	4.40E-01	2.56E+06	.210	.248	.384	.517	.298	.610
4.5	5.38	52.5	9.51E+10	7.99E-04	3.66E+01	4.07E-01	2.60E+06	.206	.245	.381	.514	.295	.607
5.0	5.55	52.9	8.96E+10	7.59E-04	3.39E+01	3.78E-01	2.64E+06	.202	.243	.378	.511	.292	.604
6.0	5.86	53.7	7.98E+10	6.86E-04	2.95E+01	3.28E-01	2.71E+06	.196	.238	.372	.505	.287	.597
7.0	6.14	54.4	7.14E+10	6.23E-04	2.58E+01	2.88E-01	2.76E+06	.190	.235	.368	.500	.282	.591
8.0	6.39	55.1	6.42E+10	5.66E-04	2.28E+01	2.54E-01	2.82E+06	.184	.232	.364	.496	.277	.586
9.0	6.62	55.7	5.79E+10	5.16E-04	2.02E+01	2.25E-01	2.86E+06	.180	.229	.360	.492	.273	.581
10.0	6.83	56.3	5.23E+10	4.72E-04	1.80E+01	2.01E-01	2.90E+06	.176	.227	.357	.489	.270	.576
12.0	7.21	57.3	4.31E+10	3.96E-04	1.45E+01	1.62E-01	2.97E+06	.169	.223	.352	.483	.263	.566
14.0	7.53	58.2	3.58E+10	3.34E-04	1.18E+01	1.32E-01	3.02E+06	.163	.220	.348	.478	.257	.558
16.0	7.82	59.1	2.99E+10	2.83E-04	9.75E+00	1.09E-01	3.07E+06	.158	.217	.344	.474	.252	.550
18.0	8.07	59.8	2.51E+10	2.41E-04	8.09E+00	9.01E-02	3.11E+06	.154	.215	.341	.471	.247	.543
20.0	8.29	60.5	2.12E+10	2.06E-04	6.76E+00	7.53E-02	3.14E+06	.150	.213	.339	.468	.242	.537
22.0	8.49	61.1	1.80E+10	1.76E-04	5.68E+00	6.33E-02	3.16E+06	.147	.211	.336	.465	.238	.531
25.0	8.76	62.0	1.41E+10	1.40E-04	4.41E+00	4.92E-02	3.20E+06	.143	.209	.333	.461	.233	.522
30.0	9.14	63.2	9.56E+09	9.68E-05	2.95E+00	3.29E-02	3.24E+06	.138	.206	.329	.456	.225	.509
35.0	9.44	64.3	6.56E+09	6.76E-05	2.01E+00	2.24E-02	3.27E+06	.134	.203	.325	.451	.218	.498
40.0	9.70	65.3	4.55E+09	4.76E-05	1.38E+00	1.54E-02	3.29E+06	.131	.201	.321	.447	.212	.488
45.0	9.93	66.2	3.18E+09	3.37E-05	9.65E-01	1.07E-02	3.30E+06	.128	.199	.318	.443	.207	.480
50.0	10.12	66.9	2.24E+09	2.41E-05	6.78E-01	7.55E-03	3.31E+06	.126	.196	.315	.439	.202	.472
55.0	10.30	67.7	1.59E+09	1.73E-05	4.80E-01	5.35E-03	3.32E+06	.125	.194	.312	.435	.198	.464
60.0	10.45	68.3	1.14E+09	1.24E-05	3.42E-01	3.81E-03	3.32E+06	.123	.192	.309	.431	.194	.458

Absorber material: Cu

Thick-ness mm	HVL mm Al	\<E\> keV	Φ/It 1/As/cm²	Ψ/It J/As/cm²	Ka/It mGy/As	Relative dose	Φ/Ka photons /cm²/µGy	68 CaWO4	68 GOS	118 GOS	181 GOS	80 SPF	180 CsI
0.00	3.23	46.7	1.82E+11	1.36E-03	8.98E+01	1.00E+00	2.03E+06	.265	.287	.430	.562	.339	.649
.02	3.67	48.0	1.65E+11	1.27E-03	7.62E+01	8.49E-01	2.16E+06	.251	.277	.418	.550	.328	.640
.04	4.07	49.1	1.51E+11	1.18E-03	6.63E+01	7.39E-01	2.27E+06	.239	.268	.408	.541	.320	.632
.06	4.43	50.0	1.39E+11	1.11E-03	5.87E+01	6.54E-01	2.37E+06	.229	.261	.400	.533	.312	.625
.08	4.75	50.9	1.29E+11	1.05E-03	5.27E+01	5.87E-01	2.45E+06	.221	.255	.393	.526	.306	.619
.10	5.05	51.6	1.20E+11	9.96E-04	4.77E+01	5.32E-01	2.52E+06	.214	.250	.387	.520	.301	.613
.20	6.21	54.7	8.95E+10	7.84E-04	3.22E+01	3.59E-01	2.78E+06	.188	.234	.366	.499	.280	.589
.30	7.03	56.9	7.02E+10	6.40E-04	2.39E+01	2.66E-01	2.94E+06	.172	.225	.355	.486	.266	.570
.40	7.64	58.6	5.69E+10	5.34E-04	1.87E+01	2.08E-01	3.04E+06	.161	.219	.347	.477	.255	.555
.50	8.13	60.0	4.71E+10	4.52E-04	1.51E+01	1.68E-01	3.11E+06	.153	.214	.341	.470	.246	.542
.60	8.51	61.2	3.95E+10	3.87E-04	1.25E+01	1.39E-01	3.17E+06	.147	.211	.336	.465	.238	.530
.70	8.84	62.2	3.36E+10	3.35E-04	1.05E+01	1.17E-01	3.21E+06	.142	.209	.333	.460	.231	.520
.80	9.11	63.1	2.88E+10	2.91E-04	8.90E+00	9.92E-02	3.24E+06	.138	.206	.329	.457	.225	.510
.90	9.35	64.0	2.49E+10	2.55E-04	7.64E+00	8.51E-02	3.26E+06	.135	.204	.326	.453	.220	.502
1.00	9.55	64.7	2.16E+10	2.24E-04	6.60E+00	7.36E-02	3.28E+06	.132	.203	.324	.450	.216	.495
1.20	9.90	66.0	1.66E+10	1.76E-04	5.03E+00	5.60E-02	3.30E+06	.128	.199	.319	.444	.208	.481
1.40	10.18	67.2	1.30E+10	1.39E-04	3.91E+00	4.35E-02	3.32E+06	.126	.196	.314	.438	.201	.470
1.60	10.42	68.2	1.02E+10	1.12E-04	3.08E+00	3.43E-02	3.32E+06	.123	.193	.310	.433	.195	.460
1.80	10.62	69.0	8.17E+09	9.04E-05	2.46E+00	2.74E-02	3.33E+06	.122	.190	.306	.428	.190	.451
2.00	10.80	69.9	6.58E+09	7.37E-05	1.98E+00	2.20E-02	3.33E+06	.120	.188	.302	.423	.186	.443
2.50	11.17	71.6	3.95E+09	4.53E-05	1.19E+00	1.32E-02	3.33E+06	.118	.182	.294	.412	.176	.426
3.00	11.46	73.1	2.46E+09	2.87E-05	7.40E-01	8.24E-03	3.32E+06	.117	.177	.286	.403	.169	.412
3.50	11.69	74.3	1.57E+09	1.86E-05	4.73E-01	5.27E-03	3.31E+06	.117	.173	.280	.395	.163	.400
4.00	11.89	75.4	1.02E+09	1.23E-05	3.09E-01	3.45E-03	3.29E+06	.116	.169	.274	.388	.158	.391
4.50	12.05	76.3	6.75E+08	8.24E-06	2.06E-01	2.29E-03	3.28E+06	.116	.166	.270	.382	.154	.382
5.00	12.19	77.1	4.53E+08	5.59E-06	1.38E-01	1.54E-03	3.27E+06	.116	.163	.265	.376	.150	.375
6.00	12.42	78.4	2.11E+08	2.65E-06	6.49E-02	7.23E-04	3.25E+06	.116	.159	.259	.367	.145	.364
7.00	12.59	79.5	1.01E+08	1.29E-06	3.14E-02	3.50E-04	3.23E+06	.116	.155	.253	.360	.140	.355
8.00	12.73	80.4	5.01E+07	6.46E-07	1.56E-02	1.74E-04	3.21E+06	.116	.152	.249	.355	.137	.347

Reference focal distance: 100 cm   X-ray tube voltage: 90 kV

Absorber material:		Water						Relative energy absorption in the image detector (coverage mg/cm²)					
Thick-ness mm	HVL mm Al	\<E\> keV	Φ/It 1/As/cm²	Ψ/It J/As/cm²	Ka/It mGy/As	Relative dose	Φ/Ka photons /cm²/µGy	68 CaWO4	68 GOS	118 GOS	181 GOS	80 SPF	180 CsI

No added filter

0.0	3.23	46.7	1.82E+11	1.36E-03	8.98E+01	1.00E+00	2.03E+06	.265	.287	.430	.562	.339	.649
2.5	4.28	49.7	9.25E+10	7.36E-04	3.97E+01	4.42E-01	2.33E+06	.233	.264	.403	.536	.315	.627
5.0	5.09	51.9	4.93E+10	4.10E-04	1.95E+01	2.17E-01	2.53E+06	.212	.249	.385	.519	.299	.610
7.5	5.75	53.6	2.71E+10	2.33E-04	1.01E+01	1.12E-01	2.68E+06	.197	.239	.373	.506	.287	.596
10.0	6.30	55.1	1.51E+10	1.34E-04	5.41E+00	6.03E-02	2.79E+06	.185	.232	.364	.496	.277	.584
12.5	6.77	56.4	8.58E+09	7.75E-05	2.97E+00	3.31E-02	2.88E+06	.176	.227	.357	.488	.268	.573
15.0	7.17	57.5	4.92E+09	4.53E-05	1.66E+00	1.85E-02	2.96E+06	.169	.222	.351	.482	.261	.562
17.5	7.52	58.6	2.84E+09	2.67E-05	9.43E-01	1.05E-02	3.01E+06	.162	.219	.347	.477	.254	.553
20.0	7.84	59.5	1.65E+09	1.58E-05	5.40E-01	6.02E-03	3.06E+06	.157	.216	.343	.472	.248	.544
22.5	8.12	60.3	9.69E+08	9.36E-06	3.12E-01	3.48E-03	3.10E+06	.153	.213	.339	.468	.243	.536
25.0	8.37	61.1	5.70E+08	5.58E-06	1.82E-01	2.03E-03	3.14E+06	.149	.211	.336	.464	.238	.529
27.5	8.59	61.8	3.37E+08	3.34E-06	1.07E-01	1.19E-03	3.16E+06	.145	.209	.333	.461	.233	.522
30.0	8.80	62.5	2.00E+08	2.01E-06	6.28E-02	7.00E-04	3.19E+06	.143	.207	.330	.458	.229	.515
32.5	8.99	63.1	1.19E+08	1.21E-06	3.72E-02	4.15E-04	3.21E+06	.140	.206	.328	.455	.225	.509
35.0	9.16	63.7	7.13E+07	7.28E-07	2.21E-02	2.46E-04	3.23E+06	.138	.204	.326	.452	.221	.503

Added filter 0.5 mm Al

0.0	3.56	47.6	1.67E+11	1.27E-03	7.84E+01	1.00E+00	2.13E+06	.255	.279	.421	.554	.331	.642
2.5	4.52	50.3	8.61E+10	6.94E-04	3.60E+01	4.59E-01	2.39E+06	.227	.259	.397	.531	.310	.622
5.0	5.28	52.4	4.63E+10	3.89E-04	1.80E+01	2.29E-01	2.58E+06	.208	.246	.382	.515	.295	.606
7.5	5.90	54.0	2.56E+10	2.21E-04	9.42E+00	1.20E-01	2.71E+06	.194	.237	.371	.503	.284	.593
10.0	6.43	55.5	1.44E+10	1.28E-04	5.09E+00	6.49E-02	2.82E+06	.183	.231	.362	.494	.274	.581
12.5	6.88	56.7	8.16E+09	7.41E-05	2.81E+00	3.59E-02	2.90E+06	.174	.225	.355	.487	.266	.570
15.0	7.26	57.8	4.69E+09	4.34E-05	1.58E+00	2.01E-02	2.97E+06	.167	.221	.350	.481	.259	.560
17.5	7.61	58.8	2.72E+09	2.56E-05	8.97E-01	1.14E-02	3.03E+06	.161	.218	.346	.475	.253	.551
20.0	7.91	59.7	1.58E+09	1.51E-05	5.15E-01	6.58E-03	3.07E+06	.156	.215	.342	.471	.247	.543
22.5	8.18	60.5	9.29E+08	9.00E-06	2.99E-01	3.81E-03	3.11E+06	.152	.213	.338	.467	.242	.535
25.0	8.42	61.3	5.47E+08	5.37E-06	1.74E-01	2.22E-03	3.14E+06	.148	.211	.335	.463	.237	.527
27.5	8.64	62.0	3.24E+08	3.22E-06	1.02E-01	1.30E-03	3.17E+06	.145	.209	.332	.460	.232	.520
30.0	8.85	62.6	1.93E+08	1.93E-06	6.03E-02	7.70E-04	3.19E+06	.142	.207	.330	.457	.228	.514
32.5	9.03	63.3	1.15E+08	1.16E-06	3.57E-02	4.56E-04	3.21E+06	.139	.205	.328	.454	.224	.508
35.0	9.20	63.8	6.87E+07	7.02E-07	2.13E-02	2.71E-04	3.23E+06	.137	.204	.325	.451	.221	.502

Added filter 0.1 mm Cu

0.0	5.05	51.6	1.20E+11	9.96E-04	4.77E+01	1.00E+00	2.52E+06	.214	.250	.387	.520	.301	.613
2.5	5.71	53.4	6.59E+10	5.64E-04	2.47E+01	5.17E-01	2.67E+06	.198	.240	.374	.507	.288	.599
5.0	6.25	54.9	3.68E+10	3.24E-04	1.32E+01	2.77E-01	2.79E+06	.187	.233	.365	.498	.278	.586
7.5	6.72	56.2	2.08E+10	1.87E-04	7.24E+00	1.52E-01	2.88E+06	.177	.227	.358	.490	.270	.575
10.0	7.12	57.3	1.19E+10	1.09E-04	4.04E+00	8.47E-02	2.95E+06	.170	.223	.352	.483	.262	.565
12.5	7.48	58.3	6.88E+09	6.43E-05	2.29E+00	4.80E-02	3.01E+06	.163	.220	.348	.478	.256	.556
15.0	7.79	59.2	4.00E+09	3.80E-05	1.31E+00	2.74E-02	3.06E+06	.158	.217	.344	.473	.250	.547
17.5	8.07	60.1	2.34E+09	2.26E-05	7.56E-01	1.59E-02	3.10E+06	.153	.214	.340	.469	.244	.539
20.0	8.32	60.9	1.38E+09	1.34E-05	4.40E-01	9.22E-03	3.13E+06	.150	.212	.337	.465	.239	.531
22.5	8.54	61.6	8.14E+08	8.03E-06	2.58E-01	5.40E-03	3.16E+06	.146	.210	.334	.462	.235	.524
25.0	8.75	62.3	4.83E+08	4.82E-06	1.52E-01	3.18E-03	3.18E+06	.143	.208	.331	.459	.231	.518
27.5	8.94	62.9	2.87E+08	2.90E-06	8.97E-02	1.88E-03	3.20E+06	.141	.206	.329	.456	.227	.511
30.0	9.12	63.5	1.72E+08	1.75E-06	5.33E-02	1.12E-03	3.22E+06	.138	.205	.327	.453	.223	.505
32.5	9.28	64.1	1.03E+08	1.06E-06	3.18E-02	6.66E-04	3.24E+06	.136	.203	.325	.451	.219	.500
35.0	9.43	64.6	6.18E+07	6.39E-07	1.90E-02	3.98E-04	3.25E+06	.134	.202	.323	.448	.216	.494

Added filter 0.2 mm Cu

0.0	6.21	54.7	8.95E+10	7.84E-04	3.22E+01	1.00E+00	2.78E+06	.188	.234	.366	.499	.280	.589
2.5	6.67	56.0	5.06E+10	4.54E-04	1.76E+01	5.48E-01	2.87E+06	.178	.228	.359	.491	.272	.578
5.0	7.07	57.1	2.89E+10	2.65E-04	9.83E+00	3.05E-01	2.94E+06	.171	.224	.353	.485	.264	.568
7.5	7.43	58.1	1.67E+10	1.55E-04	5.56E+00	1.73E-01	3.00E+06	.164	.220	.349	.479	.257	.558
10.0	7.74	59.0	9.69E+09	9.17E-05	3.18E+00	9.86E-02	3.05E+06	.159	.217	.345	.474	.251	.550
12.5	8.02	59.9	5.67E+09	5.43E-05	1.83E+00	5.69E-02	3.09E+06	.154	.215	.341	.470	.246	.542
15.0	8.27	60.6	3.33E+09	3.23E-05	1.06E+00	3.31E-02	3.13E+06	.150	.212	.338	.467	.241	.534
17.5	8.50	61.4	1.96E+09	1.93E-05	6.23E-01	1.93E-02	3.16E+06	.147	.210	.335	.463	.236	.527
20.0	8.70	62.0	1.16E+09	1.16E-05	3.66E-01	1.14E-02	3.18E+06	.144	.209	.332	.460	.232	.520
22.5	8.89	62.7	6.93E+08	6.95E-06	2.16E-01	6.72E-03	3.20E+06	.141	.207	.330	.457	.228	.514
25.0	9.07	63.3	4.13E+08	4.19E-06	1.28E-01	3.98E-03	3.22E+06	.139	.205	.328	.454	.224	.508
27.5	9.23	63.8	2.47E+08	2.53E-06	7.64E-02	2.37E-03	3.24E+06	.137	.204	.326	.452	.221	.502
30.0	9.38	64.4	1.48E+08	1.53E-06	4.57E-02	1.42E-03	3.25E+06	.135	.202	.324	.449	.217	.497
32.5	9.52	64.9	8.93E+07	9.28E-07	2.74E-02	8.50E-04	3.26E+06	.133	.201	.322	.447	.214	.491
35.0	9.65	65.4	5.38E+07	5.64E-07	1.65E-02	5.11E-04	3.27E+06	.132	.200	.320	.445	.211	.486

Table IV.3.7

Characteristic X-ray beam quantities without scattered radiation and relative energy absorption in typical image detectors at variable absorber or phantom materials for diagnostic X-ray tube assemblies with W/Re anode at 10° target angle and an inherent (quality equivalent) filtration of 2.5 mm Al

Reference focal distance: 100 cm   X-ray tube voltage: 100 kV

Absorber material: Al

Thickness mm	HVL mm Al	<E> keV	Φ/It 1/As/cm²	Ψ/It J/As/cm²	Ka/It mGy/As	Relative dose	Φ/Ka photons /cm²/µGy	Relative energy absorption in the image detector (coverage mg/cm²) 68 CaWO4	68 GOS	118 GOS	181 GOS	80 SPF	180 CsI
0.0	3.59	49.6	2.27E+11	1.80E-03	1.06E+02	1.00E+00	2.14E+06	.240	.268	.405	.535	.312	.612
.5	3.95	50.5	2.10E+11	1.70E-03	9.39E+01	8.83E-01	2.23E+06	.231	.261	.397	.527	.305	.606
1.0	4.27	51.3	1.94E+11	1.60E-03	8.40E+01	7.90E-01	2.32E+06	.224	.255	.391	.521	.299	.600
1.5	4.56	52.1	1.81E+11	1.51E-03	7.59E+01	7.13E-01	2.39E+06	.217	.250	.385	.515	.294	.595
2.0	4.83	52.7	1.69E+11	1.43E-03	6.91E+01	6.49E-01	2.45E+06	.211	.246	.380	.510	.289	.590
2.5	5.08	53.4	1.59E+11	1.36E-03	6.33E+01	5.95E-01	2.51E+06	.206	.242	.375	.505	.285	.585
3.0	5.31	53.9	1.49E+11	1.29E-03	5.83E+01	5.48E-01	2.56E+06	.201	.239	.371	.501	.281	.581
3.5	5.52	54.5	1.40E+11	1.22E-03	5.39E+01	5.07E-01	2.60E+06	.197	.236	.367	.497	.278	.577
4.0	5.72	55.0	1.32E+11	1.17E-03	5.00E+01	4.70E-01	2.65E+06	.193	.233	.364	.494	.274	.573
4.5	5.91	55.5	1.25E+11	1.11E-03	4.66E+01	4.38E-01	2.68E+06	.189	.231	.361	.490	.271	.570
5.0	6.08	55.9	1.18E+11	1.06E-03	4.35E+01	4.09E-01	2.72E+06	.186	.229	.358	.487	.269	.566
6.0	6.41	56.7	1.06E+11	9.66E-04	3.82E+01	3.59E-01	2.78E+06	.180	.225	.353	.482	.263	.560
7.0	6.70	57.5	9.58E+10	8.83E-04	3.38E+01	3.18E-01	2.83E+06	.175	.222	.349	.477	.259	.553
8.0	6.96	58.2	8.67E+10	8.08E-04	3.01E+01	2.83E-01	2.88E+06	.170	.219	.345	.473	.254	.548
9.0	7.20	58.8	7.88E+10	7.42E-04	2.70E+01	2.54E-01	2.92E+06	.166	.216	.342	.470	.251	.542
10.0	7.42	59.4	7.17E+10	6.82E-04	2.43E+01	2.28E-01	2.95E+06	.163	.214	.339	.466	.247	.537
12.0	7.80	60.5	5.98E+10	5.80E-04	1.99E+01	1.87E-01	3.01E+06	.157	.210	.334	.460	.240	.528
14.0	8.14	61.5	5.02E+10	4.95E-04	1.64E+01	1.55E-01	3.06E+06	.152	.207	.330	.456	.235	.519
16.0	8.43	62.3	4.25E+10	4.24E-04	1.37E+01	1.29E-01	3.09E+06	.148	.205	.326	.451	.229	.511
18.0	8.69	63.1	3.61E+10	3.65E-04	1.15E+01	1.09E-01	3.12E+06	.144	.202	.323	.447	.225	.504
20.0	8.92	63.9	3.08E+10	3.15E-04	9.77E+00	9.18E-02	3.15E+06	.141	.200	.320	.444	.220	.497
22.0	9.12	64.5	2.63E+10	2.72E-04	8.31E+00	7.81E-02	3.17E+06	.138	.198	.317	.441	.216	.490
25.0	9.40	65.5	2.10E+10	2.20E-04	6.56E+00	6.17E-02	3.19E+06	.135	.196	.314	.436	.211	.482
30.0	9.78	66.9	1.45E+10	1.55E-04	4.51E+00	4.24E-02	3.22E+06	.131	.192	.308	.430	.203	.468
35.0	10.10	68.1	1.02E+10	1.11E-04	3.14E+00	2.96E-02	3.24E+06	.127	.189	.303	.424	.196	.457
40.0	10.37	69.2	7.21E+09	7.99E-05	2.22E+00	2.09E-02	3.25E+06	.125	.186	.299	.418	.190	.446
45.0	10.60	70.2	5.15E+09	5.78E-05	1.58E+00	1.49E-02	3.25E+06	.122	.183	.295	.413	.185	.437
50.0	10.81	71.1	3.70E+09	4.21E-05	1.14E+00	1.07E-02	3.26E+06	.121	.181	.291	.408	.180	.428
55.0	11.00	71.9	2.67E+09	3.08E-05	8.22E-01	7.73E-03	3.26E+06	.119	.178	.287	.403	.176	.420
60.0	11.17	72.7	1.94E+09	2.27E-05	5.98E-01	5.62E-03	3.25E+06	.118	.176	.284	.398	.172	.413

Absorber material: Cu

Thickness mm	HVL mm Al	<E> keV	Φ/It 1/As/cm²	Ψ/It J/As/cm²	Ka/It mGy/As	Relative dose	Φ/Ka photons /cm²/µGy	Relative energy absorption in the image detector (coverage mg/cm²) 68 CaWO4	68 GOS	118 GOS	181 GOS	80 SPF	180 CsI
0.00	3.59	49.6	2.27E+11	1.80E-03	1.06E+02	1.00E+00	2.14E+06	.240	.268	.405	.535	.312	.612
.02	4.08	50.9	2.08E+11	1.69E-03	9.17E+01	8.62E-01	2.27E+06	.228	.259	.394	.524	.302	.603
.04	4.51	52.0	1.92E+11	1.60E-03	8.08E+01	7.60E-01	2.38E+06	.218	.251	.385	.516	.295	.595
.06	4.90	53.0	1.79E+11	1.52E-03	7.24E+01	6.81E-01	2.47E+06	.209	.245	.378	.508	.288	.588
.08	5.24	53.8	1.67E+11	1.44E-03	6.57E+01	6.18E-01	2.54E+06	.202	.240	.372	.502	.282	.582
.10	5.55	54.6	1.57E+11	1.37E-03	6.01E+01	5.65E-01	2.61E+06	.196	.235	.366	.496	.277	.576
.20	6.77	57.7	1.20E+11	1.11E-03	4.23E+01	3.98E-01	2.85E+06	.173	.221	.348	.476	.257	.551
.30	7.62	60.0	9.69E+10	9.31E-04	3.25E+01	3.05E-01	2.98E+06	.159	.212	.336	.463	.243	.532
.40	8.25	61.8	8.02E+10	7.94E-04	2.61E+01	2.45E-01	3.07E+06	.150	.206	.328	.454	.233	.516
.50	8.74	63.3	6.76E+10	6.86E-04	2.16E+01	2.03E-01	3.13E+06	.143	.202	.322	.447	.224	.502
.60	9.14	64.6	5.78E+10	5.98E-04	1.82E+01	1.71E-01	3.17E+06	.138	.199	.317	.441	.216	.490
.70	9.47	65.7	5.00E+10	5.26E-04	1.56E+01	1.47E-01	3.20E+06	.134	.196	.313	.435	.210	.480
.80	9.75	66.7	4.36E+10	4.65E-04	1.35E+01	1.27E-01	3.22E+06	.131	.193	.309	.431	.204	.470
.90	9.99	67.6	3.82E+10	4.14E-04	1.18E+01	1.11E-01	3.24E+06	.128	.191	.306	.426	.199	.461
1.00	10.20	68.4	3.37E+10	3.70E-04	1.04E+01	9.77E-02	3.25E+06	.126	.188	.302	.422	.194	.453
1.20	10.56	69.9	2.66E+10	2.98E-04	8.18E+00	7.69E-02	3.26E+06	.123	.184	.296	.414	.186	.439
1.40	10.86	71.2	2.14E+10	2.44E-04	6.55E+00	6.16E-02	3.26E+06	.120	.180	.291	.407	.179	.427
1.60	11.11	72.4	1.73E+10	2.01E-04	5.32E+00	5.00E-02	3.26E+06	.118	.177	.285	.401	.173	.416
1.80	11.33	73.4	1.42E+10	1.67E-04	4.37E+00	4.11E-02	3.25E+06	.117	.174	.281	.395	.168	.407
2.00	11.52	74.4	1.17E+10	1.40E-04	3.62E+00	3.40E-02	3.25E+06	.116	.171	.276	.389	.164	.398
2.50	11.93	76.5	7.51E+09	9.21E-05	2.33E+00	2.19E-02	3.23E+06	.114	.164	.266	.376	.154	.379
3.00	12.24	78.3	4.97E+09	6.24E-05	1.55E+00	1.46E-02	3.20E+06	.112	.158	.257	.365	.147	.364
3.50	12.50	79.8	3.38E+09	4.32E-05	1.06E+00	9.99E-03	3.18E+06	.112	.153	.250	.356	.140	.352
4.00	12.72	81.1	2.34E+09	3.04E-05	7.41E-01	6.97E-03	3.16E+06	.111	.149	.244	.348	.135	.342
4.50	12.90	82.2	1.65E+09	2.17E-05	5.25E-01	4.94E-03	3.14E+06	.110	.146	.239	.341	.131	.333
5.00	13.05	83.2	1.17E+09	1.57E-05	3.77E-01	3.54E-03	3.12E+06	.110	.143	.234	.335	.128	.326
6.00	13.29	84.8	6.17E+08	8.38E-06	2.00E-01	1.88E-03	3.08E+06	.108	.138	.227	.325	.122	.314
7.00	13.47	86.1	3.34E+08	4.61E-06	1.09E-01	1.03E-03	3.05E+06	.107	.134	.221	.318	.118	.305
8.00	13.61	87.1	1.86E+08	2.59E-06	6.13E-02	5.76E-04	3.03E+06	.106	.131	.216	.312	.115	.297

Reference focal distance: 100 cm   X-ray tube voltage: 100 kV

Absorber material:		Water					Φ/Ka	Relative energy absorption in the image detector (coverage mg/cm²)					
Thick-													
ness	HVL	<E>	Φ/It	Ψ/It	Ka/It	Relative	photons	68	68	118	181	80	180
mm	mm Al	keV	1/As/cm²	J/As/cm²	mGy/As	dose	/cm²/µGy	CaWO4	GOS	GOS	GOS	SPF	CsI

No added filter

0.0	3.59	49.6	2.27E+11	1.80E-03	1.06E+02	1.00E+00	2.14E+06	.240	.268	.405	.535	.312	.612
2.5	4.74	52.6	1.19E+11	1.00E-03	4.88E+01	4.59E-01	2.43E+06	.212	.247	.381	.510	.290	.590
5.0	5.62	54.9	6.46E+10	5.69E-04	2.46E+01	2.32E-01	2.62E+06	.194	.234	.365	.494	.274	.572
7.5	6.32	56.8	3.61E+10	3.29E-04	1.31E+01	1.23E-01	2.76E+06	.180	.225	.353	.482	.262	.557
10.0	6.90	58.4	2.05E+10	1.92E-04	7.18E+00	6.75E-02	2.86E+06	.170	.218	.344	.472	.253	.544
12.5	7.39	59.7	1.18E+10	1.13E-04	4.03E+00	3.79E-02	2.94E+06	.162	.213	.337	.464	.244	.532
15.0	7.82	61.0	6.89E+09	6.73E-05	2.29E+00	2.16E-02	3.00E+06	.156	.209	.331	.457	.237	.521
17.5	8.19	62.1	4.04E+09	4.02E-05	1.32E+00	1.25E-02	3.05E+06	.150	.205	.326	.451	.230	.511
20.0	8.52	63.1	2.38E+09	2.41E-05	7.72E-01	7.26E-03	3.09E+06	.146	.202	.322	.446	.224	.502
22.5	8.81	64.0	1.42E+09	1.45E-05	4.54E-01	4.27E-03	3.12E+06	.142	.199	.318	.441	.219	.493
25.0	9.07	64.9	8.45E+08	8.78E-06	2.69E-01	2.53E-03	3.14E+06	.139	.197	.315	.437	.214	.485
27.5	9.31	65.7	5.06E+08	5.33E-06	1.60E-01	1.50E-03	3.16E+06	.136	.194	.311	.433	.209	.477
30.0	9.53	66.5	3.05E+08	3.24E-06	9.57E-02	9.00E-04	3.18E+06	.134	.192	.308	.429	.205	.470
32.5	9.73	67.2	1.84E+08	1.98E-06	5.75E-02	5.41E-04	3.19E+06	.131	.190	.305	.425	.201	.463
35.0	9.91	67.9	1.11E+08	1.21E-06	3.47E-02	3.26E-04	3.20E+06	.129	.188	.302	.422	.197	.457

Added filter   0.5 mm Al

0.0	3.95	50.5	2.10E+11	1.70E-03	9.39E+01	1.00E+00	2.23E+06	.231	.261	.397	.527	.305	.606
2.5	5.00	53.3	1.11E+11	9.48E-04	4.46E+01	4.75E-01	2.49E+06	.207	.243	.376	.506	.285	.585
5.0	5.82	55.4	6.10E+10	5.42E-04	2.29E+01	2.44E-01	2.66E+06	.190	.231	.361	.491	.271	.568
7.5	6.48	57.2	3.43E+10	3.14E-04	1.23E+01	1.31E-01	2.79E+06	.178	.223	.351	.479	.260	.554
10.0	7.04	58.7	1.96E+10	1.84E-04	6.78E+00	7.23E-02	2.88E+06	.168	.217	.342	.470	.251	.541
12.5	7.51	60.0	1.13E+10	1.09E-04	3.82E+00	4.07E-02	2.96E+06	.161	.212	.336	.462	.242	.529
15.0	7.92	61.2	6.59E+09	6.46E-05	2.19E+00	2.33E-02	3.01E+06	.154	.208	.330	.456	.235	.519
17.5	8.27	62.3	3.87E+09	3.86E-05	1.27E+00	1.35E-02	3.06E+06	.149	.204	.325	.450	.229	.509
20.0	8.59	63.3	2.29E+09	2.32E-05	7.39E-01	7.87E-03	3.10E+06	.145	.201	.321	.445	.223	.500
22.5	8.87	64.2	1.36E+09	1.40E-05	4.35E-01	4.64E-03	3.13E+06	.141	.199	.317	.440	.218	.491
25.0	9.13	65.1	8.13E+08	8.47E-06	2.58E-01	2.75E-03	3.15E+06	.138	.196	.314	.436	.213	.483
27.5	9.36	65.9	4.88E+08	5.14E-06	1.54E-01	1.64E-03	3.17E+06	.135	.194	.311	.432	.208	.476
30.0	9.57	66.6	2.93E+08	3.13E-06	9.21E-02	9.82E-04	3.18E+06	.133	.192	.307	.428	.204	.469
32.5	9.77	67.3	1.77E+08	1.91E-06	5.54E-02	5.91E-04	3.20E+06	.131	.190	.305	.425	.200	.462
35.0	9.95	68.0	1.07E+08	1.17E-06	3.35E-02	3.57E-04	3.21E+06	.129	.188	.302	.421	.196	.456

Added filter   0.1 mm Cu

0.0	5.55	54.6	1.57E+11	1.37E-03	6.01E+01	1.00E+00	2.61E+06	.196	.235	.366	.496	.277	.576
2.5	6.26	56.5	8.76E+10	7.92E-04	3.18E+01	5.30E-01	2.75E+06	.182	.226	.355	.484	.265	.561
5.0	6.84	58.0	4.97E+10	4.62E-04	1.74E+01	2.90E-01	2.85E+06	.172	.219	.346	.474	.255	.548
7.5	7.33	59.4	2.86E+10	2.72E-04	9.75E+00	1.62E-01	2.93E+06	.164	.214	.339	.466	.246	.536
10.0	7.75	60.6	1.66E+10	1.61E-04	5.55E+00	9.23E-02	3.00E+06	.157	.210	.333	.459	.239	.525
12.5	8.12	61.7	9.73E+09	9.63E-05	3.20E+00	5.32E-02	3.04E+06	.152	.206	.328	.453	.232	.515
15.0	8.45	62.8	5.74E+09	5.77E-05	1.86E+00	3.09E-02	3.08E+06	.147	.203	.324	.448	.226	.505
17.5	8.74	63.7	3.40E+09	3.47E-05	1.09E+00	1.82E-02	3.12E+06	.143	.200	.320	.443	.221	.497
20.0	9.00	64.6	2.03E+09	2.10E-05	6.45E-01	1.07E-02	3.14E+06	.140	.198	.316	.439	.216	.489
22.5	9.24	65.4	1.21E+09	1.27E-05	3.84E-01	6.38E-03	3.16E+06	.137	.196	.313	.435	.211	.481
25.0	9.46	66.1	7.29E+08	7.72E-06	2.29E-01	3.81E-03	3.18E+06	.134	.193	.310	.431	.207	.474
27.5	9.66	66.9	4.40E+08	4.71E-06	1.38E-01	2.29E-03	3.19E+06	.132	.191	.307	.427	.203	.467
30.0	9.84	67.5	2.66E+08	2.88E-06	8.29E-02	1.38E-03	3.20E+06	.130	.189	.304	.424	.199	.460
32.5	10.02	68.2	1.61E+08	1.76E-06	5.02E-02	8.34E-04	3.21E+06	.128	.188	.301	.421	.195	.454
35.0	10.17	68.8	9.80E+07	1.08E-06	3.04E-02	5.06E-04	3.22E+06	.127	.186	.299	.417	.192	.448

Added filter   0.2 mm Cu

0.0	6.77	57.7	1.20E+11	1.11E-03	4.23E+01	1.00E+00	2.85E+06	.173	.221	.348	.476	.257	.551
2.5	7.26	59.1	6.92E+10	6.55E-04	2.36E+01	5.58E-01	2.93E+06	.165	.215	.341	.468	.249	.539
5.0	7.69	60.3	4.01E+10	3.88E-04	1.34E+01	3.17E-01	2.99E+06	.158	.211	.335	.461	.241	.528
7.5	8.05	61.4	2.35E+10	2.31E-04	7.72E+00	1.82E-01	3.04E+06	.153	.207	.330	.455	.234	.518
10.0	8.38	62.4	1.38E+10	1.38E-04	4.48E+00	1.06E-01	3.08E+06	.148	.204	.325	.450	.228	.509
12.5	8.67	63.4	8.18E+09	8.30E-05	2.63E+00	6.21E-02	3.11E+06	.144	.201	.321	.445	.223	.500
15.0	8.94	64.2	4.87E+09	5.01E-05	1.55E+00	3.67E-02	3.14E+06	.141	.199	.318	.441	.218	.492
17.5	9.17	65.0	2.91E+09	3.03E-05	9.21E-01	2.18E-02	3.16E+06	.138	.197	.314	.437	.213	.484
20.0	9.39	65.8	1.75E+09	1.84E-05	5.50E-01	1.30E-02	3.18E+06	.135	.194	.311	.433	.209	.477
22.5	9.59	66.5	1.05E+09	1.12E-05	3.29E-01	7.79E-03	3.19E+06	.133	.192	.308	.429	.205	.470
25.0	9.78	67.2	6.35E+08	6.84E-06	1.98E-01	4.69E-03	3.20E+06	.131	.190	.305	.426	.201	.463
27.5	9.95	67.9	3.85E+08	4.19E-06	1.20E-01	2.83E-03	3.21E+06	.129	.189	.303	.422	.197	.457
30.0	10.11	68.5	2.34E+08	2.57E-06	7.26E-02	1.72E-03	3.22E+06	.127	.187	.300	.419	.194	.451
32.5	10.26	69.1	1.42E+08	1.58E-06	4.41E-02	1.04E-03	3.23E+06	.126	.185	.298	.416	.190	.445
35.0	10.40	69.7	8.69E+07	9.70E-07	2.69E-02	6.36E-04	3.23E+06	.125	.183	.295	.413	.187	.440

Table IV.3.8

Characteristic X-ray beam quantities without scattered radiation and relative energy absorption in typical image detectors at variable absorber or phantom materials for diagnostic X-ray tube assemblies with W/Re anode at 10° target angle and an inherent (quality equivalent) filtration of 2.5 mm Al

Reference focal distance: 100 cm   X-ray tube voltage: 110 kV

Absorber material: Al

Thick-ness mm	HVL mm Al	<E> keV	Φ/It 1/As/cm²	Ψ/It J/As/cm²	Ka/It mGy/As	Relative dose	Φ/Ka photons /cm²/μGy	68 CaWO4	68 GOS	118 GOS	181 GOS	80 SPF	180 CsI
0.0	3.98	52.3	2.76E+11	2.31E-03	1.24E+02	1.00E+00	2.23E+06	.221	.251	.383	.509	.289	.579
.5	4.36	53.3	2.56E+11	2.19E-03	1.10E+02	8.91E-01	2.32E+06	.213	.245	.376	.502	.283	.572
1.0	4.71	54.1	2.39E+11	2.07E-03	9.96E+01	8.04E-01	2.40E+06	.206	.240	.370	.496	.277	.567
1.5	5.01	54.8	2.24E+11	1.97E-03	9.07E+01	7.32E-01	2.47E+06	.200	.236	.365	.491	.273	.561
2.0	5.29	55.5	2.10E+11	1.87E-03	8.32E+01	6.71E-01	2.52E+06	.195	.232	.360	.486	.268	.557
2.5	5.55	56.1	1.98E+11	1.78E-03	7.67E+01	6.19E-01	2.58E+06	.190	.229	.356	.482	.264	.552
3.0	5.79	56.7	1.87E+11	1.70E-03	7.11E+01	5.74E-01	2.62E+06	.186	.226	.352	.478	.261	.548
3.5	6.01	57.3	1.76E+11	1.62E-03	6.62E+01	5.34E-01	2.67E+06	.182	.223	.349	.474	.257	.544
4.0	6.22	57.8	1.67E+11	1.55E-03	6.18E+01	4.98E-01	2.70E+06	.179	.220	.345	.471	.254	.540
4.5	6.41	58.3	1.58E+11	1.48E-03	5.78E+01	4.67E-01	2.74E+06	.176	.218	.343	.468	.251	.536
5.0	6.59	58.7	1.50E+11	1.41E-03	5.43E+01	4.38E-01	2.77E+06	.173	.216	.340	.465	.249	.532
6.0	6.93	59.6	1.36E+11	1.30E-03	4.81E+01	3.88E-01	2.82E+06	.167	.213	.335	.460	.244	.526
7.0	7.23	60.4	1.23E+11	1.19E-03	4.29E+01	3.47E-01	2.87E+06	.163	.209	.331	.455	.239	.519
8.0	7.49	61.1	1.12E+11	1.10E-03	3.86E+01	3.11E-01	2.91E+06	.159	.207	.328	.451	.235	.514
9.0	7.74	61.8	1.02E+11	1.01E-03	3.48E+01	2.81E-01	2.94E+06	.155	.204	.324	.448	.231	.508
10.0	7.96	62.4	9.37E+10	9.37E-04	3.15E+01	2.54E-01	2.97E+06	.152	.202	.321	.444	.228	.503
12.0	8.36	63.5	7.89E+10	8.03E-04	2.61E+01	2.11E-01	3.02E+06	.147	.199	.316	.438	.221	.493
14.0	8.69	64.5	6.69E+10	6.92E-04	2.19E+01	1.77E-01	3.06E+06	.143	.195	.312	.433	.216	.484
16.0	8.99	65.5	5.71E+10	5.98E-04	1.85E+01	1.49E-01	3.09E+06	.139	.193	.308	.428	.211	.476
18.0	9.25	66.3	4.89E+10	5.19E-04	1.57E+01	1.27E-01	3.11E+06	.136	.190	.305	.424	.206	.469
20.0	9.48	67.1	4.20E+10	4.51E-04	1.34E+01	1.08E-01	3.13E+06	.133	.188	.301	.420	.202	.461
22.0	9.69	67.8	3.62E+10	3.93E-04	1.15E+01	9.30E-02	3.14E+06	.131	.186	.299	.416	.198	.455
25.0	9.97	68.8	2.91E+10	3.21E-04	9.23E+00	7.45E-02	3.16E+06	.128	.183	.294	.411	.192	.446
30.0	10.37	70.4	2.05E+10	2.32E-04	6.47E+00	5.22E-02	3.17E+06	.124	.179	.288	.404	.184	.432
35.0	10.70	71.7	1.46E+10	1.68E-04	4.60E+00	3.72E-02	3.18E+06	.121	.175	.283	.396	.178	.419
40.0	10.98	73.0	1.05E+10	1.23E-04	3.31E+00	2.67E-02	3.18E+06	.118	.172	.278	.390	.172	.408
45.0	11.23	74.1	7.64E+09	9.07E-05	2.40E+00	1.94E-02	3.18E+06	.116	.169	.273	.384	.166	.398
50.0	11.45	75.2	5.58E+09	6.72E-05	1.76E+00	1.42E-02	3.17E+06	.115	.166	.268	.378	.161	.389
55.0	11.65	76.2	4.10E+09	5.00E-05	1.29E+00	1.04E-02	3.16E+06	.113	.163	.264	.372	.157	.381
60.0	11.83	77.1	3.02E+09	3.74E-05	9.58E-01	7.73E-03	3.15E+06	.112	.160	.260	.367	.153	.373

Absorber material: Cu

Thick-ness mm	HVL mm Al	<E> keV	Φ/It 1/As/cm²	Ψ/It J/As/cm²	Ka/It mGy/As	Relative dose	Φ/Ka photons /cm²/μGy	68 CaWO4	68 GOS	118 GOS	181 GOS	80 SPF	180 CsI
0.00	3.98	52.3	2.76E+11	2.31E-03	1.24E+02	1.00E+00	2.23E+06	.221	.251	.383	.509	.289	.579
.02	4.50	53.6	2.55E+11	2.19E-03	1.08E+02	8.74E-01	2.35E+06	.210	.243	.373	.500	.280	.570
.04	4.96	54.7	2.37E+11	2.08E-03	9.66E+01	7.79E-01	2.45E+06	.201	.236	.365	.492	.273	.562
.06	5.36	55.7	2.22E+11	1.98E-03	8.74E+01	7.06E-01	2.54E+06	.193	.231	.358	.485	.267	.555
.08	5.73	56.6	2.09E+11	1.90E-03	8.00E+01	6.46E-01	2.61E+06	.187	.226	.353	.479	.261	.548
.10	6.05	57.4	1.97E+11	1.82E-03	7.39E+01	5.96E-01	2.67E+06	.181	.222	.348	.474	.256	.542
.20	7.31	60.6	1.55E+11	1.51E-03	5.38E+01	4.34E-01	2.88E+06	.162	.209	.330	.454	.238	.518
.30	8.17	63.0	1.27E+11	1.28E-03	4.24E+01	3.42E-01	3.00E+06	.149	.200	.319	.441	.224	.498
.40	8.80	64.8	1.07E+11	1.11E-03	3.49E+01	2.81E-01	3.07E+06	.141	.195	.311	.432	.214	.482
.50	9.30	66.4	9.17E+10	9.76E-04	2.94E+01	2.37E-01	3.12E+06	.135	.190	.304	.424	.205	.468
.60	9.70	67.8	7.95E+10	8.63E-04	2.53E+01	2.04E-01	3.15E+06	.131	.186	.299	.417	.198	.455
.70	10.03	69.0	6.96E+10	7.69E-04	2.20E+01	1.77E-01	3.16E+06	.127	.183	.294	.411	.192	.444
.80	10.32	70.1	6.14E+10	6.89E-04	1.93E+01	1.56E-01	3.18E+06	.124	.180	.290	.405	.186	.434
.90	10.56	71.1	5.45E+10	6.20E-04	1.71E+01	1.38E-01	3.18E+06	.122	.177	.286	.400	.181	.425
1.00	10.78	72.0	4.86E+10	5.61E-04	1.53E+01	1.23E-01	3.19E+06	.120	.175	.282	.395	.176	.417
1.20	11.16	73.7	3.92E+10	4.63E-04	1.23E+01	9.94E-02	3.18E+06	.117	.170	.275	.387	.168	.402
1.40	11.47	75.2	3.21E+10	3.87E-04	1.01E+01	8.16E-02	3.18E+06	.115	.166	.268	.378	.161	.389
1.60	11.74	76.6	2.66E+10	3.26E-04	8.41E+00	6.78E-02	3.17E+06	.113	.162	.262	.371	.155	.378
1.80	11.98	77.8	2.23E+10	2.77E-04	7.06E+00	5.70E-02	3.15E+06	.111	.158	.257	.364	.150	.368
2.00	12.18	78.9	1.88E+10	2.37E-04	5.98E+00	4.83E-02	3.14E+06	.110	.155	.252	.357	.145	.358
2.50	12.62	81.4	1.26E+10	1.65E-04	4.07E+00	3.28E-02	3.10E+06	.108	.148	.241	.343	.136	.339
3.00	12.96	83.5	8.77E+09	1.17E-04	2.86E+00	2.31E-02	3.07E+06	.106	.141	.231	.331	.128	.323
3.50	13.23	85.2	6.25E+09	8.53E-05	2.06E+00	1.66E-02	3.03E+06	.105	.136	.224	.320	.122	.311
4.00	13.46	86.7	4.54E+09	6.30E-05	1.51E+00	1.22E-02	3.00E+06	.103	.132	.217	.312	.117	.300
4.50	13.64	88.0	3.35E+09	4.72E-05	1.12E+00	9.08E-03	2.97E+06	.102	.129	.212	.305	.113	.292
5.00	13.80	89.1	2.50E+09	3.57E-05	8.47E-01	6.84E-03	2.95E+06	.101	.126	.207	.299	.110	.284
6.00	14.05	91.0	1.43E+09	2.09E-05	4.94E-01	3.98E-03	2.91E+06	.100	.121	.200	.289	.104	.272
7.00	14.23	92.5	8.48E+08	1.26E-05	2.95E-01	2.38E-03	2.87E+06	.098	.117	.194	.281	.100	.263
8.00	14.38	93.7	5.12E+08	7.69E-06	1.80E-01	1.45E-03	2.84E+06	.097	.114	.189	.275	.097	.256

Reference focal distance: 100 cm   X-ray tube voltage: 110 kV

Absorber material:		Water						Relative energy absorption in the					
Thick-							Φ/Ka	image detector (coverage mg/cm²)					
ness	HVL	\<E\>	Φ/It	Ψ/It	Ka/It	Relative	photons	68	68	118	181	80	180
mm	mm Al	keV	1/As/cm²	J/As/cm²	mGy/As	dose	/cm²/μGy	CaWO4	GOS	GOS	GOS	SPF	CsI

No added filter

0.0	3.98	52.3	2.76E+11	2.31E-03	1.24E+02	1.00E+00	2.23E+06	.221	.251	.383	.509	.289	.579
2.5	5.21	55.5	1.47E+11	1.31E-03	5.87E+01	4.74E-01	2.51E+06	.196	.232	.360	.486	.268	.556
5.0	6.14	57.9	8.16E+10	7.56E-04	3.04E+01	2.45E-01	2.68E+06	.179	.220	.345	.470	.253	.537
7.5	6.87	59.8	4.63E+10	4.44E-04	1.65E+01	1.33E-01	2.81E+06	.167	.212	.334	.458	.242	.522
10.0	7.48	61.5	2.67E+10	2.63E-04	9.23E+00	7.45E-02	2.89E+06	.158	.205	.325	.448	.232	.508
12.5	7.99	63.0	1.56E+10	1.57E-04	5.27E+00	4.25E-02	2.96E+06	.151	.200	.318	.440	.224	.495
15.0	8.43	64.3	9.18E+09	9.46E-05	3.05E+00	2.46E-02	3.01E+06	.145	.196	.312	.433	.216	.484
17.5	8.81	65.5	5.45E+09	5.72E-05	1.79E+00	1.44E-02	3.05E+06	.141	.192	.307	.426	.210	.473
20.0	9.15	66.6	3.26E+09	3.47E-05	1.06E+00	8.54E-03	3.08E+06	.137	.189	.302	.420	.204	.463
22.5	9.46	67.7	1.96E+09	2.12E-05	6.31E-01	5.09E-03	3.10E+06	.133	.186	.298	.415	.198	.454
25.0	9.73	68.6	1.18E+09	1.30E-05	3.79E-01	3.06E-03	3.12E+06	.130	.183	.293	.410	.193	.445
27.5	9.98	69.6	7.15E+08	7.96E-06	2.28E-01	1.84E-03	3.13E+06	.128	.180	.290	.405	.188	.437
30.0	10.21	70.4	4.35E+08	4.90E-06	1.39E-01	1.12E-03	3.14E+06	.125	.178	.286	.400	.184	.429
32.5	10.42	71.3	2.65E+08	3.03E-06	8.44E-02	6.81E-04	3.14E+06	.123	.175	.282	.396	.180	.422
35.0	10.61	72.1	1.62E+08	1.87E-06	5.16E-02	4.16E-04	3.15E+06	.122	.173	.279	.392	.176	.415

Added filter   0.5 mm Al

0.0	4.36	53.3	2.56E+11	2.19E-03	1.10E+02	1.00E+00	2.32E+06	.213	.245	.376	.502	.283	.572
2.5	5.48	56.1	1.38E+11	1.24E-03	5.41E+01	4.89E-01	2.56E+06	.191	.229	.356	.482	.264	.551
5.0	6.34	58.4	7.73E+10	7.23E-04	2.84E+01	2.57E-01	2.72E+06	.176	.218	.342	.467	.250	.534
7.5	7.04	60.2	4.41E+10	4.25E-04	1.56E+01	1.41E-01	2.83E+06	.165	.210	.332	.456	.239	.518
10.0	7.61	61.8	2.55E+10	2.53E-04	8.76E+00	7.93E-02	2.91E+06	.156	.204	.324	.446	.230	.505
12.5	8.10	63.3	1.49E+10	1.51E-04	5.02E+00	4.54E-02	2.97E+06	.150	.199	.317	.438	.222	.493
15.0	8.53	64.6	8.81E+09	9.11E-05	2.92E+00	2.64E-02	3.02E+06	.144	.195	.311	.431	.215	.481
17.5	8.90	65.7	5.24E+09	5.52E-05	1.71E+00	1.55E-02	3.06E+06	.140	.191	.306	.425	.208	.471
20.0	9.23	66.8	3.13E+09	3.35E-05	1.02E+00	9.20E-03	3.08E+06	.136	.188	.301	.419	.203	.461
22.5	9.52	67.9	1.88E+09	2.05E-05	6.07E-01	5.49E-03	3.10E+06	.133	.185	.297	.414	.197	.452
25.0	9.79	68.8	1.14E+09	1.25E-05	3.65E-01	3.30E-03	3.12E+06	.130	.182	.293	.409	.192	.444
27.5	10.03	69.7	6.90E+08	7.70E-06	2.20E-01	1.99E-03	3.13E+06	.127	.180	.289	.404	.188	.435
30.0	10.25	70.6	4.20E+08	4.75E-06	1.34E-01	1.21E-03	3.14E+06	.125	.177	.285	.399	.183	.428
32.5	10.46	71.4	2.56E+08	2.93E-06	8.15E-02	7.38E-04	3.14E+06	.123	.175	.282	.395	.179	.420
35.0	10.65	72.2	1.57E+08	1.81E-06	4.99E-02	4.51E-04	3.15E+06	.121	.173	.278	.391	.175	.414

Added filter   0.1 mm Cu

0.0	6.05	57.4	1.97E+11	1.82E-03	7.39E+01	1.00E+00	2.67E+06	.181	.222	.348	.474	.256	.542
2.5	6.79	59.4	1.12E+11	1.06E-03	4.00E+01	5.41E-01	2.80E+06	.169	.214	.337	.461	.245	.527
5.0	7.39	61.0	6.44E+10	6.29E-04	2.23E+01	3.02E-01	2.89E+06	.160	.207	.328	.451	.235	.513
7.5	7.90	62.5	3.75E+10	3.75E-04	1.27E+01	1.72E-01	2.96E+06	.153	.202	.321	.443	.227	.500
10.0	8.34	63.8	2.20E+10	2.25E-04	7.33E+00	9.92E-02	3.01E+06	.147	.197	.314	.436	.219	.489
12.5	8.72	65.0	1.31E+10	1.36E-04	4.29E+00	5.80E-02	3.05E+06	.142	.194	.309	.429	.212	.478
15.0	9.06	66.2	7.79E+09	8.25E-05	2.53E+00	3.43E-02	3.08E+06	.138	.190	.304	.423	.206	.468
17.5	9.37	67.2	4.67E+09	5.03E-05	1.51E+00	2.04E-02	3.10E+06	.134	.187	.300	.418	.201	.459
20.0	9.64	68.2	2.81E+09	3.07E-05	9.03E-01	1.22E-02	3.12E+06	.131	.184	.296	.413	.196	.450
22.5	9.89	69.1	1.70E+09	1.88E-05	5.44E-01	7.36E-03	3.13E+06	.129	.182	.292	.408	.191	.442
25.0	10.12	70.0	1.03E+09	1.16E-05	3.29E-01	4.46E-03	3.14E+06	.126	.179	.288	.403	.186	.434
27.5	10.33	70.8	6.29E+08	7.14E-06	2.00E-01	2.71E-03	3.14E+06	.124	.177	.285	.399	.182	.426
30.0	10.53	71.6	3.85E+08	4.41E-06	1.22E-01	1.65E-03	3.15E+06	.122	.175	.281	.394	.178	.419
32.5	10.71	72.4	2.36E+08	2.73E-06	7.48E-02	1.01E-03	3.15E+06	.121	.172	.278	.390	.174	.412
35.0	10.88	73.1	1.45E+08	1.70E-06	4.60E-02	6.22E-04	3.15E+06	.119	.170	.275	.386	.171	.406

Added filter   0.2 mm Cu

0.0	7.31	60.6	1.55E+11	1.51E-03	5.38E+01	1.00E+00	2.88E+06	.162	.209	.330	.454	.238	.518
2.5	7.81	62.1	9.02E+10	8.97E-04	3.06E+01	5.68E-01	2.95E+06	.154	.203	.323	.446	.229	.505
5.0	8.25	63.4	5.29E+10	5.38E-04	1.76E+01	3.27E-01	3.00E+06	.148	.199	.317	.438	.222	.493
7.5	8.63	64.6	3.13E+10	3.24E-04	1.03E+01	1.91E-01	3.04E+06	.143	.195	.311	.432	.215	.483
10.0	8.97	65.7	1.86E+10	1.96E-04	6.06E+00	1.13E-01	3.07E+06	.139	.192	.306	.426	.209	.473
12.5	9.28	66.7	1.12E+10	1.19E-04	3.60E+00	6.69E-02	3.10E+06	.135	.189	.302	.421	.203	.463
15.0	9.55	67.7	6.71E+09	7.28E-05	2.15E+00	4.00E-02	3.12E+06	.132	.186	.298	.415	.198	.454
17.5	9.80	68.6	4.06E+09	4.46E-05	1.30E+00	2.41E-02	3.13E+06	.130	.183	.294	.411	.193	.446
20.0	10.03	69.5	2.46E+09	2.74E-05	7.83E-01	1.45E-02	3.14E+06	.127	.181	.290	.406	.189	.438
22.5	10.25	70.4	1.50E+09	1.69E-05	4.75E-01	8.83E-03	3.15E+06	.125	.178	.287	.401	.184	.430
25.0	10.44	71.2	9.13E+08	1.04E-05	2.90E-01	5.38E-03	3.15E+06	.123	.176	.283	.397	.180	.423
27.5	10.63	71.9	5.59E+08	6.44E-06	1.77E-01	3.29E-03	3.15E+06	.122	.174	.280	.393	.177	.416
30.0	10.80	72.7	3.43E+08	3.99E-06	1.09E-01	2.02E-03	3.16E+06	.120	.172	.277	.389	.173	.410
32.5	10.96	73.4	2.11E+08	2.48E-06	6.69E-02	1.24E-03	3.15E+06	.119	.170	.274	.385	.170	.404
35.0	11.11	74.1	1.30E+08	1.54E-06	4.13E-02	7.66E-04	3.15E+06	.117	.168	.271	.381	.166	.398

Table IV.3.10

Characteristic X-ray beam quantities without scattered radiation and relative energy absorption in typical image detectors at variable absorber or phantom materials for diagnostic X-ray tube assemblies with W/Re anode at 10° target angle and an inherent (quality equivalent) filtration of 2.5 mm Al

Reference focal distance: 100 cm   X-ray tube voltage: 150 kV

Absorber material: Al

Thickness mm	HVL mm Al	$\langle E \rangle$ keV	Φ/It 1/As/cm²	Ψ/It J/As/cm²	Ka/It mGy/As	Relative dose	Φ/Ka photons /cm²/µGy	68 CaWO4	68 GOS	118 GOS	181 GOS	80 SPF	180 CsI
0.0	5.54	62.1	4.86E+11	4.83E-03	2.01E+02	1.00E+00	2.41E+06	.169	.202	.314	.426	.225	.472
.5	5.99	63.1	4.58E+11	4.62E-03	1.84E+02	9.16E-01	2.48E+06	.164	.198	.309	.421	.220	.466
1.0	6.37	63.9	4.33E+11	4.43E-03	1.71E+02	8.47E-01	2.54E+06	.160	.194	.304	.416	.215	.461
1.5	6.72	64.7	4.10E+11	4.25E-03	1.59E+02	7.88E-01	2.58E+06	.156	.191	.300	.411	.212	.456
2.0	7.02	65.4	3.89E+11	4.08E-03	1.48E+02	7.37E-01	2.62E+06	.152	.188	.297	.407	.208	.451
2.5	7.30	66.1	3.70E+11	3.92E-03	1.39E+02	6.92E-01	2.66E+06	.149	.186	.293	.404	.205	.447
3.0	7.56	66.7	3.53E+11	3.77E-03	1.31E+02	6.52E-01	2.69E+06	.146	.183	.290	.400	.202	.442
3.5	7.80	67.3	3.37E+11	3.63E-03	1.24E+02	6.16E-01	2.71E+06	.144	.181	.287	.397	.199	.438
4.0	8.01	67.9	3.22E+11	3.50E-03	1.18E+02	5.84E-01	2.74E+06	.141	.179	.285	.394	.197	.434
4.5	8.22	68.4	3.08E+11	3.37E-03	1.12E+02	5.54E-01	2.76E+06	.139	.178	.282	.391	.194	.431
5.0	8.41	69.0	2.95E+11	3.25E-03	1.06E+02	5.27E-01	2.78E+06	.137	.176	.280	.388	.192	.427
6.0	8.75	69.9	2.70E+11	3.03E-03	9.64E+01	4.79E-01	2.81E+06	.134	.173	.276	.383	.188	.420
7.0	9.06	70.8	2.49E+11	2.83E-03	8.80E+01	4.37E-01	2.83E+06	.130	.170	.272	.379	.184	.414
8.0	9.34	71.7	2.30E+11	2.64E-03	8.07E+01	4.01E-01	2.85E+06	.128	.168	.269	.374	.180	.408
9.0	9.59	72.4	2.13E+11	2.47E-03	7.43E+01	3.69E-01	2.86E+06	.125	.165	.265	.370	.177	.402
10.0	9.81	73.2	1.97E+11	2.31E-03	6.86E+01	3.40E-01	2.87E+06	.123	.163	.262	.367	.173	.397
12.0	10.22	74.6	1.70E+11	2.03E-03	5.88E+01	2.92E-01	2.89E+06	.119	.160	.257	.360	.168	.387
14.0	10.57	75.9	1.47E+11	1.79E-03	5.08E+01	2.52E-01	2.90E+06	.116	.156	.252	.354	.162	.378
16.0	10.87	77.1	1.28E+11	1.58E-03	4.42E+01	2.19E-01	2.90E+06	.113	.153	.247	.348	.158	.369
18.0	11.14	78.2	1.12E+11	1.40E-03	3.86E+01	1.92E-01	2.90E+06	.111	.151	.243	.342	.153	.361
20.0	11.39	79.2	9.82E+10	1.25E-03	3.39E+01	1.68E-01	2.90E+06	.108	.148	.239	.337	.149	.353
22.0	11.61	80.2	8.62E+10	1.11E-03	2.98E+01	1.48E-01	2.89E+06	.106	.145	.236	.333	.146	.346
25.0	11.91	81.7	7.13E+10	9.33E-04	2.47E+01	1.23E-01	2.88E+06	.104	.142	.230	.326	.140	.336
30.0	12.35	83.9	5.25E+10	7.05E-04	1.84E+01	9.11E-02	2.86E+06	.101	.137	.222	.315	.133	.321
35.0	12.72	85.9	3.90E+10	5.37E-04	1.38E+01	6.84E-02	2.83E+06	.098	.132	.215	.306	.126	.308
40.0	13.04	87.8	2.93E+10	4.11E-04	1.04E+01	5.18E-02	2.81E+06	.095	.128	.208	.297	.120	.296
45.0	13.33	89.5	2.21E+10	3.17E-04	7.95E+00	3.95E-02	2.78E+06	.093	.124	.202	.289	.115	.285
50.0	13.59	91.2	1.68E+10	2.45E-04	6.10E+00	3.03E-02	2.75E+06	.091	.120	.197	.281	.110	.275
55.0	13.82	92.7	1.28E+10	1.91E-04	4.71E+00	2.34E-02	2.72E+06	.090	.117	.191	.274	.106	.267
60.0	14.03	94.2	9.85E+09	1.49E-04	3.65E+00	1.81E-02	2.69E+06	.088	.113	.186	.268	.102	.258

Absorber material: Cu

Thickness mm	HVL mm Al	$\langle E \rangle$ keV	Φ/It 1/As/cm²	Ψ/It J/As/cm²	Ka/It mGy/As	Relative dose	Φ/Ka photons /cm²/µGy	68 CaWO4	68 GOS	118 GOS	181 GOS	80 SPF	180 CsI
0.00	5.54	62.1	4.86E+11	4.83E-03	2.01E+02	1.00E+00	2.41E+06	.169	.202	.314	.426	.225	.472
.02	6.15	63.4	4.58E+11	4.65E-03	1.83E+02	9.07E-01	2.50E+06	.162	.196	.307	.419	.218	.464
.04	6.66	64.6	4.34E+11	4.49E-03	1.68E+02	8.36E-01	2.58E+06	.156	.192	.301	.412	.212	.456
.06	7.11	65.6	4.13E+11	4.34E-03	1.57E+02	7.78E-01	2.63E+06	.151	.188	.296	.406	.207	.450
.08	7.50	66.6	3.94E+11	4.21E-03	1.47E+02	7.30E-01	2.68E+06	.147	.184	.291	.401	.203	.443
.10	7.84	67.5	3.78E+11	4.08E-03	1.39E+02	6.90E-01	2.72E+06	.143	.181	.287	.396	.199	.438
.20	9.13	71.0	3.14E+11	3.58E-03	1.11E+02	5.50E-01	2.84E+06	.130	.170	.272	.378	.183	.413
.30	10.00	73.7	2.70E+11	3.19E-03	9.35E+01	4.64E-01	2.89E+06	.121	.162	.261	.365	.171	.393
.40	10.64	76.0	2.36E+11	2.88E-03	8.12E+01	4.03E-01	2.91E+06	.115	.156	.252	.354	.162	.377
.50	11.14	78.0	2.09E+11	2.62E-03	7.19E+01	3.57E-01	2.91E+06	.111	.151	.244	.344	.154	.363
.60	11.56	79.8	1.87E+11	2.40E-03	6.45E+01	3.20E-01	2.91E+06	.107	.147	.238	.336	.147	.350
.70	11.91	81.4	1.69E+11	2.21E-03	5.84E+01	2.90E-01	2.89E+06	.104	.143	.232	.328	.141	.338
.80	12.21	82.9	1.54E+11	2.04E-03	5.33E+01	2.65E-01	2.88E+06	.102	.139	.226	.321	.136	.328
.90	12.48	84.3	1.40E+11	1.89E-03	4.89E+01	2.43E-01	2.86E+06	.100	.136	.221	.314	.131	.319
1.00	12.72	85.6	1.28E+11	1.76E-03	4.51E+01	2.24E-01	2.85E+06	.098	.133	.217	.308	.127	.310
1.20	13.14	88.1	1.09E+11	1.54E-03	3.89E+01	1.93E-01	2.81E+06	.095	.127	.208	.296	.119	.294
1.40	13.49	90.3	9.41E+10	1.36E-03	3.39E+01	1.68E-01	2.77E+06	.092	.122	.200	.286	.112	.281
1.60	13.79	92.3	8.18E+10	1.21E-03	2.99E+01	1.48E-01	2.74E+06	.090	.118	.193	.277	.107	.269
1.80	14.06	94.1	7.18E+10	1.08E-03	2.65E+01	1.32E-01	2.71E+06	.088	.114	.187	.269	.102	.259
2.00	14.29	95.8	6.35E+10	9.74E-04	2.37E+01	1.18E-01	2.67E+06	.087	.110	.181	.261	.098	.250
2.50	14.76	99.4	4.79E+10	7.63E-04	1.84E+01	9.13E-02	2.60E+06	.084	.103	.170	.246	.089	.231
3.00	15.12	102.4	3.72E+10	6.10E-04	1.46E+01	7.26E-02	2.54E+06	.081	.097	.161	.233	.083	.217
3.50	15.40	104.8	2.95E+10	4.95E-04	1.18E+01	5.87E-02	2.49E+06	.079	.092	.153	.223	.078	.205
4.00	15.62	106.9	2.37E+10	4.06E-04	9.67E+00	4.80E-02	2.45E+06	.077	.088	.147	.215	.074	.196
4.50	15.80	108.7	1.93E+10	3.36E-04	8.00E+00	3.97E-02	2.41E+06	.075	.085	.143	.209	.071	.189
5.00	15.95	110.2	1.59E+10	2.80E-04	6.67E+00	3.31E-02	2.38E+06	.074	.083	.138	.203	.068	.183
6.00	16.18	112.8	1.10E+10	1.98E-04	4.71E+00	2.34E-02	2.33E+06	.071	.079	.132	.194	.064	.173
7.00	16.36	114.9	7.74E+09	1.42E-04	3.39E+00	1.68E-02	2.29E+06	.069	.075	.127	.187	.061	.166
8.00	16.51	116.6	5.54E+09	1.04E-04	2.46E+00	1.22E-02	2.25E+06	.067	.073	.123	.181	.058	.160

Reference focal distance: 100 cm   X-ray tube voltage: 150 kV

Absorber material:		Water					Φ/Ka	Relative energy absorption in the image detector (coverage mg/cm²)					
Thick-ness mm	HVL mm Al	\<E\> keV	Φ/It 1/As/cm²	Ψ/It J/As/cm²	Ka/It mGy/As	Relative dose	photons /cm²/µGy	68 CaWO4	68 GOS	118 GOS	181 GOS	80 SPF	180 CsI

No added filter

0.0	5.54	62.1	4.86E+11	4.83E-03	2.01E+02	1.00E+00	2.41E+06	.169	.202	.314	.426	.225	.472
2.5	6.99	65.7	2.74E+11	2.88E-03	1.05E+02	5.21E-01	2.61E+06	.152	.188	.295	.405	.207	.448
5.0	8.03	68.5	1.59E+11	1.74E-03	5.84E+01	2.90E-01	2.72E+06	.140	.177	.282	.389	.193	.428
7.5	8.85	70.9	9.38E+10	1.07E-03	3.37E+01	1.67E-01	2.78E+06	.131	.169	.271	.376	.183	.410
10.0	9.52	73.1	5.61E+10	6.58E-04	1.99E+01	9.89E-02	2.82E+06	.125	.163	.261	.365	.173	.394
12.5	10.08	75.2	3.39E+10	4.09E-04	1.20E+01	5.94E-02	2.84E+06	.119	.157	.253	.354	.165	.380
15.0	10.57	77.1	2.07E+10	2.55E-04	7.26E+00	3.61E-02	2.85E+06	.115	.152	.245	.345	.158	.366
17.5	11.01	78.8	1.27E+10	1.60E-04	4.46E+00	2.21E-02	2.85E+06	.111	.148	.239	.336	.151	.354
20.0	11.39	80.5	7.83E+09	1.01E-04	2.76E+00	1.37E-02	2.84E+06	.107	.143	.232	.328	.145	.342
22.5	11.74	82.2	4.85E+09	6.39E-05	1.72E+00	8.52E-03	2.83E+06	.104	.139	.226	.320	.139	.331
25.0	12.06	83.7	3.02E+09	4.06E-05	1.07E+00	5.34E-03	2.81E+06	.102	.136	.220	.312	.134	.321
27.5	12.36	85.2	1.89E+09	2.58E-05	6.77E-01	3.36E-03	2.80E+06	.099	.132	.215	.305	.129	.311
30.0	12.63	86.7	1.19E+09	1.65E-05	4.28E-01	2.12E-03	2.78E+06	.097	.129	.210	.298	.124	.302
32.5	12.88	88.1	7.49E+08	1.06E-05	2.71E-01	1.35E-03	2.76E+06	.095	.126	.205	.292	.120	.293
35.0	13.12	89.5	4.74E+08	6.79E-06	1.73E-01	8.59E-04	2.74E+06	.094	.123	.200	.286	.116	.285

Added filter   0.5 mm Al

0.0	5.99	63.1	4.58E+11	4.62E-03	1.84E+02	1.00E+00	2.48E+06	.164	.198	.309	.421	.220	.466
2.5	7.28	66.4	2.60E+11	2.77E-03	9.84E+01	5.33E-01	2.65E+06	.149	.185	.292	.402	.204	.444
5.0	8.25	69.1	1.52E+11	1.68E-03	5.54E+01	3.00E-01	2.74E+06	.138	.175	.279	.386	.191	.424
7.5	9.02	71.4	9.00E+10	1.03E-03	3.22E+01	1.75E-01	2.80E+06	.130	.168	.268	.374	.181	.407
10.0	9.65	73.6	5.40E+10	6.37E-04	1.91E+01	1.04E-01	2.83E+06	.123	.162	.259	.362	.172	.391
12.5	10.20	75.5	3.27E+10	3.96E-04	1.15E+01	6.24E-02	2.84E+06	.118	.156	.251	.352	.163	.377
15.0	10.67	77.4	2.00E+10	2.48E-04	7.01E+00	3.80E-02	2.85E+06	.114	.151	.244	.343	.156	.364
17.5	11.09	79.2	1.23E+10	1.55E-04	4.31E+00	2.34E-02	2.85E+06	.110	.147	.237	.334	.150	.352
20.0	11.47	80.8	7.57E+09	9.81E-05	2.67E+00	1.45E-02	2.84E+06	.107	.143	.231	.326	.144	.340
22.5	11.81	82.4	4.70E+09	6.21E-05	1.66E+00	9.02E-03	2.83E+06	.104	.139	.225	.318	.138	.329
25.0	12.12	84.0	2.93E+09	3.94E-05	1.04E+00	5.65E-03	2.81E+06	.101	.135	.220	.311	.133	.319
27.5	12.41	85.5	1.84E+09	2.51E-05	6.57E-01	3.56E-03	2.79E+06	.099	.132	.214	.304	.128	.310
30.0	12.68	86.9	1.15E+09	1.61E-05	4.16E-01	2.25E-03	2.78E+06	.097	.128	.209	.297	.123	.300
32.5	12.92	88.3	7.28E+08	1.03E-05	2.64E-01	1.43E-03	2.76E+06	.095	.125	.204	.291	.119	.292
35.0	13.16	89.7	4.60E+08	6.61E-06	1.68E-01	9.12E-04	2.74E+06	.093	.122	.200	.285	.115	.284

Added filter   0.1 mm Cu

0.0	7.84	67.5	3.78E+11	4.08E-03	1.39E+02	1.00E+00	2.72E+06	.143	.181	.287	.396	.199	.438
2.5	8.65	69.9	2.23E+11	2.49E-03	7.98E+01	5.74E-01	2.79E+06	.134	.173	.276	.383	.188	.420
5.0	9.32	72.0	1.33E+11	1.53E-03	4.69E+01	3.38E-01	2.83E+06	.127	.166	.266	.371	.178	.404
7.5	9.89	74.0	8.00E+10	9.49E-04	2.81E+01	2.02E-01	2.85E+06	.122	.161	.258	.361	.170	.389
10.0	10.38	75.9	4.86E+10	5.91E-04	1.70E+01	1.22E-01	2.86E+06	.117	.156	.251	.351	.162	.375
12.5	10.81	77.7	2.97E+10	3.70E-04	1.04E+01	7.47E-02	2.86E+06	.113	.151	.244	.342	.155	.363
15.0	11.20	79.4	1.83E+10	2.32E-04	6.40E+00	4.61E-02	2.86E+06	.109	.147	.237	.334	.149	.351
17.5	11.55	81.0	1.13E+10	1.47E-04	3.97E+00	2.86E-02	2.85E+06	.106	.143	.231	.326	.143	.340
20.0	11.87	82.5	7.03E+09	9.29E-05	2.48E+00	1.78E-02	2.83E+06	.104	.139	.225	.319	.138	.329
22.5	12.17	84.0	4.39E+09	5.91E-05	1.56E+00	1.12E-02	2.82E+06	.101	.135	.220	.311	.133	.319
25.0	12.45	85.5	2.75E+09	3.76E-05	9.81E-01	7.06E-03	2.80E+06	.099	.132	.214	.305	.128	.310
27.5	12.70	86.9	1.73E+09	2.41E-05	6.21E-01	4.47E-03	2.78E+06	.097	.129	.209	.298	.123	.301
30.0	12.94	88.3	1.09E+09	1.54E-05	3.95E-01	2.84E-03	2.76E+06	.095	.125	.205	.292	.119	.292
32.5	13.17	89.6	6.90E+08	9.90E-06	2.52E-01	1.81E-03	2.74E+06	.093	.122	.200	.286	.115	.284
35.0	13.38	90.9	4.38E+08	6.38E-06	1.61E-01	1.16E-03	2.72E+06	.092	.120	.196	.280	.112	.277

Added filter   0.2 mm Cu

0.0	9.13	71.0	3.14E+11	3.58E-03	1.11E+02	1.00E+00	2.84E+06	.130	.170	.272	.378	.183	.413
2.5	9.70	73.0	1.89E+11	2.21E-03	6.60E+01	5.95E-01	2.86E+06	.124	.164	.263	.368	.174	.398
5.0	10.19	74.8	1.14E+11	1.37E-03	3.98E+01	3.59E-01	2.87E+06	.119	.159	.256	.358	.167	.384
7.5	10.62	76.5	6.98E+10	8.56E-04	2.42E+01	2.19E-01	2.88E+06	.115	.154	.249	.349	.160	.372
10.0	11.01	78.2	4.28E+10	5.36E-04	1.49E+01	1.34E-01	2.87E+06	.111	.150	.242	.341	.153	.360
12.5	11.36	79.8	2.64E+10	3.38E-04	9.21E+00	8.32E-02	2.87E+06	.108	.146	.236	.333	.147	.348
15.0	11.69	81.3	1.64E+10	2.13E-04	5.74E+00	5.18E-02	2.85E+06	.105	.142	.230	.325	.142	.337
17.5	11.98	82.8	1.02E+10	1.35E-04	3.59E+00	3.24E-02	2.84E+06	.103	.138	.224	.318	.136	.327
20.0	12.26	84.3	6.37E+09	8.60E-05	2.26E+00	2.04E-02	2.82E+06	.101	.135	.219	.311	.132	.318
22.5	12.52	85.7	3.99E+09	5.48E-05	1.42E+00	1.28E-02	2.81E+06	.098	.131	.214	.304	.127	.309
25.0	12.77	87.1	2.51E+09	3.51E-05	9.02E-01	8.14E-03	2.79E+06	.097	.128	.209	.298	.123	.300
27.5	12.99	88.4	1.59E+09	2.25E-05	5.74E-01	5.18E-03	2.77E+06	.095	.125	.204	.291	.119	.292
30.0	13.21	89.7	1.00E+09	1.44E-05	3.66E-01	3.30E-03	2.75E+06	.093	.122	.200	.285	.115	.284
32.5	13.41	91.0	6.38E+08	9.30E-06	2.34E-01	2.11E-03	2.72E+06	.092	.120	.196	.280	.111	.276
35.0	13.60	92.2	4.06E+08	6.00E-06	1.50E-01	1.36E-03	2.70E+06	.090	.117	.192	.274	.108	.269

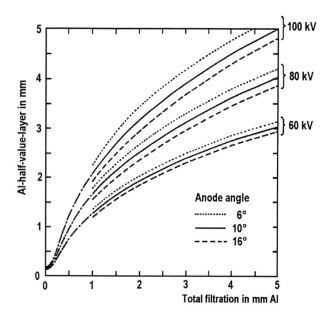

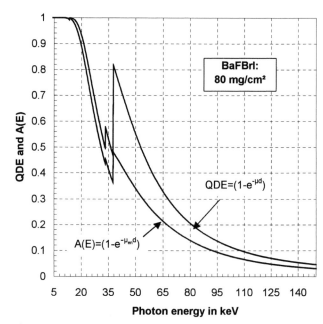

Fig. IV.3.1 Relation between the Al quality equivalent total filtration and the Al half value layer at various X-ray tube voltages and anode angles (DIN 1990)

Fig. IV.3.3 Quantum detection efficiency (QDE) and absorbed energy [A(E)] in dependence on photon energy in a storage phosphor screen with a cover thickness of 80 mg/cm^2

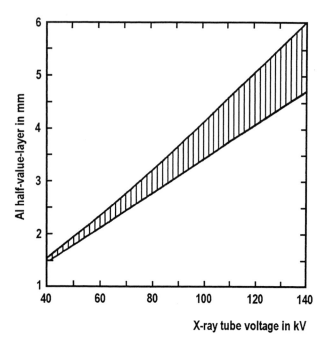

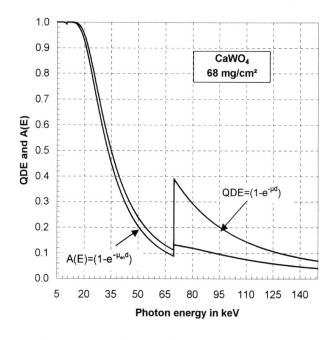

Fig. IV.3.2 Possible spread of Al half value layer in dependence on X-ray tube voltage for a brand-new W-anode, angles varying from 6° to 16°. Radiation quality: DC- or 12-pulse generator, filtration 2.5 mm Al

Fig. IV.3.4 Quantum detection efficiency (QDE) and absorbed energy [A(E)] in dependence on photon energy in an intensifying screen with a cover thickness of 68 mg/cm^2

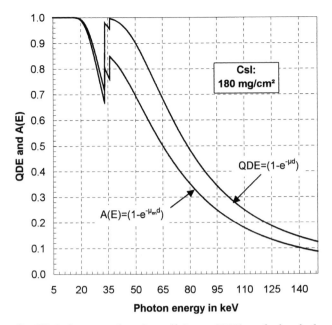

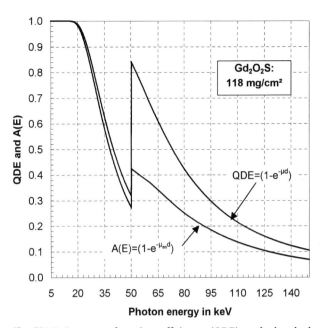

Fig. IV.3.5 Quantum detection efficiency (QDE) and absorbed energy [A(E)] in dependence on photon energy in an image intensifier input screen with a cover thickness of 180 mg/cm²

Fig. IV.3.7 Quantum detection efficiency (QDE) and absorbed energy [A(E)] in dependence on photon energy in an intensifying screen with a cover thickness of 118 mg/cm² (sensitivity class 300)

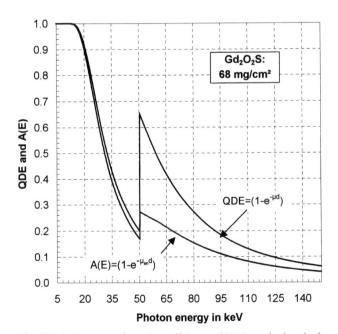

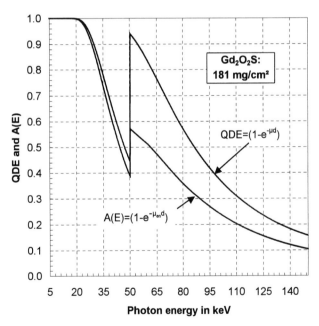

Fig. IV.3.6 Quantum detection efficiency (QDE) and absorbed energy [A(E)] in dependence on photon energy in an intensifying screen with a cover thickness of 68 mg/cm² (sensitivity class 100)

Fig. IV.3.8 Quantum detection efficiency (QDE) and absorbed energy [A(E)] in dependence on photon energy in an intensifying screen with a cover thickness of 181 mg/cm² (sensitivity class 600)

IV.3.2 Mammography

Table IV.3.11

Characteristic X-ray beam quantities without scattered radiation and relative energy absorption in typical image detectors at variable absorber or phantom materials for mammographic X-ray beam qualities with the indicated anode-filter combination:

Anode Mo / 0.04 mm Nb   Reference focal distance 60 cm

Absorber material: PMMA

Thick-ness cm	HVL mm Al	<E> keV	Φ/It 1/As/cm²	Ψ/It J/As/cm²	Ka/It mGy/As	Relative dose	Φ/Ka photons /cm²/µGy	34 CaWO4	34 GOS	40 SPF	60 SPF	60 aSe	80 CsI
X-ray tube voltage: 22 kV													
0.0	.24	14.9	1.82E+10	4.34E-05	6.60E+01	1.00E+00	2.75E+05	.900	.882	.855	.938	.934	.969
1.0	.35	16.2	4.80E+09	1.25E-05	1.38E+01	2.08E-01	3.49E+05	.871	.845	.819	.918	.939	.956
2.0	.42	16.9	1.61E+09	4.35E-06	4.16E+00	6.31E-02	3.87E+05	.852	.823	.798	.905	.937	.947
3.0	.46	17.3	5.95E+08	1.65E-06	1.45E+00	2.19E-02	4.11E+05	.839	.807	.783	.895	.932	.940
4.0	.49	17.5	2.32E+08	6.53E-07	5.43E-01	8.22E-03	4.28E+05	.829	.795	.771	.887	.927	.934
5.0	.51	17.8	9.39E+07	2.67E-07	2.13E-01	3.22E-03	4.41E+05	.820	.785	.761	.881	.922	.929
6.0	.53	18.0	3.90E+07	1.12E-07	8.62E-02	1.31E-03	4.52E+05	.812	.777	.753	.874	.918	.924
7.0	.54	18.1	1.66E+07	4.82E-08	3.58E-02	5.43E-04	4.62E+05	.805	.768	.745	.868	.913	.919
8.0	.56	18.3	7.17E+06	2.10E-08	1.52E-02	2.30E-04	4.72E+05	.798	.760	.737	.862	.909	.914
X-ray tube voltage: 24 kV													
0.0	.26	15.4	2.62E+10	6.48E-05	8.83E+01	1.00E+00	2.96E+05	.883	.861	.835	.926	.931	.960
1.0	.38	16.7	7.66E+09	2.05E-05	2.06E+01	2.34E-01	3.71E+05	.854	.825	.800	.905	.933	.946
2.0	.45	17.3	2.74E+09	7.59E-06	6.74E+00	7.63E-02	4.07E+05	.835	.803	.779	.891	.927	.936
3.0	.48	17.7	1.07E+09	3.02E-06	2.48E+00	2.81E-02	4.30E+05	.821	.787	.763	.880	.920	.927
4.0	.51	18.0	4.35E+08	1.25E-06	9.70E-01	1.10E-02	4.48E+05	.807	.772	.749	.869	.913	.918
5.0	.54	18.3	1.84E+08	5.38E-07	3.96E-01	4.48E-03	4.64E+05	.794	.758	.735	.858	.905	.909
6.0	.56	18.6	7.98E+07	2.37E-07	1.66E-01	1.89E-03	4.79E+05	.781	.743	.721	.847	.897	.899
7.0	.58	18.9	3.56E+07	1.08E-07	7.20E-02	8.15E-04	4.95E+05	.768	.729	.706	.835	.888	.889
8.0	.61	19.2	1.63E+07	5.01E-08	3.19E-02	3.61E-04	5.11E+05	.753	.713	.691	.822	.878	.877
X-ray tube voltage: 26 kV													
0.0	.28	15.9	3.59E+10	9.13E-05	1.14E+02	1.00E+00	3.14E+05	.867	.843	.817	.913	.926	.950
1.0	.41	17.1	1.13E+10	3.09E-05	2.90E+01	2.54E-01	3.89E+05	.836	.805	.781	.890	.923	.933
2.0	.47	17.7	4.24E+09	1.20E-05	9.97E+00	8.73E-02	4.26E+05	.813	.780	.756	.872	.913	.919
3.0	.51	18.2	1.72E+09	5.02E-06	3.81E+00	3.34E-02	4.52E+05	.793	.757	.734	.855	.901	.904
4.0	.54	18.7	7.35E+08	2.20E-06	1.55E+00	1.36E-02	4.75E+05	.772	.735	.712	.837	.888	.888
5.0	.58	19.1	3.27E+08	1.00E-06	6.55E-01	5.74E-03	4.99E+05	.750	.711	.690	.817	.873	.871
6.0	.61	19.6	1.51E+08	4.74E-07	2.88E-01	2.52E-03	5.24E+05	.728	.687	.666	.796	.857	.852
7.0	.65	20.1	7.22E+07	2.33E-07	1.31E-01	1.15E-03	5.52E+05	.704	.662	.642	.775	.840	.832
8.0	.70	20.7	3.58E+07	1.19E-07	6.15E-02	5.39E-04	5.83E+05	.682	.638	.619	.755	.824	.813
X-ray tube voltage: 28 kV													
0.0	.30	16.3	4.73E+10	1.23E-04	1.43E+02	1.00E+00	3.30E+05	.851	.825	.800	.900	.919	.939
1.0	.43	17.5	1.57E+10	4.40E-05	3.88E+01	2.71E-01	4.05E+05	.816	.784	.760	.872	.910	.917
2.0	.49	18.2	6.14E+09	1.79E-05	1.38E+01	9.66E-02	4.44E+05	.788	.753	.730	.848	.894	.897
3.0	.53	18.8	2.59E+09	7.82E-06	5.45E+00	3.81E-02	4.75E+05	.759	.722	.700	.823	.875	.874
4.0	.57	19.5	1.16E+09	3.61E-06	2.28E+00	1.59E-02	5.07E+05	.729	.690	.669	.796	.854	.849
5.0	.62	20.2	5.41E+08	1.75E-06	9.99E-01	6.97E-03	5.41E+05	.698	.657	.637	.767	.831	.822
6.0	.67	20.9	2.65E+08	8.89E-07	4.57E-01	3.19E-03	5.81E+05	.667	.623	.605	.738	.807	.794
7.0	.74	21.6	1.36E+08	4.72E-07	2.18E-01	1.52E-03	6.24E+05	.637	.592	.575	.710	.785	.768
8.0	.81	22.3	7.29E+07	2.60E-07	1.09E-01	7.60E-04	6.70E+05	.610	.565	.548	.685	.765	.744
X-ray tube voltage: 30 kV													
0.0	.32	16.7	5.94E+10	1.59E-04	1.73E+02	1.00E+00	3.43E+05	.833	.805	.781	.884	.908	.925
1.0	.44	18.0	2.06E+10	5.93E-05	4.90E+01	2.83E-01	4.21E+05	.791	.757	.734	.849	.891	.895
2.0	.51	18.8	8.38E+09	2.53E-05	1.80E+01	1.04E-01	4.66E+05	.753	.716	.695	.815	.867	.865
3.0	.56	19.7	3.69E+09	1.17E-05	7.29E+00	4.21E-02	5.06E+05	.714	.675	.655	.779	.838	.831
4.0	.62	20.6	1.73E+09	5.72E-06	3.15E+00	1.82E-02	5.51E+05	.674	.633	.614	.742	.808	.795
5.0	.68	21.6	8.61E+08	2.97E-06	1.43E+00	8.28E-03	6.01E+05	.634	.591	.574	.704	.777	.759
6.0	.76	22.5	4.52E+08	1.63E-06	6.87E-01	3.97E-03	6.58E+05	.598	.553	.537	.669	.748	.726
7.0	.86	23.3	2.49E+08	9.32E-07	3.47E-01	2.00E-03	7.18E+05	.566	.521	.506	.639	.723	.696
8.0	.96	24.1	1.43E+08	5.52E-07	1.84E-01	1.06E-03	7.78E+05	.541	.494	.480	.614	.702	.672
X-ray tube voltage: 32 kV													
0.0	.33	17.0	7.35E+10	2.00E-04	2.06E+02	1.00E+00	3.57E+05	.816	.787	.764	.869	.897	.911
1.0	.46	18.4	2.65E+10	7.80E-05	6.08E+01	2.95E-01	4.35E+05	.768	.734	.712	.827	.873	.874
2.0	.53	19.4	1.11E+10	3.45E-05	2.29E+01	1.11E-01	4.85E+05	.723	.686	.665	.786	.841	.836
3.0	.58	20.5	5.05E+09	1.66E-05	9.48E+00	4.60E-02	5.33E+05	.676	.637	.618	.742	.805	.794
4.0	.65	21.6	2.46E+09	8.52E-06	4.19E+00	2.03E-02	5.88E+05	.630	.588	.571	.697	.768	.750
5.0	.73	22.7	1.28E+09	4.65E-06	1.96E+00	9.50E-03	6.53E+05	.586	.542	.527	.655	.732	.709
6.0	.83	23.8	7.02E+08	2.67E-06	9.70E-01	4.71E-03	7.24E+05	.547	.503	.489	.618	.701	.673
7.0	.96	24.7	4.05E+08	1.60E-06	5.09E-01	2.47E-03	7.97E+05	.516	.470	.458	.588	.675	.642
8.0	1.10	25.5	2.44E+08	9.93E-07	2.81E-01	1.36E-03	8.67E+05	.492	.445	.434	.564	.654	.619

Table IV.3.12

Characteristic X-ray beam quantities without scattered radiation and relative energy absorption in typical image detectors at variable absorber or phantom materials for mammographic X-ray beam qualities with the indicated anode-filter combination:

Anode Mo / 0.03 mm Mo   Reference focal distance 60 cm

Absorber material: PMMA

Thickness cm	HVL mm Al	<E> keV	Φ/It 1/As/cm²	Ψ/It J/As/cm²	Ka/It mGy/As	Relative dose	Φ/Ka photons /cm²/µGy	34 CaWO4	34 GOS	40 SPF	60 SPF	60 aSe	80 CsI

Relative energy absorption in the image detector (coverage mg/cm²)

X-ray tube voltage: 22 kV

0.0	.24	15.2	2.06E+10	5.00E-05	7.31E+01	1.00E+00	2.82E+05	.887	.867	.841	.928	.928	.961
1.0	.37	16.6	5.71E+09	1.52E-05	1.57E+01	2.14E-01	3.65E+05	.853	.825	.800	.904	.931	.945
2.0	.44	17.3	2.02E+09	5.60E-06	4.96E+00	6.78E-02	4.07E+05	.832	.800	.776	.889	.926	.934
3.0	.49	17.8	7.88E+08	2.24E-06	1.81E+00	2.48E-02	4.35E+05	.817	.783	.759	.878	.919	.926
4.0	.52	18.1	3.25E+08	9.41E-07	7.16E-01	9.79E-03	4.54E+05	.806	.770	.746	.869	.913	.919
5.0	.55	18.3	1.39E+08	4.08E-07	2.96E-01	4.05E-03	4.70E+05	.796	.759	.736	.861	.908	.913
6.0	.57	18.5	6.11E+07	1.81E-07	1.27E-01	1.73E-03	4.82E+05	.789	.750	.727	.855	.904	.908
7.0	.59	18.7	2.74E+07	8.19E-08	5.55E-02	7.60E-04	4.93E+05	.782	.743	.720	.849	.900	.904
8.0	.61	18.8	1.25E+07	3.76E-08	2.48E-02	3.40E-04	5.02E+05	.776	.736	.713	.844	.896	.899

X-ray tube voltage: 24 kV

0.0	.27	15.8	3.03E+10	7.67E-05	9.90E+01	1.00E+00	3.06E+05	.868	.844	.819	.914	.925	.951
1.0	.41	17.1	9.37E+09	2.57E-05	2.41E+01	2.43E-01	3.89E+05	.835	.804	.780	.890	.923	.935
2.0	.48	17.7	3.54E+09	1.01E-05	8.25E+00	8.33E-02	4.29E+05	.816	.781	.758	.876	.917	.924
3.0	.52	18.1	1.45E+09	4.21E-06	3.19E+00	3.22E-02	4.54E+05	.801	.765	.742	.864	.910	.914
4.0	.55	18.4	6.21E+08	1.83E-06	1.31E+00	1.32E-02	4.73E+05	.789	.751	.728	.854	.903	.906
5.0	.58	18.7	2.74E+08	8.22E-07	5.62E-01	5.67E-03	4.89E+05	.778	.740	.717	.845	.896	.899
6.0	.60	18.9	1.24E+08	3.77E-07	2.48E-01	2.50E-03	5.02E+05	.768	.729	.706	.837	.890	.892
7.0	.62	19.1	5.76E+07	1.77E-07	1.12E-01	1.13E-03	5.15E+05	.759	.718	.696	.828	.884	.884
8.0	.64	19.4	2.72E+07	8.43E-08	5.15E-02	5.20E-04	5.28E+05	.749	.708	.686	.820	.878	.877

X-ray tube voltage: 26 kV

0.0	.29	16.2	4.21E+10	1.09E-04	1.29E+02	1.00E+00	3.26E+05	.852	.826	.801	.902	.920	.941
1.0	.43	17.5	1.39E+10	3.90E-05	3.42E+01	2.65E-01	4.08E+05	.820	.787	.763	.877	.915	.923
2.0	.50	18.1	5.50E+09	1.59E-05	1.23E+01	9.55E-02	4.46E+05	.798	.763	.739	.860	.905	.910
3.0	.54	18.5	2.33E+09	6.91E-06	4.93E+00	3.82E-02	4.72E+05	.781	.743	.721	.846	.895	.898
4.0	.58	18.9	1.03E+09	3.12E-06	2.09E+00	1.62E-02	4.94E+05	.764	.725	.703	.831	.885	.885
5.0	.61	19.3	4.72E+08	1.46E-06	9.19E-01	7.13E-03	5.14E+05	.748	.708	.686	.817	.874	.872
6.0	.64	19.6	2.23E+08	7.00E-07	4.17E-01	3.24E-03	5.34E+05	.732	.691	.669	.803	.863	.859
7.0	.67	20.0	1.08E+08	3.46E-07	1.95E-01	1.51E-03	5.54E+05	.716	.673	.653	.788	.851	.846
8.0	.70	20.4	5.36E+07	1.75E-07	9.31E-02	7.22E-04	5.75E+05	.699	.656	.636	.773	.839	.832

X-ray tube voltage: 28 kV

0.0	.31	16.6	5.58E+10	1.48E-04	1.63E+02	1.00E+00	3.43E+05	.838	.809	.785	.890	.913	.931
1.0	.45	17.8	1.94E+10	5.55E-05	4.60E+01	2.82E-01	4.23E+05	.804	.769	.746	.863	.904	.910
2.0	.52	18.5	7.92E+09	2.34E-05	1.71E+01	1.05E-01	4.62E+05	.779	.742	.719	.842	.891	.893
3.0	.56	19.0	3.46E+09	1.05E-05	7.04E+00	4.32E-02	4.91E+05	.756	.717	.695	.822	.876	.875
4.0	.60	19.5	1.58E+09	4.94E-06	3.05E+00	1.87E-02	5.18E+05	.733	.692	.671	.801	.859	.856
5.0	.64	20.0	7.51E+08	2.41E-06	1.38E+00	8.45E-03	5.45E+05	.709	.667	.647	.779	.842	.835
6.0	.68	20.6	3.70E+08	1.22E-06	6.43E-01	3.95E-03	5.75E+05	.685	.642	.623	.757	.824	.814
7.0	.73	21.1	1.88E+08	6.37E-07	3.10E-01	1.91E-03	6.06E+05	.661	.617	.599	.734	.806	.793
8.0	.78	21.7	9.89E+07	3.43E-07	1.55E-01	9.50E-04	6.40E+05	.639	.593	.576	.713	.789	.772

X-ray tube voltage: 30 kV

0.0	.33	17.0	7.02E+10	1.91E-04	1.97E+02	1.00E+00	3.56E+05	.822	.792	.768	.876	.904	.919
1.0	.47	18.2	2.54E+10	7.43E-05	5.82E+01	2.95E-01	4.37E+05	.783	.748	.725	.843	.888	.892
2.0	.54	19.0	1.07E+10	3.26E-05	2.23E+01	1.13E-01	4.80E+05	.751	.713	.692	.815	.869	.867
3.0	.59	19.7	4.82E+09	1.52E-05	9.35E+00	4.74E-02	5.16E+05	.720	.680	.660	.787	.846	.840
4.0	.64	20.4	2.29E+09	7.49E-06	4.15E+00	2.10E-02	5.52E+05	.688	.646	.627	.757	.822	.812
5.0	.69	21.2	1.14E+09	3.86E-06	1.93E+00	9.76E-03	5.92E+05	.656	.613	.595	.727	.797	.783
6.0	.75	21.9	5.91E+08	2.08E-06	9.32E-01	4.72E-03	6.35E+05	.624	.580	.563	.697	.772	.753
7.0	.82	22.6	3.19E+08	1.16E-06	4.69E-01	2.38E-03	6.82E+05	.595	.550	.534	.668	.748	.726
8.0	.90	23.3	1.79E+08	6.69E-07	2.45E-01	1.24E-03	7.30E+05	.569	.523	.508	.643	.726	.701

X-ray tube voltage: 32 kV

0.0	.35	17.3	8.70E+10	2.41E-04	2.36E+02	1.00E+00	3.69E+05	.807	.776	.753	.862	.894	.906
1.0	.48	18.6	3.26E+10	9.71E-05	7.24E+01	3.07E-01	4.50E+05	.764	.728	.706	.826	.873	.874
2.0	.55	19.5	1.41E+10	4.38E-05	2.83E+01	1.20E-01	4.96E+05	.727	.688	.667	.791	.847	.843
3.0	.61	20.3	6.51E+09	2.12E-05	1.21E+01	5.13E-02	5.38E+05	.688	.648	.629	.756	.818	.809
4.0	.66	21.2	3.18E+09	1.08E-05	5.46E+00	2.32E-02	5.83E+05	.650	.608	.590	.719	.787	.773
5.0	.73	22.1	1.64E+09	5.81E-06	2.59E+00	1.10E-02	6.33E+05	.612	.568	.552	.682	.756	.737
6.0	.80	23.1	8.85E+08	3.27E-06	1.28E+00	5.45E-03	6.89E+05	.576	.531	.516	.647	.727	.702
7.0	.90	23.9	4.99E+08	1.91E-06	6.66E-01	2.83E-03	7.49E+05	.544	.499	.485	.616	.700	.672
8.0	1.00	24.7	2.92E+08	1.16E-06	3.61E-01	1.53E-03	8.09E+05	.518	.471	.459	.590	.677	.645

Table IV.3.13

Characteristic X-ray beam quantities without scattered radiation and relative energy absorption in typical image detectors at variable absorber or phantom materials for mammographic X-ray beam qualities with the indicated anode-filter combination:

Anode Mo / 0.025 mm Rh   Reference focal distance 60 cm

Absorber material: PMMA

Thick-ness cm	HVL mm Al	\<E\> keV	Φ/It 1/As/cm²	Ψ/It J/As/cm²	Ka/It mGy/As	Relative dose	Φ/Ka photons /cm²/µGy	34 CaWO4	34 GOS	40 SPF	60 SPF	60 aSe	80 CsI
\multicolumn{14}{l}{X-ray tube voltage: 24 kV}
0.0	.31	16.6	2.69E+10	7.16E-05	7.86E+01	1.00E+00	3.43E+05	.837	.809	.784	.889	.913	.931
1.0	.45	17.9	9.35E+09	2.67E-05	2.21E+01	2.81E-01	4.23E+05	.802	.767	.744	.862	.904	.911
2.0	.53	18.5	3.83E+09	1.14E-05	8.20E+00	1.04E-01	4.67E+05	.779	.741	.719	.844	.894	.896
3.0	.58	19.0	1.69E+09	5.13E-06	3.40E+00	4.32E-02	4.97E+05	.761	.721	.699	.829	.883	.883
4.0	.62	19.4	7.77E+08	2.41E-06	1.49E+00	1.90E-02	5.21E+05	.746	.705	.683	.816	.874	.873
5.0	.66	19.7	3.70E+08	1.17E-06	6.85E-01	8.71E-03	5.41E+05	.733	.691	.669	.805	.866	.863
6.0	.69	19.9	1.81E+08	5.79E-07	3.24E-01	4.12E-03	5.59E+05	.721	.678	.657	.795	.858	.854
7.0	.72	20.2	9.04E+07	2.93E-07	1.57E-01	2.00E-03	5.76E+05	.711	.667	.646	.786	.851	.846
8.0	.74	20.4	4.60E+07	1.50E-07	7.78E-02	9.91E-04	5.91E+05	.702	.657	.637	.777	.845	.838
\multicolumn{14}{l}{X-ray tube voltage: 26 kV}
0.0	.34	17.1	3.88E+10	1.06E-04	1.06E+02	1.00E+00	3.67E+05	.818	.787	.763	.874	.905	.919
1.0	.48	18.3	1.44E+10	4.22E-05	3.24E+01	3.06E-01	4.45E+05	.784	.748	.725	.847	.894	.898
2.0	.56	18.9	6.15E+09	1.86E-05	1.26E+01	1.19E-01	4.87E+05	.761	.722	.700	.828	.881	.881
3.0	.61	19.4	2.80E+09	8.71E-06	5.41E+00	5.11E-02	5.18E+05	.742	.701	.679	.811	.870	.867
4.0	.65	19.8	1.33E+09	4.22E-06	2.45E+00	2.32E-02	5.43E+05	.725	.683	.662	.797	.859	.855
5.0	.69	20.1	6.53E+08	2.11E-06	1.15E+00	1.09E-02	5.66E+05	.710	.667	.647	.784	.849	.843
6.0	.73	20.5	3.29E+08	1.08E-06	5.61E-01	5.30E-03	5.87E+05	.697	.653	.633	.772	.840	.832
7.0	.76	20.7	1.69E+08	5.63E-07	2.79E-01	2.64E-03	6.06E+05	.685	.640	.620	.761	.832	.822
8.0	.79	21.0	8.88E+07	2.99E-07	1.42E-01	1.34E-03	6.24E+05	.675	.629	.609	.751	.824	.813
\multicolumn{14}{l}{X-ray tube voltage: 28 kV}
0.0	.37	17.4	5.22E+10	1.46E-04	1.36E+02	1.00E+00	3.83E+05	.806	.774	.750	.864	.899	.910
1.0	.50	18.5	2.01E+10	5.97E-05	4.39E+01	3.22E-01	4.58E+05	.772	.735	.713	.836	.886	.888
2.0	.57	19.2	8.79E+09	2.70E-05	1.76E+01	1.29E-01	5.00E+05	.748	.708	.686	.815	.872	.870
3.0	.62	19.7	4.08E+09	1.29E-05	7.68E+00	5.64E-02	5.31E+05	.726	.685	.664	.797	.858	.853
4.0	.67	20.2	1.98E+09	6.38E-06	3.53E+00	2.60E-02	5.59E+05	.707	.664	.644	.779	.844	.838
5.0	.71	20.6	9.88E+08	3.26E-06	1.69E+00	1.24E-02	5.85E+05	.689	.645	.626	.763	.832	.823
6.0	.76	21.0	5.09E+08	1.71E-06	8.33E-01	6.12E-03	6.10E+05	.673	.628	.609	.748	.820	.808
7.0	.80	21.3	2.68E+08	9.16E-07	4.22E-01	3.10E-03	6.34E+05	.658	.612	.593	.734	.808	.795
8.0	.84	21.7	1.44E+08	5.01E-07	2.19E-01	1.61E-03	6.58E+05	.644	.597	.579	.720	.797	.782
\multicolumn{14}{l}{X-ray tube voltage: 30 kV}
0.0	.38	17.7	6.61E+10	1.88E-04	1.67E+02	1.00E+00	3.96E+05	.793	.760	.737	.852	.891	.899
1.0	.52	18.8	2.63E+10	7.92E-05	5.58E+01	3.34E-01	4.71E+05	.757	.719	.697	.822	.874	.874
2.0	.59	19.5	1.17E+10	3.67E-05	2.28E+01	1.37E-01	5.14E+05	.729	.689	.668	.797	.856	.852
3.0	.64	20.2	5.56E+09	1.80E-05	1.01E+01	6.06E-02	5.49E+05	.703	.661	.641	.774	.838	.831
4.0	.69	20.7	2.75E+09	9.14E-06	4.73E+00	2.83E-02	5.82E+05	.679	.636	.617	.752	.820	.810
5.0	.75	21.3	1.41E+09	4.82E-06	2.30E+00	1.38E-02	6.15E+05	.656	.612	.594	.730	.803	.789
6.0	.80	21.8	7.50E+08	2.62E-06	1.16E+00	6.93E-03	6.48E+05	.635	.589	.572	.710	.786	.769
7.0	.85	22.3	4.09E+08	1.46E-06	6.01E-01	3.60E-03	6.81E+05	.615	.568	.552	.690	.769	.750
8.0	.91	22.8	2.29E+08	8.33E-07	3.21E-01	1.92E-03	7.13E+05	.596	.549	.533	.672	.754	.732
\multicolumn{14}{l}{X-ray tube voltage: 32 kV}
0.0	.40	18.0	8.20E+10	2.36E-04	2.01E+02	1.00E+00	4.07E+05	.782	.748	.726	.842	.883	.890
1.0	.53	19.1	3.33E+10	1.02E-04	6.92E+01	3.44E-01	4.81E+05	.744	.706	.684	.809	.863	.861
2.0	.60	19.9	1.51E+10	4.81E-05	2.87E+01	1.43E-01	5.25E+05	.712	.672	.651	.781	.841	.835
3.0	.66	20.6	7.27E+09	2.40E-05	1.29E+01	6.41E-02	5.64E+05	.682	.640	.621	.752	.819	.808
4.0	.71	21.3	3.67E+09	1.25E-05	6.08E+00	3.02E-02	6.03E+05	.653	.609	.591	.725	.795	.781
5.0	.77	22.0	1.93E+09	6.77E-06	2.99E+00	1.49E-02	6.43E+05	.625	.580	.563	.697	.773	.755
6.0	.84	22.6	1.05E+09	3.79E-06	1.53E+00	7.60E-03	6.85E+05	.598	.552	.536	.672	.751	.729
7.0	.90	23.3	5.87E+08	2.19E-06	8.07E-01	4.01E-03	7.27E+05	.574	.527	.512	.647	.730	.705
8.0	.98	23.9	3.39E+08	1.29E-06	4.40E-01	2.19E-03	7.70E+05	.551	.505	.491	.625	.710	.683
\multicolumn{14}{l}{X-ray tube voltage: 34 kV}
0.0	.41	18.3	9.85E+10	2.88E-04	2.36E+02	1.00E+00	4.18E+05	.769	.734	.712	.829	.873	.877
1.0	.54	19.4	4.09E+10	1.27E-04	8.31E+01	3.52E-01	4.92E+05	.726	.688	.667	.792	.847	.844
2.0	.61	20.3	1.89E+10	6.15E-05	3.49E+01	1.48E-01	5.40E+05	.689	.648	.629	.757	.820	.811
3.0	.67	21.2	9.28E+09	3.15E-05	1.58E+01	6.72E-02	5.85E+05	.653	.610	.592	.722	.791	.778
4.0	.74	22.1	4.79E+09	1.69E-05	7.57E+00	3.21E-02	6.33E+05	.617	.573	.557	.688	.761	.744
5.0	.81	22.9	2.58E+09	9.48E-06	3.78E+00	1.60E-02	6.83E+05	.583	.539	.524	.655	.733	.712
6.0	.89	23.7	1.45E+09	5.51E-06	1.97E+00	8.34E-03	7.36E+05	.552	.507	.493	.624	.706	.681
7.0	.98	24.5	8.41E+08	3.31E-06	1.06E+00	4.51E-03	7.91E+05	.525	.479	.466	.596	.681	.653
8.0	1.07	25.3	5.03E+08	2.04E-06	5.95E-01	2.52E-03	8.46E+05	.500	.454	.443	.572	.659	.628

Table IV.3.14

Characteristic X-ray beam quantities without scattered radiation and relative energy absorption in typical image detectors at variable absorber or phantom materials for mammographic X-ray beam qualities with the indicated anode-filter combination:

Anode Rh / 0.025 mm Rh   Reference focal distance   60 cm

Absorber material:			PMMA				Φ/K_a	Relative energy absorption in the image detector (coverage mg/cm²)					
Thickness cm	HVL mm Al	\<E\> keV	Φ/It 1/As/cm²	Ψ/It J/As/cm²	K_a/It mGy/As	Relative dose	photons /cm²/µGy	34 CaWO4	34 GOS	40 SPF	60 SPF	60 aSe	80 CsI

X-ray tube voltage: 24 kV

0.0	.30	16.7	1.92E+10	5.13E-05	5.66E+01	1.00E+00	3.39E+05	.826	.798	.774	.879	.904	.921
1.0	.45	18.2	6.67E+09	1.94E-05	1.54E+01	2.72E-01	4.33E+05	.783	.747	.725	.845	.891	.896
2.0	.55	19.0	2.80E+09	8.52E-06	5.76E+00	1.02E-01	4.87E+05	.756	.716	.694	.823	.878	.877
3.0	.62	19.5	1.28E+09	4.01E-06	2.44E+00	4.32E-02	5.24E+05	.736	.694	.673	.806	.866	.863
4.0	.67	19.9	6.16E+08	1.96E-06	1.11E+00	1.97E-02	5.53E+05	.720	.677	.656	.793	.857	.852
5.0	.71	20.2	3.06E+08	9.91E-07	5.31E-01	9.39E-03	5.76E+05	.708	.664	.643	.782	.849	.843
6.0	.75	20.5	1.56E+08	5.11E-07	2.62E-01	4.63E-03	5.94E+05	.698	.653	.633	.773	.842	.834
7.0	.78	20.7	8.07E+07	2.68E-07	1.32E-01	2.34E-03	6.10E+05	.689	.643	.624	.766	.836	.828
8.0	.80	20.9	4.25E+07	1.42E-07	6.81E-02	1.20E-03	6.23E+05	.682	.636	.616	.759	.831	.821

X-ray tube voltage: 26 kV

0.0	.33	17.4	2.67E+10	7.42E-05	7.27E+01	1.00E+00	3.67E+05	.798	.766	.743	.855	.890	.901
1.0	.50	18.8	1.01E+10	3.04E-05	2.16E+01	2.97E-01	4.66E+05	.755	.716	.695	.821	.874	.874
2.0	.60	19.6	4.50E+09	1.41E-05	8.62E+00	1.18E-01	5.22E+05	.728	.686	.666	.799	.859	.855
3.0	.67	20.1	2.16E+09	6.97E-06	3.86E+00	5.31E-02	5.60E+05	.709	.665	.645	.782	.848	.841
4.0	.73	20.5	1.08E+09	3.56E-06	1.84E+00	2.53E-02	5.88E+05	.694	.650	.630	.769	.838	.830
5.0	.77	20.8	5.60E+08	1.87E-06	9.16E-01	1.26E-02	6.11E+05	.682	.637	.617	.759	.830	.820
6.0	.81	21.1	2.95E+08	9.95E-07	4.69E-01	6.44E-03	6.29E+05	.672	.626	.607	.750	.823	.812
7.0	.84	21.3	1.58E+08	5.39E-07	2.45E-01	3.37E-03	6.45E+05	.664	.617	.598	.742	.817	.805
8.0	.86	21.5	8.58E+07	2.95E-07	1.30E-01	1.79E-03	6.59E+05	.656	.609	.591	.735	.811	.798

X-ray tube voltage: 28 kV

0.0	.35	17.9	3.58E+10	1.03E-04	9.19E+01	1.00E+00	3.89E+05	.776	.741	.719	.836	.877	.884
1.0	.53	19.4	1.43E+10	4.43E-05	2.90E+01	3.16E-01	4.92E+05	.732	.692	.671	.801	.858	.855
2.0	.64	20.1	6.64E+09	2.14E-05	1.21E+01	1.32E-01	5.49E+05	.705	.663	.643	.778	.843	.836
3.0	.71	20.6	3.30E+09	1.09E-05	5.62E+00	6.11E-02	5.88E+05	.686	.641	.622	.760	.830	.820
4.0	.77	21.0	1.71E+09	5.76E-06	2.76E+00	3.01E-02	6.18E+05	.670	.624	.606	.746	.818	.807
5.0	.82	21.4	9.07E+08	3.11E-06	1.41E+00	1.54E-02	6.42E+05	.657	.610	.592	.734	.809	.795
6.0	.85	21.7	4.91E+08	1.71E-06	7.41E-01	8.07E-03	6.63E+05	.645	.598	.580	.722	.800	.785
7.0	.89	21.9	2.71E+08	9.51E-07	3.97E-01	4.32E-03	6.81E+05	.634	.587	.569	.712	.791	.775
8.0	.92	22.2	1.51E+08	5.37E-07	2.16E-01	2.36E-03	6.99E+05	.624	.577	.560	.703	.784	.766

X-ray tube voltage: 30 kV

0.0	.38	18.4	4.56E+10	1.35E-04	1.11E+02	1.00E+00	4.12E+05	.754	.718	.696	.817	.863	.867
1.0	.57	19.8	1.91E+10	6.08E-05	3.70E+01	3.34E-01	5.17E+05	.711	.669	.649	.780	.842	.836
2.0	.68	20.6	9.21E+09	3.04E-05	1.60E+01	1.45E-01	5.75E+05	.683	.639	.620	.756	.824	.814
3.0	.75	21.2	4.71E+09	1.60E-05	7.65E+00	6.91E-02	6.16E+05	.661	.616	.598	.736	.809	.796
4.0	.81	21.6	2.50E+09	8.66E-06	3.86E+00	3.49E-02	6.48E+05	.644	.598	.580	.719	.795	.780
5.0	.86	22.0	1.36E+09	4.81E-06	2.02E+00	1.82E-02	6.75E+05	.628	.581	.564	.704	.783	.765
6.0	.90	22.4	7.58E+08	2.72E-06	1.08E+00	9.78E-03	7.00E+05	.613	.566	.549	.690	.770	.751
7.0	.95	22.7	4.29E+08	1.56E-06	5.93E-01	5.35E-03	7.24E+05	.599	.552	.535	.676	.759	.737
8.0	.99	23.1	2.47E+08	9.11E-07	3.30E-01	2.98E-03	7.47E+05	.586	.538	.522	.663	.747	.724

X-ray tube voltage: 32 kV

0.0	.40	18.9	5.61E+10	1.70E-04	1.30E+02	1.00E+00	4.32E+05	.734	.696	.675	.798	.849	.849
1.0	.60	20.3	2.45E+10	7.97E-05	4.54E+01	3.50E-01	5.39E+05	.690	.647	.628	.760	.825	.816
2.0	.71	21.1	1.21E+10	4.10E-05	2.02E+01	1.56E-01	5.99E+05	.660	.616	.597	.733	.804	.791
3.0	.79	21.7	6.34E+09	2.21E-05	9.88E+00	7.61E-02	6.42E+05	.636	.591	.573	.711	.786	.770
4.0	.85	22.3	3.45E+09	1.23E-05	5.08E+00	3.92E-02	6.78E+05	.615	.569	.552	.690	.768	.750
5.0	.90	22.8	1.92E+09	7.01E-06	2.71E+00	2.08E-02	7.10E+05	.596	.549	.533	.671	.752	.731
6.0	.95	23.2	1.10E+09	4.08E-06	1.48E+00	1.14E-02	7.42E+05	.577	.530	.515	.653	.736	.712
7.0	1.01	23.7	6.38E+08	2.42E-06	8.26E-01	6.36E-03	7.72E+05	.560	.513	.498	.635	.720	.694
8.0	1.06	24.1	3.77E+08	1.46E-06	4.70E-01	3.62E-03	8.03E+05	.543	.496	.482	.618	.705	.677

X-ray tube voltage: 34 kV

0.0	.43	19.3	6.79E+10	2.11E-04	1.50E+02	1.00E+00	4.52E+05	.715	.676	.656	.780	.834	.833
1.0	.62	20.8	3.06E+10	1.02E-04	5.47E+01	3.63E-01	5.60E+05	.670	.627	.608	.740	.807	.797
2.0	.74	21.6	1.55E+10	5.37E-05	2.50E+01	1.66E-01	6.21E+05	.638	.594	.576	.711	.784	.769
3.0	.82	22.3	8.29E+09	2.96E-05	1.24E+01	8.26E-02	6.67E+05	.612	.566	.550	.686	.762	.744
4.0	.88	22.9	4.59E+09	1.69E-05	6.49E+00	4.32E-02	7.07E+05	.588	.542	.526	.662	.742	.721
5.0	.94	23.5	2.62E+09	9.86E-06	3.51E+00	2.33E-02	7.45E+05	.565	.519	.504	.639	.721	.698
6.0	1.00	24.1	1.53E+09	5.89E-06	1.95E+00	1.30E-02	7.83E+05	.544	.498	.484	.618	.702	.676
7.0	1.06	24.6	9.08E+08	3.58E-06	1.11E+00	7.35E-03	8.21E+05	.524	.477	.464	.597	.683	.655
8.0	1.13	25.2	5.50E+08	2.22E-06	6.40E-01	4.26E-03	8.59E+05	.505	.458	.446	.578	.665	.635

Table IV.3.15

Characteristic X-ray beam quantities without scattered radiation and relative energy absorption in typical image detectors at variable absorber or phantom materials for mammographic X-ray beam qualities with the indicated anode-filter combination:

Anode W / 0.08 mm Nb

Reference focal distance 60 cm

Absorber material: PMMA

Thickness cm	HVL mm Al	<E> keV	Φ/It 1/As/cm²	Ψ/It J/As/cm²	Ka/It mGy/As	Relative dose	Φ/Ka photons /cm²/µGy	34 CaWO4	34 GOS	40 SPF	60 SPF	60 aSe	80 CsI
X-ray tube voltage: 22 kV													
0.0	.31	15.8	9.69E+09	2.45E-05	3.00E+01	1.00E+00	3.23E+05	.882	.859	.833	.926	.940	.962
1.0	.39	16.6	3.02E+09	8.04E-06	8.10E+00	2.70E-01	3.73E+05	.862	.833	.809	.912	.940	.952
2.0	.44	17.0	1.08E+09	2.94E-06	2.70E+00	8.98E-02	3.99E+05	.848	.818	.793	.903	.936	.946
3.0	.47	17.3	4.09E+08	1.14E-06	9.80E-01	3.26E-02	4.17E+05	.838	.806	.782	.895	.932	.941
4.0	.49	17.6	1.62E+08	4.54E-07	3.75E-01	1.25E-02	4.30E+05	.830	.797	.773	.889	.928	.936
5.0	.51	17.7	6.56E+07	1.86E-07	1.49E-01	4.96E-03	4.41E+05	.824	.790	.765	.884	.925	.933
6.0	.53	17.9	2.72E+07	7.78E-08	6.05E-02	2.01E-03	4.49E+05	.818	.783	.759	.880	.922	.929
7.0	.54	18.0	1.15E+07	3.30E-08	2.51E-02	8.36E-04	4.57E+05	.813	.778	.754	.876	.919	.926
8.0	.55	18.1	4.89E+06	1.42E-08	1.06E-02	3.52E-04	4.63E+05	.809	.773	.749	.872	.917	.923
X-ray tube voltage: 24 kV													
0.0	.32	16.0	1.24E+10	3.16E-05	3.75E+01	1.00E+00	3.30E+05	.875	.851	.826	.921	.938	.958
1.0	.40	16.8	3.97E+09	1.07E-05	1.04E+01	2.78E-01	3.81E+05	.854	.825	.800	.906	.936	.948
2.0	.45	17.2	1.45E+09	3.99E-06	3.55E+00	9.45E-02	4.08E+05	.840	.808	.784	.895	.931	.940
3.0	.48	17.5	5.61E+08	1.58E-06	1.31E+00	3.50E-02	4.27E+05	.828	.794	.770	.886	.925	.933
4.0	.51	17.8	2.26E+08	6.46E-07	5.13E-01	1.37E-02	4.41E+05	.817	.783	.759	.878	.920	.926
5.0	.53	18.0	9.42E+07	2.72E-07	2.07E-01	5.52E-03	4.54E+05	.807	.772	.748	.870	.914	.919
6.0	.55	18.3	4.02E+07	1.18E-07	8.61E-02	2.30E-03	4.66E+05	.797	.761	.737	.861	.908	.912
7.0	.56	18.5	1.75E+07	5.18E-08	3.66E-02	9.76E-04	4.78E+05	.787	.749	.726	.852	.901	.905
8.0	.58	18.7	7.78E+06	2.34E-08	1.59E-02	4.23E-04	4.90E+05	.776	.737	.715	.843	.894	.896
X-ray tube voltage: 26 kV													
0.0	.32	16.2	1.51E+10	3.91E-05	4.48E+01	1.00E+00	3.37E+05	.866	.840	.815	.913	.933	.951
1.0	.41	17.0	4.97E+09	1.36E-05	1.27E+01	2.84E-01	3.90E+05	.841	.810	.786	.894	.927	.937
2.0	.46	17.6	1.86E+09	5.25E-06	4.43E+00	9.88E-02	4.21E+05	.820	.787	.764	.878	.918	.924
3.0	.50	18.0	7.46E+08	2.15E-06	1.68E+00	3.75E-02	4.44E+05	.801	.766	.743	.862	.907	.911
4.0	.53	18.5	3.13E+08	9.27E-07	6.73E-01	1.50E-02	4.66E+05	.782	.745	.722	.845	.894	.896
5.0	.56	18.9	1.37E+08	4.15E-07	2.81E-01	6.26E-03	4.88E+05	.761	.722	.700	.826	.880	.879
6.0	.60	19.4	6.22E+07	1.93E-07	1.22E-01	2.71E-03	5.12E+05	.739	.699	.677	.806	.864	.861
7.0	.63	19.9	2.93E+07	9.34E-08	5.44E-02	1.21E-03	5.38E+05	.715	.674	.653	.785	.848	.841
8.0	.67	20.4	1.43E+07	4.68E-08	2.52E-02	5.62E-04	5.67E+05	.692	.649	.629	.763	.830	.821
X-ray tube voltage: 28 kV													
0.0	.33	16.4	1.79E+10	4.72E-05	5.19E+01	1.00E+00	3.45E+05	.850	.824	.799	.899	.923	.939
1.0	.43	17.5	6.09E+09	1.70E-05	1.51E+01	2.91E-01	4.04E+05	.817	.785	.761	.872	.910	.916
2.0	.48	18.2	2.37E+09	6.90E-06	5.37E+00	1.03E-01	4.41E+05	.786	.751	.728	.846	.892	.893
3.0	.53	18.9	9.94E+08	3.01E-06	2.09E+00	4.03E-02	4.75E+05	.754	.716	.695	.817	.870	.868
4.0	.57	19.6	4.44E+08	1.40E-06	8.69E-01	1.67E-02	5.10E+05	.720	.680	.660	.786	.846	.839
5.0	.62	20.4	2.09E+08	6.84E-07	3.80E-01	7.32E-03	5.50E+05	.685	.644	.625	.754	.820	.809
6.0	.68	21.2	1.04E+08	3.53E-07	1.75E-01	3.36E-03	5.94E+05	.652	.608	.591	.723	.795	.779
7.0	.76	22.0	5.41E+07	1.90E-07	8.42E-02	1.62E-03	6.42E+05	.621	.576	.559	.694	.771	.751
8.0	.84	22.7	2.94E+07	1.07E-07	4.25E-02	8.18E-04	6.92E+05	.594	.547	.532	.668	.750	.727
X-ray tube voltage: 30 kV													
0.0	.34	16.9	2.13E+10	5.77E-05	5.95E+01	1.00E+00	3.57E+05	.822	.793	.770	.873	.902	.913
1.0	.44	18.3	7.57E+09	2.21E-05	1.77E+01	2.98E-01	4.26E+05	.772	.738	.716	.830	.875	.876
2.0	.51	19.4	3.12E+09	9.67E-06	6.51E+00	1.09E-01	4.78E+05	.724	.686	.666	.786	.842	.836
3.0	.58	20.5	1.41E+09	4.62E-06	2.65E+00	4.45E-02	5.32E+05	.675	.635	.617	.741	.806	.793
4.0	.65	21.6	6.87E+08	2.38E-06	1.16E+00	1.95E-02	5.92E+05	.629	.587	.570	.698	.770	.751
5.0	.74	22.7	3.58E+08	1.30E-06	5.44E-01	9.14E-03	6.58E+05	.588	.543	.528	.658	.737	.713
6.0	.85	23.7	1.98E+08	7.49E-07	2.71E-01	4.56E-03	7.28E+05	.553	.507	.493	.624	.708	.680
7.0	.98	24.5	1.14E+08	4.48E-07	1.43E-01	2.41E-03	7.97E+05	.525	.478	.465	.597	.686	.654
8.0	1.11	25.1	6.86E+07	2.77E-07	7.98E-02	1.34E-03	8.60E+05	.503	.456	.444	.577	.668	.633
X-ray tube voltage: 32 kV													
0.0	.35	17.5	2.46E+10	6.89E-05	6.65E+01	1.00E+00	3.70E+05	.790	.761	.738	.843	.878	.885
1.0	.46	19.1	9.15E+09	2.80E-05	2.03E+01	3.05E-01	4.51E+05	.727	.691	.671	.786	.838	.833
2.0	.54	20.5	3.98E+09	1.31E-05	7.66E+00	1.15E-01	5.19E+05	.666	.627	.609	.730	.793	.779
3.0	.63	22.0	1.91E+09	6.74E-06	3.23E+00	4.86E-02	5.92E+05	.610	.568	.552	.676	.748	.728
4.0	.74	23.3	1.00E+09	3.74E-06	1.49E+00	2.23E-02	6.74E+05	.561	.517	.503	.629	.708	.682
5.0	.87	24.5	5.60E+08	2.20E-06	7.37E-01	1.11E-02	7.60E+05	.521	.476	.463	.591	.675	.644
6.0	1.02	25.4	3.30E+08	1.35E-06	3.91E-01	5.88E-03	8.45E+05	.490	.445	.433	.561	.650	.614
7.0	1.19	26.2	2.03E+08	8.52E-07	2.20E-01	3.31E-03	9.21E+05	.468	.422	.411	.539	.630	.592
8.0	1.34	26.8	1.28E+08	5.51E-07	1.30E-01	1.96E-03	9.86E+05	.451	.405	.395	.522	.616	.576

Table IV.3.16

Characteristic X-ray beam quantities without scattered radiation and relative energy absorption in typical image detectors at variable absorber or phantom materials for mammographic X-ray beam qualities with the indicated anode-filter combination:

Anode W / 0.06 mm Mo   Reference focal distance 60 cm

Absorber material: PMMA

Thick-ness cm	HVL mm Al	\<E\> keV	Φ/It 1/As/cm²	Ψ/It J/As/cm²	Ka/It mGy/As	Relative dose	Φ/Ka photons /cm²/µGy	34 CaWO4	34 GOS	40 SPF	60 SPF	60 aSe	80 CsI
							Relative energy absorption in the image detector (coverage mg/cm²)						

X-ray tube voltage: 22 kV

0.0	.32	16.2	1.21E+10	3.15E-05	3.61E+01	1.00E+00	3.36E+05	.862	.836	.811	.911	.931	.950
1.0	.42	17.1	4.03E+09	1.11E-05	1.02E+01	2.82E-01	3.96E+05	.838	.806	.782	.893	.927	.937
2.0	.48	17.6	1.53E+09	4.34E-06	3.59E+00	9.95E-02	4.28E+05	.821	.787	.763	.881	.921	.928
3.0	.52	18.0	6.25E+08	1.80E-06	1.39E+00	3.85E-02	4.49E+05	.809	.773	.750	.871	.915	.921
4.0	.55	18.2	2.65E+08	7.74E-07	5.69E-01	1.58E-02	4.66E+05	.800	.763	.739	.864	.910	.916
5.0	.57	18.4	1.16E+08	3.41E-07	2.42E-01	6.70E-03	4.78E+05	.792	.754	.731	.858	.906	.911
6.0	.59	18.6	5.15E+07	1.53E-07	1.05E-01	2.92E-03	4.89E+05	.786	.747	.724	.853	.903	.907
7.0	.60	18.7	2.33E+07	6.99E-08	4.69E-02	1.30E-03	4.97E+05	.781	.741	.718	.849	.900	.903
8.0	.62	18.8	1.07E+07	3.23E-08	2.12E-02	5.87E-04	5.05E+05	.776	.736	.713	.845	.897	.900

X-ray tube voltage: 24 kV

0.0	.33	16.5	1.60E+10	4.20E-05	4.59E+01	1.00E+00	3.47E+05	.853	.826	.801	.904	.927	.944
1.0	.44	17.4	5.49E+09	1.53E-05	1.35E+01	2.93E-01	4.07E+05	.828	.795	.771	.885	.922	.931
2.0	.49	17.9	2.15E+09	6.16E-06	4.90E+00	1.07E-01	4.39E+05	.811	.776	.753	.873	.915	.922
3.0	.53	18.2	8.97E+08	2.62E-06	1.95E+00	4.24E-02	4.61E+05	.799	.762	.739	.863	.909	.914
4.0	.56	18.5	3.89E+08	1.15E-06	8.14E-01	1.77E-02	4.77E+05	.789	.751	.728	.855	.904	.908
5.0	.59	18.7	1.73E+08	5.18E-07	3.53E-01	7.67E-03	4.91E+05	.781	.742	.719	.848	.899	.902
6.0	.61	18.9	7.86E+07	2.38E-07	1.57E-01	3.41E-03	5.02E+05	.773	.733	.711	.842	.894	.897
7.0	.62	19.0	3.64E+07	1.11E-07	7.10E-02	1.55E-03	5.12E+05	.766	.726	.703	.836	.890	.892
8.0	.64	19.2	1.71E+07	5.24E-08	3.27E-02	7.13E-04	5.21E+05	.760	.719	.697	.830	.886	.887

X-ray tube voltage: 26 kV

0.0	.34	16.7	1.97E+10	5.27E-05	5.55E+01	1.00E+00	3.56E+05	.843	.815	.790	.896	.922	.938
1.0	.45	17.6	6.98E+09	1.97E-05	1.67E+01	3.02E-01	4.18E+05	.817	.783	.759	.875	.915	.922
2.0	.51	18.1	2.80E+09	8.15E-06	6.22E+00	1.12E-01	4.51E+05	.798	.761	.738	.860	.906	.910
3.0	.55	18.5	1.20E+09	3.56E-06	2.52E+00	4.54E-02	4.75E+05	.782	.744	.721	.847	.897	.900
4.0	.58	18.9	5.32E+08	1.61E-06	1.08E+00	1.94E-02	4.95E+05	.768	.729	.706	.835	.888	.889
5.0	.61	19.2	2.44E+08	7.50E-07	4.76E-01	8.59E-03	5.12E+05	.754	.714	.692	.823	.879	.879
6.0	.64	19.5	1.15E+08	3.58E-07	2.17E-01	3.91E-03	5.29E+05	.741	.700	.678	.812	.870	.868
7.0	.66	19.8	5.52E+07	1.75E-07	1.01E-01	1.82E-03	5.45E+05	.728	.685	.664	.799	.860	.857
8.0	.69	20.1	2.71E+07	8.72E-08	4.82E-02	8.69E-04	5.62E+05	.714	.671	.650	.786	.850	.845

X-ray tube voltage: 28 kV

0.0	.35	16.9	2.36E+10	6.39E-05	6.47E+01	1.00E+00	3.64E+05	.831	.801	.777	.884	.914	.927
1.0	.46	17.9	8.57E+09	2.46E-05	2.00E+01	3.08E-01	4.29E+05	.799	.764	.741	.859	.902	.907
2.0	.53	18.6	3.54E+09	1.05E-05	7.57E+00	1.17E-01	4.67E+05	.774	.737	.714	.838	.888	.889
3.0	.57	19.1	1.56E+09	4.77E-06	3.14E+00	4.85E-02	4.97E+05	.751	.712	.690	.818	.873	.871
4.0	.61	19.6	7.19E+08	2.26E-06	1.37E+00	2.12E-02	5.24E+05	.729	.688	.667	.798	.857	.853
5.0	.65	20.1	3.44E+08	1.11E-06	6.25E-01	9.65E-03	5.51E+05	.706	.664	.644	.776	.840	.833
6.0	.69	20.7	1.71E+08	5.65E-07	2.95E-01	4.55E-03	5.80E+05	.683	.639	.620	.755	.822	.812
7.0	.74	21.2	8.74E+07	2.97E-07	1.43E-01	2.21E-03	6.10E+05	.659	.615	.597	.732	.804	.791
8.0	.78	21.7	4.61E+07	1.60E-07	7.18E-02	1.11E-03	6.42E+05	.637	.591	.574	.711	.786	.770

X-ray tube voltage: 30 kV

0.0	.36	17.3	2.80E+10	7.77E-05	7.44E+01	1.00E+00	3.76E+05	.808	.777	.754	.863	.897	.907
1.0	.48	18.5	1.05E+10	3.13E-05	2.35E+01	3.16E-01	4.49E+05	.766	.730	.708	.828	.876	.877
2.0	.55	19.4	4.54E+09	1.41E-05	9.15E+00	1.23E-01	4.96E+05	.729	.691	.670	.795	.851	.847
3.0	.61	20.2	2.10E+09	6.80E-06	3.90E+00	5.24E-02	5.38E+05	.694	.653	.634	.762	.825	.816
4.0	.67	21.0	1.03E+09	3.47E-06	1.77E+00	2.38E-02	5.81E+05	.659	.616	.598	.729	.798	.784
5.0	.73	21.9	5.28E+08	1.85E-06	8.43E-01	1.13E-02	6.27E+05	.625	.581	.564	.697	.771	.753
6.0	.80	22.7	2.84E+08	1.03E-06	4.19E-01	5.64E-03	6.76E+05	.594	.549	.533	.666	.745	.722
7.0	.88	23.4	1.58E+08	5.94E-07	2.18E-01	2.92E-03	7.27E+05	.565	.519	.505	.638	.721	.695
8.0	.97	24.1	9.14E+07	3.53E-07	1.17E-01	1.58E-03	7.80E+05	.540	.493	.480	.613	.700	.670

X-ray tube voltage: 32 kV

0.0	.37	17.8	3.24E+10	9.22E-05	8.34E+01	1.00E+00	3.88E+05	.783	.751	.729	.839	.878	.884
1.0	.50	19.2	1.26E+10	3.88E-05	2.69E+01	3.22E-01	4.69E+05	.731	.693	.673	.793	.846	.842
2.0	.58	20.3	5.64E+09	1.84E-05	1.07E+01	1.28E-01	5.27E+05	.684	.644	.625	.750	.812	.802
3.0	.65	21.4	2.74E+09	9.38E-06	4.69E+00	5.62E-02	5.83E+05	.639	.597	.580	.707	.776	.761
4.0	.73	22.5	1.41E+09	5.08E-06	2.20E+00	2.64E-02	6.43E+05	.597	.554	.538	.667	.742	.721
5.0	.82	23.5	7.70E+08	2.89E-06	1.09E+00	1.31E-02	7.06E+05	.560	.515	.501	.630	.710	.685
6.0	.92	24.4	4.39E+08	1.72E-06	5.69E-01	6.82E-03	7.72E+05	.527	.482	.469	.598	.683	.653
7.0	1.04	25.2	2.60E+08	1.05E-06	3.11E-01	3.73E-03	8.38E+05	.500	.454	.442	.571	.659	.625
8.0	1.16	25.9	1.59E+08	6.61E-07	1.77E-01	2.12E-03	9.01E+05	.478	.432	.421	.549	.640	.603

Table IV.3.17

Characteristic X-ray beam quantities without scattered radiation and relative energy absorption in typical image detectors at variable absorber or phantom materials for mammographic X-ray beam qualities with the indicated anode-filter combination:

Anode W / 0.05 mm Rh   Reference focal distance   60 cm

Absorber material:			PMMA					Relative energy absorption in the image detector (coverage mg/cm²)					
Thick- ness cm	HVL mm Al	\<E\> keV	Φ/It 1/As/cm²	Ψ/It J/As/cm²	Ka/It mGy/As	Relative dose	Φ/Ka photons /cm²/µGy	34 CaWO4	34 GOS	40 SPF	60 SPF	60 aSe	80 CsI

X-ray tube voltage: 24 kV

0.0	.43	18.0	1.49E+10	4.32E-05	3.54E+01	1.00E+00	4.22E+05	.787	.752	.729	.849	.893	.898
1.0	.54	18.9	6.18E+09	1.87E-05	1.28E+01	3.62E-01	4.82E+05	.759	.720	.698	.826	.880	.880
2.0	.61	19.5	2.81E+09	8.76E-06	5.40E+00	1.52E-01	5.21E+05	.739	.697	.676	.809	.869	.866
3.0	.66	19.9	1.34E+09	4.27E-06	2.45E+00	6.91E-02	5.49E+05	.723	.681	.660	.796	.859	.855
4.0	.70	20.2	6.64E+08	2.14E-06	1.16E+00	3.28E-02	5.71E+05	.711	.667	.647	.785	.851	.845
5.0	.74	20.4	3.36E+08	1.10E-06	5.70E-01	1.61E-02	5.90E+05	.701	.656	.636	.776	.844	.837
6.0	.77	20.6	1.74E+08	5.74E-07	2.87E-01	8.10E-03	6.05E+05	.692	.647	.627	.769	.839	.830
7.0	.80	20.8	9.09E+07	3.03E-07	1.47E-01	4.15E-03	6.19E+05	.685	.639	.619	.762	.833	.824
8.0	.82	21.0	4.82E+07	1.62E-07	7.65E-02	2.16E-03	6.30E+05	.678	.632	.613	.756	.829	.819

X-ray tube voltage: 26 kV

0.0	.46	18.5	2.05E+10	6.09E-05	4.60E+01	1.00E+00	4.46E+05	.766	.728	.706	.830	.880	.882
1.0	.58	19.4	8.90E+09	2.76E-05	1.75E+01	3.81E-01	5.08E+05	.738	.697	.676	.807	.866	.863
2.0	.65	19.9	4.20E+09	1.34E-05	7.67E+00	1.67E-01	5.47E+05	.718	.675	.655	.790	.854	.849
3.0	.71	20.3	2.08E+09	6.76E-06	3.60E+00	7.84E-02	5.76E+05	.703	.659	.639	.777	.845	.837
4.0	.75	20.6	1.06E+09	3.50E-06	1.77E+00	3.85E-02	5.99E+05	.691	.646	.626	.767	.837	.828
5.0	.79	20.9	5.52E+08	1.84E-06	8.94E-01	1.95E-02	6.17E+05	.681	.635	.616	.758	.830	.820
6.0	.82	21.1	2.93E+08	9.87E-07	4.62E-01	1.01E-02	6.33E+05	.673	.627	.607	.751	.824	.813
7.0	.84	21.2	1.57E+08	5.35E-07	2.43E-01	5.29E-03	6.46E+05	.666	.619	.600	.744	.819	.807
8.0	.87	21.4	8.53E+07	2.93E-07	1.30E-01	2.82E-03	6.58E+05	.660	.612	.594	.738	.815	.802

X-ray tube voltage: 28 kV

0.0	.48	18.8	2.56E+10	7.73E-05	5.57E+01	1.00E+00	4.61E+05	.753	.715	.693	.819	.872	.872
1.0	.60	19.7	1.14E+10	3.60E-05	2.18E+01	3.92E-01	5.23E+05	.725	.684	.663	.796	.857	.853
2.0	.67	20.2	5.50E+09	1.78E-05	9.77E+00	1.75E-01	5.63E+05	.705	.662	.642	.779	.845	.838
3.0	.73	20.6	2.77E+09	9.13E-06	4.67E+00	8.39E-02	5.92E+05	.690	.645	.626	.765	.835	.826
4.0	.78	20.9	1.43E+09	4.81E-06	2.33E+00	4.19E-02	6.16E+05	.678	.632	.613	.754	.826	.815
5.0	.81	21.2	7.60E+08	2.58E-06	1.20E+00	2.15E-02	6.35E+05	.667	.621	.602	.744	.818	.806
6.0	.85	21.4	4.09E+08	1.40E-06	6.28E-01	1.13E-02	6.52E+05	.658	.611	.592	.736	.812	.798
7.0	.87	21.6	2.23E+08	7.73E-07	3.35E-01	6.02E-03	6.66E+05	.650	.602	.584	.728	.805	.791
8.0	.90	21.8	1.23E+08	4.30E-07	1.81E-01	3.26E-03	6.80E+05	.642	.594	.576	.721	.800	.784

X-ray tube voltage: 30 kV

0.0	.50	19.1	3.12E+10	9.56E-05	6.57E+01	1.00E+00	4.74E+05	.739	.700	.679	.806	.861	.860
1.0	.62	20.0	1.42E+10	4.55E-05	2.63E+01	4.01E-01	5.39E+05	.709	.667	.647	.780	.844	.838
2.0	.70	20.6	6.98E+09	2.30E-05	1.20E+01	1.83E-01	5.81E+05	.687	.643	.624	.761	.830	.820
3.0	.76	21.0	3.58E+09	1.21E-05	5.85E+00	8.89E-02	6.13E+05	.670	.625	.606	.745	.817	.805
4.0	.81	21.4	1.89E+09	6.50E-06	2.96E+00	4.51E-02	6.39E+05	.654	.608	.590	.731	.806	.792
5.0	.85	21.7	1.02E+09	3.57E-06	1.55E+00	2.35E-02	6.63E+05	.641	.594	.576	.718	.795	.779
6.0	.89	22.1	5.64E+08	1.99E-06	8.25E-01	1.26E-02	6.84E+05	.628	.581	.564	.705	.785	.767
7.0	.92	22.4	3.16E+08	1.13E-06	4.49E-01	6.83E-03	7.04E+05	.616	.569	.552	.694	.775	.756
8.0	.95	22.6	1.79E+08	6.50E-07	2.48E-01	3.77E-03	7.23E+05	.604	.557	.540	.682	.765	.744

X-ray tube voltage: 32 kV

0.0	.51	19.4	3.64E+10	1.13E-04	7.49E+01	1.00E+00	4.86E+05	.725	.685	.664	.792	.850	.846
1.0	.64	20.3	1.69E+10	5.50E-05	3.05E+01	4.07E-01	5.54E+05	.693	.650	.630	.764	.830	.821
2.0	.72	21.0	8.44E+09	2.84E-05	1.41E+01	1.89E-01	5.98E+05	.668	.624	.605	.742	.812	.800
3.0	.78	21.5	4.41E+09	1.52E-05	6.96E+00	9.30E-02	6.34E+05	.647	.601	.584	.722	.796	.782
4.0	.84	22.0	2.38E+09	8.38E-06	3.58E+00	4.78E-02	6.65E+05	.628	.582	.565	.704	.781	.764
5.0	.89	22.4	1.31E+09	4.72E-06	1.89E+00	2.53E-02	6.94E+05	.610	.564	.547	.686	.766	.747
6.0	.93	22.9	7.42E+08	2.71E-06	1.03E+00	1.37E-02	7.22E+05	.594	.547	.531	.670	.751	.730
7.0	.98	23.3	4.26E+08	1.59E-06	5.69E-01	7.60E-03	7.49E+05	.577	.530	.515	.653	.737	.713
8.0	1.02	23.7	2.49E+08	9.42E-07	3.20E-01	4.28E-03	7.76E+05	.562	.514	.500	.637	.723	.697

X-ray tube voltage: 34 kV

0.0	.53	19.8	4.22E+10	1.34E-04	8.44E+01	1.00E+00	5.00E+05	.705	.665	.645	.773	.833	.827
1.0	.66	20.8	2.00E+10	6.68E-05	3.49E+01	4.14E-01	5.72E+05	.669	.626	.607	.740	.808	.797
2.0	.74	21.6	1.02E+10	3.54E-05	1.64E+01	1.94E-01	6.23E+05	.639	.595	.577	.712	.785	.771
3.0	.81	22.3	5.47E+09	1.95E-05	8.22E+00	9.74E-02	6.65E+05	.613	.568	.551	.687	.764	.746
4.0	.88	22.9	3.02E+09	1.11E-05	4.29E+00	5.09E-02	7.05E+05	.589	.543	.528	.663	.743	.723
5.0	.94	23.5	1.72E+09	6.46E-06	2.31E+00	2.74E-02	7.42E+05	.567	.520	.506	.641	.723	.700
6.0	1.00	24.1	1.00E+09	3.85E-06	1.28E+00	1.52E-02	7.80E+05	.545	.499	.485	.619	.703	.678
7.0	1.06	24.6	5.94E+08	2.34E-06	7.26E-01	8.61E-03	8.18E+05	.525	.478	.465	.598	.684	.656
8.0	1.12	25.2	3.60E+08	1.45E-06	4.20E-01	4.98E-03	8.57E+05	.505	.459	.447	.578	.665	.636

Table IV.3.18

Characteristic X-ray beam quantities without scattered radiation and relative energy absorption in typical image detectors at variable absorber or phantom materials for mammographic X-ray beam qualities with the indicated anode-filter combination:

Anode W / 0.05 mm Pd   Reference focal distance 60 cm

Absorber material: PMMA

Thick-ness cm	HVL mm Al	\<E> keV	Φ/It 1/As/cm²	Ψ/It J/As/cm²	Ka/It mGy/As	Relative dose	Φ/Ka photons /cm²/µGy	34 CaWO4	34 GOS	40 SPF	60 SPF	60 aSe	80 CsI
								\multicolumn{6}{c}{Relative energy absorption in the image detector (coverage mg/cm²)}

X-ray tube voltage: 24 kV

0.0	.44	18.2	1.46E+10	4.25E-05	3.42E+01	1.00E+00	4.28E+05	.783	.747	.724	.845	.890	.895
1.0	.55	19.0	6.11E+09	1.86E-05	1.25E+01	3.67E-01	4.88E+05	.754	.714	.693	.821	.877	.876
2.0	.62	19.6	2.80E+09	8.77E-06	5.32E+00	1.56E-01	5.26E+05	.734	.692	.671	.804	.865	.861
3.0	.67	20.0	1.35E+09	4.31E-06	2.43E+00	7.11E-02	5.55E+05	.718	.675	.654	.791	.855	.850
4.0	.72	20.3	6.71E+08	2.18E-06	1.16E+00	3.40E-02	5.78E+05	.705	.661	.641	.780	.847	.840
5.0	.75	20.6	3.43E+08	1.13E-06	5.74E-01	1.68E-02	5.97E+05	.694	.649	.629	.770	.839	.831
6.0	.78	20.8	1.78E+08	5.94E-07	2.91E-01	8.51E-03	6.14E+05	.685	.640	.620	.762	.833	.824
7.0	.81	21.0	9.42E+07	3.16E-07	1.50E-01	4.39E-03	6.28E+05	.678	.631	.612	.755	.828	.818
8.0	.83	21.1	5.03E+07	1.70E-07	7.87E-02	2.30E-03	6.40E+05	.671	.624	.605	.749	.823	.812

X-ray tube voltage: 26 kV

0.0	.48	18.9	2.13E+10	6.46E-05	4.59E+01	1.00E+00	4.64E+05	.747	.708	.687	.813	.867	.867
1.0	.61	19.8	9.54E+09	3.03E-05	1.80E+01	3.93E-01	5.30E+05	.717	.675	.655	.788	.851	.845
2.0	.68	20.4	4.64E+09	1.51E-05	8.11E+00	1.77E-01	5.72E+05	.696	.652	.633	.770	.838	.829
3.0	.75	20.8	2.36E+09	7.86E-06	3.91E+00	8.51E-02	6.04E+05	.680	.635	.616	.756	.827	.817
4.0	.80	21.1	1.24E+09	4.19E-06	1.97E+00	4.28E-02	6.29E+05	.667	.621	.602	.744	.818	.806
5.0	.84	21.4	6.63E+08	2.27E-06	1.02E+00	2.22E-02	6.50E+05	.657	.610	.591	.734	.811	.797
6.0	.87	21.6	3.61E+08	1.25E-06	5.41E-01	1.18E-02	6.68E+05	.648	.600	.582	.726	.804	.789
7.0	.90	21.8	2.00E+08	6.98E-07	2.92E-01	6.37E-03	6.83E+05	.640	.592	.574	.719	.798	.782
8.0	.93	22.0	1.11E+08	3.93E-07	1.60E-01	3.49E-03	6.96E+05	.633	.585	.567	.712	.793	.776

X-ray tube voltage: 28 kV

0.0	.51	19.3	2.73E+10	8.45E-05	5.65E+01	1.00E+00	4.84E+05	.731	.691	.670	.799	.856	.853
1.0	.64	20.2	1.26E+10	4.08E-05	2.29E+01	4.06E-01	5.50E+05	.701	.658	.639	.774	.840	.832
2.0	.72	20.7	6.28E+09	2.09E-05	1.06E+01	1.88E-01	5.93E+05	.681	.636	.617	.755	.826	.815
3.0	.78	21.2	3.27E+09	1.11E-05	5.23E+00	9.25E-02	6.25E+05	.665	.619	.600	.741	.815	.802
4.0	.83	21.5	1.75E+09	6.01E-06	2.68E+00	4.75E-02	6.51E+05	.652	.605	.587	.729	.806	.792
5.0	.87	21.8	9.53E+08	3.32E-06	1.42E+00	2.51E-02	6.72E+05	.641	.594	.576	.719	.798	.782
6.0	.91	22.0	5.28E+08	1.86E-06	7.66E-01	1.36E-02	6.90E+05	.632	.584	.567	.711	.791	.774
7.0	.94	22.2	2.97E+08	1.05E-06	4.21E-01	7.44E-03	7.06E+05	.624	.576	.559	.704	.785	.767
8.0	.97	22.4	1.68E+08	6.03E-07	2.34E-01	4.14E-03	7.19E+05	.618	.569	.552	.697	.780	.761

X-ray tube voltage: 30 kV

0.0	.53	19.7	3.37E+10	1.06E-04	6.75E+01	1.00E+00	5.00E+05	.716	.675	.654	.784	.845	.840
1.0	.66	20.5	1.60E+10	5.25E-05	2.81E+01	4.16E-01	5.68E+05	.686	.642	.623	.758	.827	.817
2.0	.75	21.1	8.11E+09	2.74E-05	1.32E+01	1.96E-01	6.12E+05	.664	.619	.600	.739	.812	.799
3.0	.81	21.6	4.30E+09	1.48E-05	6.65E+00	9.85E-02	6.46E+05	.647	.601	.583	.724	.800	.785
4.0	.86	21.9	2.34E+09	8.21E-06	3.47E+00	5.14E-02	6.74E+05	.633	.586	.569	.710	.789	.772
5.0	.91	22.2	1.30E+09	4.63E-06	1.86E+00	2.76E-02	6.97E+05	.621	.573	.556	.699	.780	.761
6.0	.95	22.5	7.34E+08	2.65E-06	1.02E+00	1.52E-02	7.18E+05	.610	.562	.546	.689	.771	.751
7.0	.98	22.7	4.20E+08	1.53E-06	5.71E-01	8.46E-03	7.36E+05	.601	.552	.536	.679	.763	.742
8.0	1.02	23.0	2.43E+08	8.95E-07	3.23E-01	4.79E-03	7.53E+05	.592	.543	.527	.670	.756	.733

X-ray tube voltage: 32 kV

0.0	.55	19.9	3.97E+10	1.27E-04	7.73E+01	1.00E+00	5.13E+05	.702	.661	.641	.772	.834	.827
1.0	.68	20.9	1.91E+10	6.38E-05	3.28E+01	4.24E-01	5.83E+05	.671	.627	.608	.744	.814	.803
2.0	.77	21.5	9.86E+09	3.39E-05	1.57E+01	2.03E-01	6.29E+05	.648	.602	.584	.723	.798	.783
3.0	.84	22.0	5.30E+09	1.87E-05	7.96E+00	1.03E-01	6.66E+05	.629	.583	.565	.705	.783	.766
4.0	.89	22.4	2.93E+09	1.05E-05	4.21E+00	5.45E-02	6.96E+05	.613	.566	.549	.690	.770	.751
5.0	.94	22.7	1.66E+09	6.04E-06	2.29E+00	2.96E-02	7.23E+05	.598	.551	.535	.676	.758	.737
6.0	.99	23.1	9.52E+08	3.52E-06	1.27E+00	1.65E-02	7.48E+05	.585	.537	.522	.662	.747	.724
7.0	1.03	23.4	5.55E+08	2.08E-06	7.20E-01	9.31E-03	7.71E+05	.573	.525	.510	.650	.736	.711
8.0	1.07	23.7	3.28E+08	1.25E-06	4.13E-01	5.35E-03	7.93E+05	.561	.513	.498	.638	.725	.699

X-ray tube voltage: 34 kV

0.0	.56	20.3	4.62E+10	1.50E-04	8.75E+01	1.00E+00	5.28E+05	.685	.643	.624	.755	.819	.810
1.0	.70	21.3	2.27E+10	7.73E-05	3.77E+01	4.31E-01	6.01E+05	.651	.607	.589	.724	.796	.782
2.0	.79	22.0	1.19E+10	4.20E-05	1.83E+01	2.09E-01	6.52E+05	.625	.579	.562	.699	.776	.759
3.0	.87	22.6	6.53E+09	2.36E-05	9.42E+00	1.08E-01	6.93E+05	.603	.556	.540	.678	.758	.738
4.0	.93	23.1	3.68E+09	1.36E-05	5.05E+00	5.77E-02	7.29E+05	.583	.536	.521	.659	.741	.719
5.0	.99	23.6	2.12E+09	8.02E-06	2.79E+00	3.18E-02	7.62E+05	.565	.518	.503	.640	.725	.701
6.0	1.04	24.0	1.25E+09	4.81E-06	1.57E+00	1.80E-02	7.94E+05	.548	.501	.487	.623	.709	.683
7.0	1.10	24.5	7.47E+08	2.93E-06	9.06E-01	1.04E-02	8.24E+05	.532	.485	.471	.607	.694	.667
8.0	1.15	24.9	4.53E+08	1.81E-06	5.31E-01	6.06E-03	8.55E+05	.517	.469	.457	.591	.679	.651

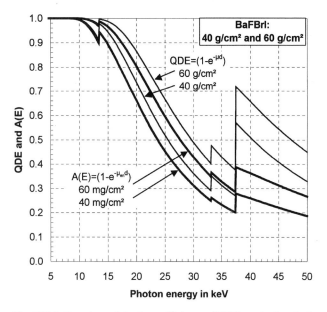

Fig. IV.3.9 Quantum detection efficiency (QDE) and absorbed energy [A(E)] in dependence on photon energy of storage phosphor screen with a cover density of 40 mg/cm² and 60 mg/cm²

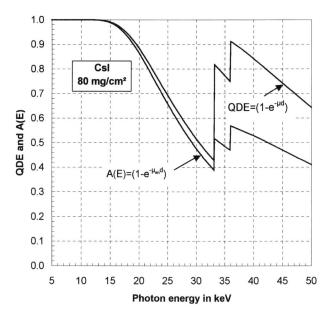

Fig. IV.3.11 Quantum detection efficiency (QDE) and absorbed energy [A(E)] in dependence on photon energy of a CsI screen with a cover density of 80 mg/cm²

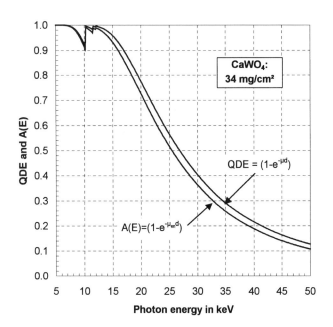

Fig. IV.3.10 Quantum detection efficiency (QDE) and absorbed energy [A(E)] in dependence on photon energy of an intensifying screen with a cover density of 34 mg/cm²

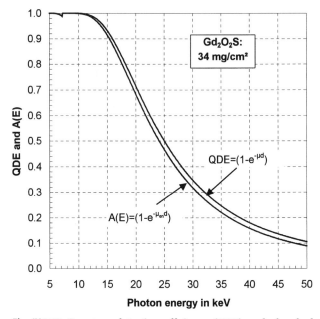

Fig. IV.3.12 Quantum detection efficiency (QDE) and absorbed energy [A(E)] in dependence on photon energy of an intensifying screen with a cover density of 34 mg/cm²

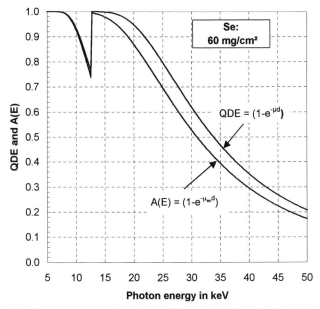

Fig. IV.3.13 Quantum detection efficiency (QDE) and absorbed energy [A(E)] in dependence on photon energy of an amorphous selenium layer with a cover density of 60 mg/cm^2

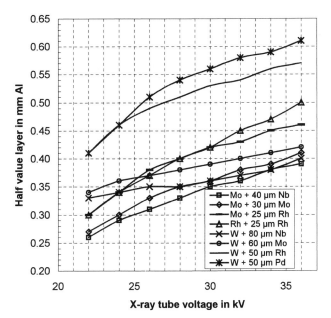

Fig. IV.3.14 Half value layer calculated in dependence on X-ray tube voltage for the most important anode-filter combinations inclusive of a 2 mm thick polycarbonate layer simulating the compression plate.

IV.4 Characteristics of the Imaging Radiation Field

IV.4.1 General X-ray Diagnostics

Table IV.4.1 (in part equivalent to table III.2.1): Typical values of the geometrical and physical characteristics of anti-scatter grids used especially in pediatrics:

Grid type	Pb 8/40			Pb 15/80		
N	40 mm^{-1}			80 mm^{-1}		
r	8			15		
d	0.072 mm			0.020 mm		
h	1.4 mm			1.6 mm		
D	0.175 mm			0.105 mm		
Measuring condition	**60 kV** **2mm Al**	**75 kV** **2 mm Al**	**100 kV** **4 mm Al**	**60 kV** **2 mm Al**	**75 kV** **2mm Al**	**100 kV** **4 mm Al**
T_p	0.61	0.63	0.65	0.68	0.70	0.74
T_s	0.076	0.088	0.125	0.062	0.080	0.137
T_t	0.17	0.18	0.22	0.17	0.19	0.24
Σ	8.0	7.1	5.2	11.0	8.7	5.4
C_{if}	3.54	3.38	2.96	3.93	3.65	3.01
B	5.80	5.37	4.56	5.77	5.21	4.07
SNR_{if}	1.50	1.46	1.39	1.63	1.60	1.49

Remarks: Geometrical data are nominal values; physical data are mean values of 20 grids
Measuring arrangement according to IEC 60627 (1978),
i.e. phantom (H_2O): field size 30 cm · 30 cm, height 20 cm;
additional filtration: 2 mm Al at 60 kV, 2 mm Al at 75 kV and
4 mm Al at 100 kV

Signs and symbols:

N strip density

r grid ratio

d thickness of lead strips

h height of lead strips

D thickness of interspace material

T_p transmission of primary radiation

T_s transmission of scattered radiation

T_t transmission of total radiation

Σ selectivity

C_{if} contrast improvement factor

B Bucky-factor

SNR_{if} improvement factor of the signal-to-noise ratio

Fig. IV.4.1

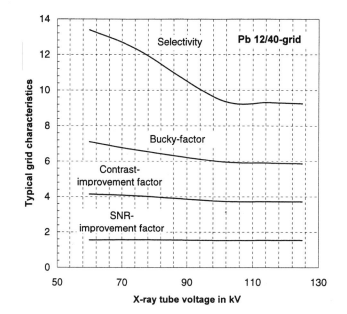

Typical geometrical and physical characteristics of anti-scatter grid Pb 12/40:

X-ray tube voltage	T_P	T_S	Σ	C_{if}	B	SNR_{if}
60 kV	0.584	0.044	13.38	4.14	7.1	1.55
75 kV	0.611	0.050	12.25	4.05	6.6	1.57
100 kV	0.626	0.066	9.47	3.75	6.0	1.53
115 kV	0.631	0.068	9.31	3.73	5.9	1.53
125 kV	0.634	0.069	9.25	3.72	5.9	1.53

Remarks: Results of Monte Carlo simulation calculation (s. chapter II.5 and III.2)
corresponding to a measuring arrangement according to IEC 60627 (1978)

Fig. IV.4.2

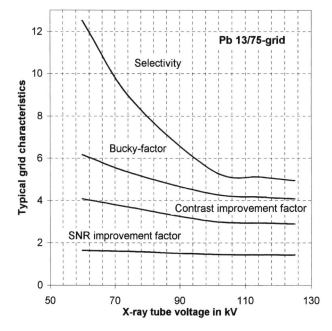

Typical geometrical and physical characteristics of anti-scatter grid Pb 13/75:

X-ray tube voltage	T_P	T_S	Σ	C_{if}	B	SNR_{if}
60 kV	0.660	0.053	12.50	4.07	6.2	2.33
75 kV	0.690	0.078	8.80	3.66	5.3	2.05
100 kV	0.700	0.129	5.43	3.02	4.3	1.63
115 kV	0.707	0.138	5.11	2.94	4.2	1.60
125 kV	0.710	0.144	4.94	2.89	4.1	1.58

Remarks: Results of Monte Carlo simulation calculation (s. chapter II.5 and III.2)
corresponding to a measuring arrangement according to IEC 60627 (1978)

Fig. IV.4.3

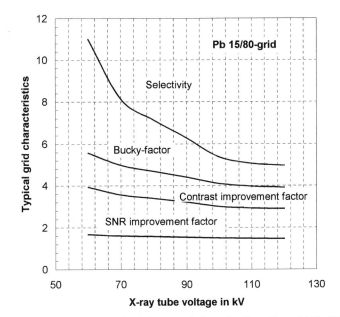

Typical geometrical and physical characteristics of anti-scatter grid Pb 15/80:

X-ray tube voltage	T_P	T_S	Σ	C_{if}	B	SNR_{if}
60 kV	0.706	0.064	11.00	3.93	5.6	1.67
70 kV	0.716	0.088	8.13	3.56	5.0	1.60
80 kV	0.724	0.102	7.13	3.39	4.7	1.57
90 kV	0.731	0.117	6.27	3.22	4.4	1.53
100 kV	0.734	0.137	5.37	3.01	4.1	1.49
110 kV	0.737	0.146	5.06	2.92	4.0	1.47
120 kV	0.741	0.149	4.96	2.90	3.9	1.46

Remarks: Results of Monte Carlo simulation calculation (s. chapter II.5 and III.2)
corresponding to a measuring arrangement according to IEC 60627 (1978)

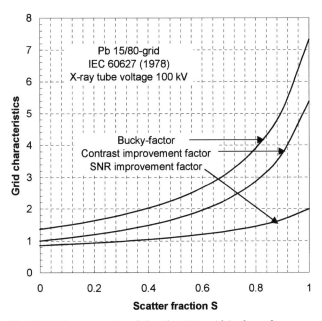

Fig. IV.4.4 Characteristics of the Pb 15/80 grid in dependence on scatter fraction S at an X-ray tube voltage of 100 kV

Fig. IV.4.5

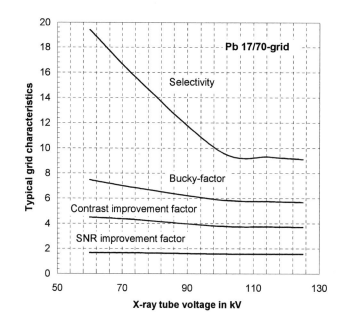

Typical geometrical and physical characteristics of anti-scatter grid Pb 17/70:

X-ray tube voltage	T_P	T_S	Σ	C_{if}	B	SNR_{if}
60 kV	0.604	0.031	19.38	4.50	7.4	2.66
75 kV	0.631	0.041	15.39	4.29	6.8	2.48
100 kV	0.644	0.066	9.74	3.79	5.9	2.01
115 kV	0.649	0.070	9.30	3.73	5.7	1.98
125 kV	0.652	0.072	9.10	3.70	5.7	1.97

Remarks: Results of Monte Carlo simulation calculation (s. chapter II.5 and III.2)
corresponding to a measuring arrangement according to IEC 60627 (1978)

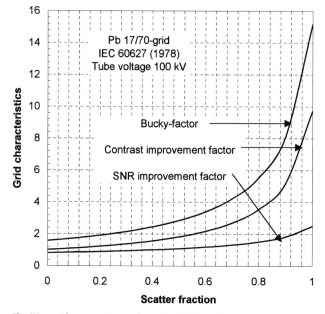

Fig. IV.4.6 Characteristics of the Pb 17/70 grid in dependence on scatter fraction S at an X-ray tube voltage of 100 kV

Fig. IV.4.7

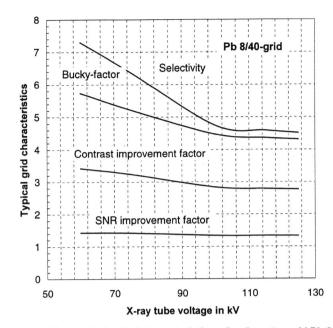

Typical geometrical and physical characteristics of anti-scatter grid Pb 8/40:

X-ray tube voltage	T_P	T_S	Σ	C_{if}	B	SNR_{if}
60 kV	0.597	0.082	7.30	3.42	5.7	1.43
75 kV	0.621	0.097	6.37	3.24	5.2	1.42
100 kV	0.634	0.134	4.74	2.83	4.5	1.34
115 kV	0.639	0.139	4.59	2.79	4.4	1.33
125 kV	0.641	0.142	4.51	2.76	4.3	1.33

Remarks: Results of Monte Carlo simulation calculation (s. chapter II.5 and III.2)
corresponding to a measuring arrangement according to IEC 60627 (1978)

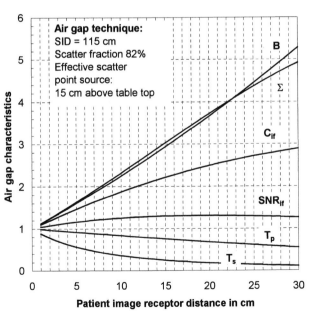

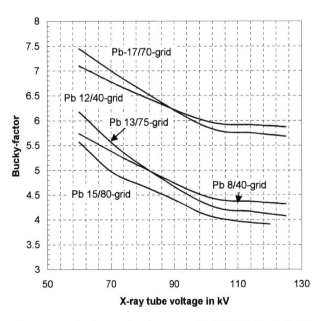

Fig. IV.4.8 Air gap characteristics (T_p, T_s, Σ, C_{if}, B and SNR_{if}) in dependence on the distance between the patient output plane and the image receptor

Fig. IV.4.9 Bucky-factor of anti-scatter grids Pb 8/40, Pb 13/75, Pb 15/80 and Pb 17/70 in dependence on X-ray tube voltage. Measuring arrangement according to IEC 60627 (1978)

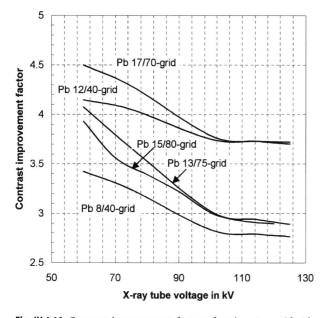

Fig. IV.4.10 Contrast improvement factor of anti-scatter grids Pb 8/40, Pb 12/40, Pb 13/75, Pb 15/80 and Pb 17/70 in dependence on X-ray tube voltage. Measuring arrangement according to IEC 60627 (1978)

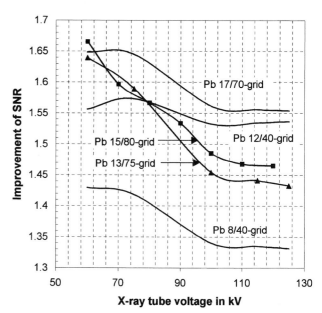

Fig. IV.4.12 Improvement factor for the signal-to-noise ratio of anti-scatter grids Pb 8/40, Pb 12/40, Pb 13/75, Pb 15/80 and Pb 17/70 in dependence on X-ray tube voltage. Measuring arrangement according to IEC 60627 (1978)

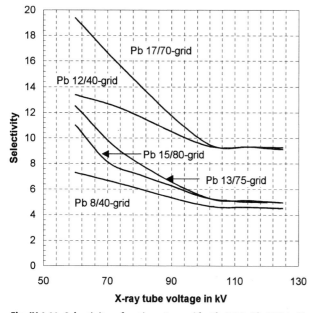

Fig. IV.4.11 Selectivity of anti-scatter grids Pb 8/40, Pb 12/40, Pb 13/75, Pb 15/80 and Pb 17/70 in dependence on X-ray tube voltage. Measuring arrangement according to IEC 60627 (1978)

Table IV.4.2 Scatter fraction S at image receptor without grid

Focal-detector-distance:		100 cm								
Air gap:		0 cm								
Tube voltage	Tp = 1 Anti scatter grid selectivity	Phantom-thickness	Beam diameter in cm / Field area in cm²								
			5	10	15	20	25	30	35	40	45
kV	Σ	cm	20	79	177	314	491	707	962	1257	1590
50	1	5	0.232	0.355	0.412	0.438	0.452	0.460	0.469	0.475	0.478
		10	0.277	0.460	0.549	0.603	0.629	0.645	0.659	0.668	0.671
		15	0.342	0.544	0.648	0.700	0.729	0.746	0.763	0.775	0.779
		20	0.402	0.613	0.715	0.771	0.800	0.815	0.829	0.838	0.843
		25	0.452	0.668	0.770	0.821	0.844	0.866	0.883	0.896	0.901
		30	0.497	0.715	0.807	0.861	0.884	0.906	0.923	0.936	0.943
		35	0.539	0.758	0.844	0.896	0.915	0.930	0.945	0.956	0.963
60	1	5	0.242	0.365	0.422	0.448	0.462	0.470	0.479	0.485	0.488
		10	0.287	0.470	0.559	0.613	0.639	0.655	0.669	0.678	0.681
		15	0.352	0.554	0.658	0.710	0.739	0.756	0.773	0.785	0.789
		20	0.412	0.623	0.725	0.781	0.810	0.825	0.839	0.848	0.853
		25	0.462	0.678	0.780	0.831	0.854	0.876	0.893	0.906	0.911
		30	0.507	0.725	0.817	0.871	0.894	0.916	0.933	0.946	0.953
		35	0.549	0.768	0.854	0.906	0.925	0.940	0.955	0.966	0.973
70	1	5	0.252	0.375	0.432	0.458	0.472	0.480	0.489	0.495	0.498
		10	0.297	0.480	0.569	0.623	0.649	0.665	0.679	0.688	0.691
		15	0.362	0.564	0.668	0.720	0.749	0.766	0.783	0.795	0.799
		20	0.422	0.633	0.735	0.791	0.820	0.835	0.849	0.858	0.863
		25	0.472	0.688	0.790	0.841	0.864	0.886	0.903	0.916	0.921
		30	0.517	0.735	0.827	0.881	0.904	0.926	0.943	0.956	0.963
		35	0.559	0.778	0.864	0.916	0.935	0.950	0.965	0.976	0.983
80	1	5	0.262	0.385	0.442	0.468	0.482	0.490	0.499	0.505	0.508
		10	0.307	0.490	0.579	0.633	0.659	0.675	0.689	0.698	0.701
		15	0.372	0.574	0.678	0.730	0.759	0.776	0.793	0.805	0.809
		20	0.432	0.643	0.745	0.801	0.830	0.845	0.859	0.868	0.873
		25	0.482	0.698	0.800	0.851	0.874	0.896	0.913	0.926	0.931
		30	0.527	0.745	0.837	0.891	0.914	0.936	0.953	0.966	0.973
		35	0.569	0.788	0.874	0.926	0.945	0.960	0.975	0.986	0.993
90	1	5	0.262	0.385	0.442	0.468	0.482	0.490	0.499	0.505	0.508
		10	0.307	0.490	0.579	0.633	0.659	0.675	0.689	0.698	0.701
		15	0.372	0.574	0.678	0.730	0.759	0.776	0.793	0.805	0.809
		20	0.432	0.643	0.745	0.801	0.830	0.845	0.859	0.868	0.873
		25	0.482	0.698	0.800	0.851	0.874	0.896	0.913	0.926	0.931
		30	0.527	0.745	0.837	0.891	0.914	0.936	0.953	0.966	0.973
		35	0.569	0.788	0.874	0.926	0.945	0.960	0.975	0.986	0.993
100	1	5	0.262	0.385	0.442	0.468	0.482	0.490	0.499	0.505	0.508
		10	0.307	0.490	0.579	0.633	0.659	0.675	0.689	0.698	0.701
		15	0.372	0.574	0.678	0.730	0.759	0.776	0.793	0.805	0.809
		20	0.432	0.643	0.745	0.801	0.830	0.845	0.859	0.868	0.873
		25	0.482	0.698	0.800	0.851	0.874	0.896	0.913	0.926	0.931
		30	0.527	0.745	0.837	0.891	0.914	0.936	0.953	0.966	0.973
		35	0.569	0.788	0.874	0.926	0.945	0.960	0.975	0.986	0.993
110	1	5	0.268	0.391	0.448	0.474	0.488	0.496	0.505	0.511	0.514
		10	0.313	0.496	0.585	0.639	0.665	0.681	0.695	0.704	0.707
		15	0.378	0.580	0.684	0.736	0.765	0.782	0.799	0.811	0.815
		20	0.438	0.649	0.751	0.807	0.836	0.851	0.865	0.874	0.879
		25	0.488	0.704	0.806	0.857	0.880	0.902	0.919	0.932	0.937
		30	0.533	0.751	0.843	0.897	0.920	0.942	0.959	0.972	0.979
		35	0.575	0.794	0.880	0.932	0.951	0.966	0.981	0.992	0.999
125	1	5	0.271	0.394	0.451	0.477	0.491	0.499	0.508	0.514	0.517
		10	0.316	0.499	0.588	0.642	0.668	0.684	0.698	0.707	0.710
		15	0.381	0.583	0.687	0.739	0.768	0.785	0.802	0.814	0.818
		20	0.441	0.652	0.754	0.810	0.839	0.854	0.868	0.877	0.882
		25	0.491	0.707	0.809	0.860	0.883	0.905	0.922	0.935	0.940
		30	0.536	0.754	0.846	0.900	0.923	0.945	0.962	0.975	0.982
		35	0.578	0.797	0.883	0.935	0.954	0.969	0.984	0.995	0.999

Table IV.4.3 Scatter fraction S at image receptor with grid Pb 12/40

Focal-detector-distance:		100 cm								
Air gap:		0 cm								

Tube voltage	Tp = 0.64 Anti scatter grid selectivity	Phantom-thickness	Beam diameter in cm / Field area in cm²								
			5	10	15	20	25	30	35	40	45
kV	Σ	cm	20	79	177	314	491	707	962	1257	1590
50	14.1	5	0.021	0.038	0.047	0.052	0.055	0.057	0.059	0.060	0.061
		10	0.026	0.057	0.079	0.097	0.107	0.114	0.120	0.125	0.127
		15	0.035	0.078	0.116	0.142	0.160	0.172	0.186	0.197	0.200
		20	0.045	0.101	0.151	0.193	0.220	0.239	0.256	0.269	0.276
		25	0.055	0.125	0.192	0.245	0.278	0.314	0.349	0.379	0.393
		30	0.065	0.151	0.229	0.305	0.352	0.405	0.460	0.511	0.542
		35	0.076	0.182	0.277	0.378	0.434	0.487	0.549	0.609	0.649
60	13.4	5	0.023	0.041	0.052	0.057	0.060	0.062	0.064	0.066	0.066
		10	0.029	0.062	0.086	0.106	0.117	0.124	0.131	0.136	0.138
		15	0.039	0.085	0.126	0.155	0.175	0.187	0.203	0.214	0.219
		20	0.050	0.110	0.165	0.210	0.241	0.261	0.280	0.295	0.303
		25	0.060	0.136	0.210	0.268	0.305	0.344	0.384	0.418	0.434
		30	0.071	0.165	0.250	0.335	0.387	0.447	0.510	0.569	0.605
		35	0.083	0.198	0.304	0.417	0.480	0.541	0.613	0.683	0.729
70	12.6	5	0.026	0.046	0.057	0.063	0.066	0.068	0.071	0.072	0.073
		10	0.032	0.068	0.095	0.116	0.128	0.136	0.144	0.149	0.151
		15	0.043	0.093	0.138	0.170	0.192	0.206	0.223	0.236	0.240
		20	0.055	0.120	0.181	0.231	0.265	0.287	0.308	0.325	0.334
		25	0.066	0.149	0.230	0.296	0.336	0.381	0.425	0.464	0.482
		30	0.078	0.181	0.276	0.370	0.429	0.497	0.568	0.635	0.677
		35	0.091	0.218	0.335	0.462	0.534	0.604	0.687	0.767	0.822
80	11.7	5	0.030	0.051	0.063	0.070	0.074	0.076	0.078	0.080	0.081
		10	0.036	0.076	0.105	0.128	0.142	0.151	0.159	0.165	0.167
		15	0.048	0.103	0.153	0.188	0.212	0.228	0.247	0.261	0.266
		20	0.061	0.133	0.200	0.256	0.294	0.319	0.342	0.361	0.371
		25	0.074	0.165	0.255	0.328	0.373	0.423	0.473	0.516	0.537
		30	0.087	0.200	0.306	0.411	0.477	0.554	0.635	0.711	0.758
		35	0.101	0.241	0.372	0.515	0.596	0.675	0.769	0.862	0.924
90	10.7	5	0.032	0.055	0.069	0.076	0.080	0.082	0.085	0.087	0.088
		10	0.040	0.082	0.114	0.139	0.153	0.163	0.171	0.178	0.180
		15	0.052	0.112	0.165	0.202	0.228	0.244	0.264	0.279	0.284
		20	0.066	0.144	0.215	0.274	0.313	0.338	0.362	0.382	0.392
		25	0.080	0.177	0.273	0.348	0.394	0.445	0.496	0.539	0.559
		30	0.094	0.215	0.325	0.433	0.500	0.576	0.655	0.729	0.774
		35	0.110	0.258	0.393	0.537	0.618	0.694	0.785	0.872	0.930
100	9.5	5	0.036	0.062	0.077	0.085	0.089	0.092	0.095	0.097	0.098
		10	0.045	0.092	0.126	0.153	0.169	0.180	0.189	0.196	0.198
		15	0.059	0.124	0.182	0.222	0.249	0.267	0.287	0.303	0.309
		20	0.074	0.159	0.236	0.298	0.339	0.365	0.390	0.410	0.421
		25	0.089	0.195	0.297	0.376	0.423	0.475	0.525	0.568	0.589
		30	0.105	0.236	0.352	0.463	0.529	0.605	0.682	0.752	0.794
		35	0.122	0.281	0.422	0.567	0.645	0.719	0.804	0.885	0.938
110	9	5	0.039	0.067	0.083	0.091	0.096	0.099	0.102	0.104	0.105
		10	0.048	0.099	0.135	0.164	0.181	0.192	0.202	0.209	0.212
		15	0.063	0.133	0.194	0.237	0.266	0.284	0.306	0.323	0.329
		20	0.080	0.170	0.251	0.317	0.361	0.389	0.415	0.436	0.447
		25	0.096	0.209	0.317	0.400	0.450	0.505	0.558	0.603	0.625
		30	0.112	0.251	0.374	0.492	0.562	0.642	0.723	0.797	0.841
		35	0.131	0.300	0.449	0.602	0.685	0.762	0.852	0.936	0.992
125	8.5	5	0.042	0.071	0.088	0.097	0.102	0.105	0.108	0.111	0.112
		10	0.052	0.105	0.144	0.174	0.191	0.203	0.214	0.221	0.224
		15	0.067	0.141	0.206	0.250	0.281	0.300	0.323	0.340	0.347
		20	0.085	0.181	0.265	0.334	0.379	0.409	0.436	0.457	0.469
		25	0.102	0.221	0.333	0.420	0.471	0.527	0.582	0.628	0.650
		30	0.120	0.265	0.393	0.514	0.586	0.667	0.749	0.824	0.868
		35	0.139	0.316	0.470	0.627	0.711	0.789	0.879	0.963	0.980

Table IV.4.4 Scatter fraction S at image receptor with grid Pb 8/40

| Focal-detector-distance: | | | 100 cm | | | | | | | | |
| Air gap: | | | 0 cm | | | | | | | | |

| Tube voltage | Tp = 0.64 Anti scatter grid selectivity | Phantom-thickness | Beam diameter in cm / Field area in cm² | | | | | | | | |
kV	Σ	cm	5 20	10 79	15 177	20 314	25 491	30 707	35 962	40 1257	45 1590
50	8.6	5	0.034	0.060	0.075	0.083	0.088	0.090	0.093	0.095	0.096
		10	0.043	0.090	0.124	0.150	0.165	0.175	0.183	0.190	0.192
		15	0.057	0.122	0.177	0.213	0.239	0.254	0.272	0.286	0.291
		20	0.072	0.156	0.226	0.282	0.317	0.339	0.360	0.376	0.385
		25	0.087	0.189	0.281	0.348	0.387	0.428	0.468	0.500	0.515
		30	0.103	0.226	0.328	0.419	0.471	0.527	0.583	0.631	0.660
		35	0.120	0.267	0.386	0.499	0.557	0.609	0.667	0.719	0.752
60	8	5	0.038	0.067	0.084	0.092	0.097	0.100	0.103	0.105	0.107
		10	0.048	0.100	0.137	0.165	0.181	0.192	0.202	0.208	0.211
		15	0.063	0.134	0.194	0.234	0.262	0.279	0.299	0.314	0.319
		20	0.080	0.171	0.248	0.309	0.347	0.372	0.394	0.412	0.421
		25	0.097	0.208	0.308	0.381	0.423	0.468	0.511	0.546	0.563
		30	0.114	0.248	0.359	0.458	0.514	0.576	0.636	0.688	0.719
		35	0.132	0.293	0.423	0.545	0.608	0.664	0.726	0.783	0.819
70	7.4	5	0.044	0.075	0.093	0.103	0.108	0.111	0.115	0.117	0.118
		10	0.054	0.111	0.151	0.182	0.200	0.212	0.222	0.229	0.232
		15	0.071	0.149	0.214	0.258	0.288	0.306	0.328	0.344	0.350
		20	0.090	0.189	0.273	0.339	0.380	0.407	0.431	0.450	0.461
		25	0.108	0.229	0.338	0.417	0.463	0.511	0.558	0.595	0.613
		30	0.126	0.273	0.393	0.500	0.561	0.627	0.692	0.748	0.781
		35	0.146	0.321	0.462	0.594	0.661	0.722	0.789	0.849	0.887
80	6.6	5	0.051	0.087	0.107	0.118	0.124	0.127	0.131	0.134	0.135
		10	0.063	0.127	0.172	0.207	0.226	0.240	0.251	0.259	0.262
		15	0.082	0.169	0.242	0.291	0.323	0.344	0.367	0.385	0.392
		20	0.103	0.214	0.307	0.379	0.424	0.453	0.479	0.500	0.511
		25	0.123	0.259	0.378	0.464	0.513	0.565	0.614	0.654	0.673
		30	0.144	0.307	0.438	0.553	0.618	0.688	0.755	0.814	0.848
		35	0.166	0.360	0.513	0.653	0.724	0.786	0.855	0.917	0.956
90	5.9	5	0.057	0.096	0.118	0.130	0.136	0.140	0.144	0.148	0.149
		10	0.070	0.140	0.189	0.226	0.247	0.261	0.273	0.281	0.285
		15	0.091	0.186	0.263	0.314	0.348	0.369	0.394	0.412	0.419
		20	0.114	0.234	0.332	0.406	0.452	0.481	0.507	0.528	0.539
		25	0.136	0.281	0.405	0.492	0.541	0.593	0.641	0.679	0.697
		30	0.159	0.332	0.466	0.581	0.644	0.711	0.775	0.830	0.861
		35	0.183	0.387	0.541	0.678	0.745	0.805	0.869	0.925	0.960
100	5.2	5	0.064	0.108	0.132	0.145	0.152	0.156	0.161	0.164	0.166
		10	0.078	0.156	0.209	0.249	0.271	0.286	0.299	0.308	0.311
		15	0.102	0.206	0.289	0.342	0.378	0.399	0.424	0.443	0.450
		20	0.127	0.257	0.360	0.437	0.483	0.513	0.539	0.559	0.570
		25	0.152	0.307	0.436	0.523	0.572	0.623	0.669	0.706	0.723
		30	0.176	0.360	0.498	0.611	0.673	0.736	0.796	0.847	0.876
		35	0.202	0.417	0.572	0.705	0.769	0.824	0.882	0.933	0.965
110	4.6	5	0.074	0.123	0.150	0.164	0.172	0.176	0.182	0.185	0.187
		10	0.090	0.176	0.234	0.278	0.301	0.317	0.331	0.341	0.344
		15	0.117	0.231	0.320	0.377	0.415	0.438	0.464	0.483	0.490
		20	0.145	0.287	0.397	0.476	0.525	0.555	0.582	0.602	0.613
		25	0.172	0.340	0.475	0.566	0.615	0.666	0.712	0.748	0.765
		30	0.199	0.397	0.539	0.654	0.715	0.778	0.836	0.885	0.912
		35	0.227	0.456	0.615	0.747	0.809	0.862	0.918	0.966	0.996
125	4.1	5	0.083	0.137	0.167	0.182	0.191	0.196	0.201	0.205	0.207
		10	0.101	0.195	0.258	0.304	0.329	0.346	0.360	0.370	0.374
		15	0.130	0.254	0.349	0.409	0.447	0.470	0.497	0.517	0.524
		20	0.161	0.314	0.428	0.510	0.559	0.589	0.615	0.636	0.647
		25	0.190	0.370	0.509	0.600	0.649	0.698	0.743	0.778	0.794
		30	0.220	0.428	0.573	0.687	0.746	0.806	0.861	0.906	0.932
		35	0.250	0.489	0.648	0.777	0.836	0.886	0.938	0.982	0.998

Table IV.4.5 Scatter fraction S at image receptor with grid Pb 17/70

| Focal-detector-distance: | | | 100 cm | | | | | | | | |
| Air gap: | | | 0 cm | | | | | | | | |

| Tube voltage | Tp = 0.64 Anti scatter grid selectivity | Phantom-thickness | Beam diameter in cm / Field area in cm² | | | | | | | | |
| | | | 5 | 10 | 15 | 20 | 25 | 30 | 35 | 40 | 45 |
kV	Σ	cm	20	79	177	314	491	707	962	1257	1590
		5	0.015	0.027	0.034	0.038	0.040	0.041	0.043	0.044	0.044
50	19.8	10	0.019	0.041	0.058	0.071	0.079	0.084	0.089	0.092	0.094
		15	0.026	0.057	0.085	0.105	0.120	0.129	0.140	0.148	0.151
		20	0.033	0.074	0.113	0.145	0.168	0.182	0.196	0.208	0.214
		25	0.040	0.092	0.145	0.188	0.215	0.245	0.276	0.303	0.316
		30	0.048	0.113	0.175	0.238	0.279	0.326	0.378	0.427	0.457
		35	0.056	0.137	0.215	0.302	0.353	0.403	0.465	0.526	0.568
		5	0.017	0.031	0.039	0.043	0.046	0.047	0.049	0.050	0.050
60	18	10	0.022	0.047	0.066	0.081	0.089	0.096	0.101	0.105	0.106
		15	0.029	0.064	0.097	0.120	0.136	0.147	0.159	0.169	0.172
		20	0.037	0.084	0.128	0.165	0.191	0.208	0.224	0.237	0.244
		25	0.045	0.105	0.165	0.215	0.246	0.281	0.317	0.348	0.364
		30	0.054	0.128	0.199	0.273	0.320	0.376	0.437	0.495	0.532
		35	0.063	0.155	0.245	0.347	0.408	0.467	0.541	0.616	0.667
		5	0.021	0.037	0.046	0.051	0.054	0.055	0.057	0.059	0.059
70	15.8	10	0.026	0.055	0.077	0.095	0.105	0.112	0.118	0.122	0.124
		15	0.035	0.076	0.113	0.140	0.159	0.171	0.186	0.197	0.201
		20	0.044	0.098	0.150	0.193	0.223	0.243	0.262	0.277	0.286
		25	0.054	0.122	0.193	0.251	0.287	0.329	0.371	0.408	0.426
		30	0.063	0.150	0.233	0.319	0.375	0.441	0.512	0.581	0.625
		35	0.074	0.182	0.287	0.407	0.478	0.549	0.636	0.724	0.786
		5	0.025	0.044	0.055	0.061	0.064	0.066	0.068	0.070	0.071
80	13.6	10	0.032	0.066	0.092	0.112	0.124	0.133	0.140	0.145	0.147
		15	0.042	0.090	0.134	0.166	0.188	0.203	0.220	0.233	0.238
		20	0.053	0.117	0.177	0.229	0.263	0.287	0.309	0.327	0.336
		25	0.064	0.145	0.228	0.296	0.339	0.387	0.436	0.479	0.500
		30	0.076	0.177	0.275	0.375	0.440	0.517	0.599	0.679	0.730
		35	0.088	0.215	0.338	0.477	0.560	0.641	0.742	0.843	0.913
		5	0.030	0.052	0.064	0.071	0.075	0.077	0.080	0.082	0.082
90	11.5	10	0.037	0.077	0.107	0.130	0.144	0.153	0.161	0.167	0.170
		15	0.049	0.105	0.155	0.190	0.215	0.231	0.250	0.265	0.270
		20	0.062	0.135	0.203	0.259	0.297	0.322	0.346	0.365	0.375
		25	0.075	0.167	0.259	0.332	0.377	0.427	0.478	0.521	0.542
		30	0.088	0.203	0.309	0.416	0.482	0.558	0.639	0.715	0.761
		35	0.103	0.244	0.376	0.519	0.600	0.679	0.772	0.864	0.926
		5	0.035	0.061	0.075	0.083	0.088	0.090	0.093	0.095	0.096
100	9.7	10	0.044	0.090	0.124	0.151	0.166	0.177	0.186	0.192	0.195
		15	0.057	0.122	0.179	0.218	0.245	0.263	0.283	0.299	0.305
		20	0.073	0.157	0.232	0.293	0.334	0.361	0.385	0.405	0.416
		25	0.087	0.192	0.293	0.371	0.418	0.469	0.520	0.563	0.583
		30	0.103	0.232	0.347	0.457	0.524	0.600	0.677	0.748	0.791
		35	0.120	0.277	0.417	0.562	0.640	0.715	0.801	0.883	0.937
		5	0.039	0.067	0.083	0.091	0.096	0.099	0.102	0.104	0.105
110	9	10	0.048	0.099	0.135	0.164	0.181	0.192	0.202	0.209	0.212
		15	0.063	0.133	0.194	0.237	0.266	0.284	0.306	0.323	0.329
		20	0.080	0.170	0.251	0.317	0.361	0.389	0.415	0.436	0.447
		25	0.096	0.209	0.317	0.400	0.450	0.505	0.558	0.603	0.625
		30	0.112	0.251	0.374	0.492	0.562	0.642	0.723	0.797	0.841
		35	0.131	0.300	0.449	0.602	0.685	0.762	0.852	0.936	0.992
		5	0.042	0.071	0.088	0.097	0.102	0.105	0.108	0.111	0.112
125	8.5	10	0.052	0.105	0.144	0.174	0.191	0.203	0.214	0.221	0.224
		15	0.067	0.141	0.206	0.250	0.281	0.300	0.323	0.340	0.347
		20	0.085	0.181	0.265	0.334	0.379	0.409	0.436	0.457	0.469
		25	0.102	0.221	0.333	0.420	0.471	0.527	0.582	0.628	0.650
		30	0.120	0.265	0.393	0.514	0.586	0.667	0.749	0.824	0.868
		35	0.139	0.316	0.470	0.627	0.711	0.789	0.879	0.963	0.980

Table IV.4.6 Scatter fraction S at image receptor with grid Pb 15/80

Focal-detector-distance:		100 cm									
Air gap:		0 cm									
	Tp = 0.74		Beam diameter in cm / Field area in cm²								
Tube voltage	Anti scatter grid selectivity	Phantom-thickness	5	10	15	20	25	30	35	40	45
kV	Σ	cm	20	79	177	314	491	707	962	1257	1590
50	12.3	5	0.024	0.043	0.054	0.060	0.063	0.065	0.067	0.069	0.069
		10	0.030	0.065	0.090	0.110	0.121	0.129	0.136	0.141	0.142
		15	0.040	0.088	0.130	0.160	0.180	0.192	0.207	0.219	0.223
		20	0.052	0.114	0.170	0.215	0.245	0.264	0.282	0.297	0.304
		25	0.063	0.140	0.214	0.272	0.306	0.344	0.381	0.412	0.426
		30	0.074	0.170	0.254	0.335	0.383	0.438	0.494	0.545	0.576
		35	0.087	0.203	0.306	0.411	0.468	0.521	0.583	0.641	0.679
60	11	5	0.028	0.050	0.062	0.069	0.072	0.075	0.077	0.079	0.080
		10	0.035	0.075	0.103	0.126	0.139	0.147	0.155	0.161	0.163
		15	0.047	0.101	0.149	0.182	0.205	0.219	0.236	0.250	0.254
		20	0.060	0.131	0.194	0.245	0.279	0.301	0.321	0.337	0.346
		25	0.072	0.160	0.244	0.309	0.348	0.390	0.432	0.467	0.483
		30	0.085	0.194	0.289	0.380	0.435	0.497	0.559	0.616	0.651
		35	0.100	0.231	0.347	0.466	0.530	0.590	0.659	0.724	0.767
70	9.5	5	0.034	0.059	0.074	0.082	0.086	0.089	0.092	0.094	0.095
		10	0.043	0.089	0.122	0.148	0.163	0.173	0.182	0.188	0.191
		15	0.056	0.120	0.175	0.213	0.239	0.256	0.275	0.290	0.296
		20	0.071	0.154	0.226	0.285	0.323	0.348	0.371	0.390	0.399
		25	0.086	0.188	0.284	0.358	0.402	0.449	0.495	0.534	0.553
		30	0.101	0.226	0.335	0.438	0.499	0.567	0.636	0.698	0.735
		35	0.118	0.270	0.401	0.533	0.603	0.669	0.744	0.814	0.859
80	7.8	5	0.044	0.074	0.092	0.101	0.107	0.110	0.113	0.116	0.117
		10	0.054	0.110	0.150	0.181	0.198	0.211	0.221	0.228	0.231
		15	0.070	0.147	0.213	0.258	0.288	0.307	0.329	0.347	0.353
		20	0.089	0.188	0.273	0.341	0.384	0.412	0.438	0.458	0.469
		25	0.106	0.228	0.340	0.423	0.472	0.524	0.574	0.616	0.635
		30	0.125	0.273	0.398	0.512	0.578	0.651	0.723	0.787	0.825
		35	0.145	0.323	0.471	0.614	0.689	0.757	0.833	0.903	0.948
90	6.5	5	0.052	0.088	0.109	0.119	0.125	0.129	0.133	0.136	0.137
		10	0.064	0.129	0.175	0.209	0.229	0.243	0.254	0.262	0.265
		15	0.083	0.172	0.245	0.294	0.327	0.347	0.371	0.389	0.395
		20	0.105	0.217	0.311	0.383	0.428	0.457	0.483	0.504	0.515
		25	0.125	0.262	0.382	0.468	0.517	0.569	0.618	0.658	0.676
		30	0.146	0.311	0.442	0.557	0.622	0.691	0.758	0.816	0.850
		35	0.169	0.364	0.516	0.657	0.727	0.789	0.857	0.918	0.957
100	5.4	5	0.062	0.104	0.128	0.140	0.147	0.151	0.156	0.159	0.161
		10	0.076	0.151	0.203	0.242	0.263	0.278	0.291	0.300	0.303
		15	0.099	0.200	0.281	0.334	0.369	0.390	0.415	0.434	0.440
		20	0.123	0.250	0.352	0.427	0.474	0.503	0.530	0.550	0.561
		25	0.147	0.299	0.426	0.514	0.563	0.614	0.661	0.698	0.716
		30	0.171	0.352	0.488	0.602	0.664	0.729	0.790	0.842	0.872
		35	0.196	0.408	0.562	0.697	0.762	0.818	0.879	0.931	0.964
110	5	5	0.068	0.114	0.140	0.153	0.160	0.165	0.169	0.173	0.175
		10	0.083	0.164	0.220	0.261	0.284	0.300	0.313	0.322	0.326
		15	0.108	0.216	0.303	0.358	0.395	0.417	0.443	0.462	0.469
		20	0.135	0.270	0.377	0.456	0.504	0.534	0.561	0.582	0.593
		25	0.160	0.322	0.455	0.545	0.596	0.647	0.695	0.732	0.750
		30	0.186	0.377	0.519	0.635	0.698	0.763	0.824	0.876	0.905
		35	0.213	0.435	0.595	0.731	0.796	0.852	0.912	0.963	0.995
125	4.8	5	0.072	0.119	0.146	0.160	0.167	0.172	0.177	0.181	0.182
		10	0.088	0.172	0.229	0.272	0.295	0.311	0.325	0.334	0.338
		15	0.113	0.225	0.314	0.371	0.409	0.431	0.458	0.477	0.484
		20	0.141	0.281	0.390	0.471	0.520	0.550	0.578	0.599	0.610
		25	0.167	0.334	0.470	0.561	0.612	0.664	0.712	0.749	0.767
		30	0.194	0.390	0.534	0.652	0.715	0.780	0.841	0.892	0.921
		35	0.222	0.450	0.611	0.748	0.813	0.869	0.928	0.979	0.990

Table IV.4.7 Scatter fraction S at image receptor with air gap

Focal-detector-distance: Air gap: Tube voltage kV	Air gap cm	Grid Σ	100 cm variable Phantom-thickness cm	Beam diameter in cm / Field area in cm²								
				5 20	10 79	15 177	20 314	25 491	30 707	35 962	40 1257	45 1590
			5	0.262	0.385	0.442	0.468	0.482	0.490	0.499	0.505	0.508
80	0	1	10	0.307	0.490	0.579	0.633	0.659	0.675	0.689	0.698	0.701
			15	0.372	0.574	0.678	0.730	0.759	0.776	0.793	0.805	0.809
			20	0.432	0.643	0.745	0.801	0.830	0.845	0.859	0.868	0.873
			25	0.482	0.698	0.800	0.851	0.874	0.896	0.913	0.926	0.931
			30	0.527	0.745	0.837	0.891	0.914	0.936	0.953	0.966	0.973
			35	0.569	0.788	0.874	0.926	0.945	0.960	0.975	0.986	0.993
			5	0.106	0.190	0.245	0.279	0.308	0.330	0.348	0.367	0.385
80	10	1	10	0.124	0.252	0.343	0.415	0.465	0.505	0.533	0.559	0.581
			15	0.157	0.311	0.439	0.516	0.576	0.621	0.652	0.683	0.708
			20	0.190	0.368	0.507	0.601	0.666	0.714	0.743	0.769	0.791
			25	0.226	0.426	0.579	0.678	0.733	0.785	0.822	0.853	0.877
			30	0.260	0.485	0.629	0.743	0.799	0.852	0.889	0.919	0.943
			35	0.294	0.541	0.688	0.806	0.859	0.904	0.932	0.957	0.977
			5	0.054	0.111	0.151	0.179	0.208	0.233	0.252	0.272	0.292
80	20	1	10	0.062	0.143	0.213	0.274	0.331	0.375	0.410	0.440	0.470
			15	0.080	0.179	0.281	0.353	0.426	0.482	0.524	0.560	0.597
			20	0.099	0.215	0.333	0.425	0.514	0.576	0.625	0.660	0.693
			25	0.122	0.260	0.394	0.503	0.595	0.652	0.711	0.756	0.795
			30	0.145	0.305	0.443	0.571	0.675	0.734	0.797	0.844	0.883
			35	0.169	0.350	0.503	0.645	0.756	0.812	0.865	0.903	0.936
			5	0.032	0.073	0.102	0.128	0.149	0.172	0.192	0.209	0.227
80	30	1	10	0.036	0.091	0.144	0.190	0.236	0.280	0.319	0.350	0.378
			15	0.047	0.114	0.186	0.256	0.310	0.368	0.418	0.457	0.492
			20	0.060	0.137	0.224	0.307	0.380	0.451	0.509	0.557	0.593
			25	0.077	0.167	0.270	0.371	0.458	0.531	0.588	0.648	0.696
			30	0.093	0.198	0.316	0.422	0.532	0.615	0.678	0.744	0.796
			35	0.110	0.229	0.369	0.488	0.616	0.707	0.766	0.823	0.868
			5	0.017	0.041	0.059	0.077	0.094	0.108	0.123	0.139	0.155
80	50	1	10	0.018	0.048	0.074	0.107	0.134	0.164	0.194	0.225	0.252
			15	0.026	0.060	0.093	0.135	0.180	0.218	0.254	0.296	0.331
			20	0.034	0.071	0.111	0.163	0.214	0.263	0.313	0.364	0.405
			25	0.045	0.085	0.137	0.198	0.261	0.323	0.382	0.438	0.475
			30	0.057	0.100	0.164	0.237	0.300	0.374	0.451	0.518	0.558
			35	0.068	0.116	0.192	0.280	0.352	0.441	0.533	0.611	0.653
			5	0.023	0.029	0.039	0.056	0.079	0.096	0.106	0.118	0.131
80	80	1	10	0.023	0.030	0.042	0.062	0.089	0.111	0.125	0.145	0.167
			15	0.035	0.042	0.055	0.077	0.107	0.133	0.150	0.173	0.198
			20	0.047	0.055	0.068	0.091	0.123	0.152	0.172	0.199	0.229
			25	0.067	0.075	0.089	0.113	0.146	0.179	0.207	0.239	0.271
			30	0.085	0.093	0.109	0.134	0.169	0.206	0.241	0.280	0.317
			35	0.104	0.113	0.129	0.156	0.194	0.235	0.277	0.322	0.366
			5	0.006	0.008	0.010	0.015	0.022	0.028	0.032	0.037	0.043
80	120	1	10	0.006	0.008	0.011	0.017	0.025	0.034	0.040	0.049	0.060
			15	0.008	0.010	0.014	0.021	0.031	0.041	0.050	0.063	0.078
			20	0.011	0.013	0.017	0.025	0.036	0.049	0.061	0.077	0.099
			25	0.016	0.019	0.023	0.032	0.045	0.062	0.080	0.103	0.132
			30	0.021	0.024	0.029	0.039	0.055	0.076	0.101	0.134	0.174
			35	0.026	0.029	0.036	0.048	0.067	0.093	0.126	0.170	0.223

Table IV.4.8 Scatter fraction S at image receptor with air gap and grids

Focal-detector-distance:				100 cm								
Air gap:			variable									
Tube voltage kV	Air gap cm	Grid Σ	Phantom-thickness cm	Beam diameter in cm / Field area in cm²								
				5 / 20	10 / 79	15 / 177	20 / 314	25 / 491	30 / 707	35 / 962	40 / 1257	45 / 1590
80	0	1	5	0.262	0.385	0.442	0.468	0.482	0.490	0.499	0.505	0.508
			10	0.307	0.490	0.579	0.633	0.659	0.675	0.689	0.698	0.701
			15	0.372	0.574	0.678	0.730	0.759	0.776	0.793	0.805	0.809
		no grid	20	0.432	0.643	0.745	0.801	0.830	0.845	0.859	0.868	0.873
			25	0.482	0.698	0.800	0.851	0.874	0.896	0.913	0.926	0.931
			30	0.527	0.745	0.837	0.891	0.914	0.936	0.953	0.966	0.973
			35	0.569	0.788	0.874	0.926	0.945	0.960	0.975	0.986	0.993
80	10	1	5	0.106	0.190	0.245	0.279	0.308	0.330	0.348	0.367	0.385
			10	0.124	0.252	0.343	0.415	0.465	0.505	0.533	0.559	0.581
			15	0.157	0.311	0.439	0.516	0.576	0.621	0.652	0.683	0.708
		no grid	20	0.190	0.368	0.507	0.601	0.666	0.714	0.743	0.769	0.791
			25	0.226	0.426	0.579	0.678	0.733	0.785	0.822	0.853	0.877
			30	0.260	0.485	0.629	0.743	0.799	0.852	0.889	0.919	0.943
			35	0.294	0.541	0.688	0.806	0.859	0.904	0.932	0.957	0.977
80	10	11.7	5	0.010	0.020	0.027	0.032	0.037	0.040	0.044	0.047	0.051
			10	0.012	0.028	0.043	0.057	0.069	0.080	0.089	0.098	0.106
			15	0.016	0.037	0.063	0.083	0.104	0.123	0.138	0.156	0.172
		Pb 12/40	20	0.020	0.047	0.081	0.114	0.145	0.176	0.199	0.222	0.244
			25	0.024	0.060	0.105	0.153	0.190	0.238	0.283	0.331	0.378
			30	0.029	0.074	0.127	0.198	0.253	0.331	0.407	0.493	0.584
			35	0.034	0.091	0.159	0.263	0.343	0.446	0.539	0.658	0.785
80	10	6.6	5	0.018	0.034	0.047	0.055	0.063	0.069	0.075	0.081	0.087
			10	0.021	0.049	0.073	0.097	0.116	0.134	0.147	0.161	0.173
			15	0.027	0.064	0.106	0.139	0.171	0.199	0.221	0.246	0.269
		Pb 8/40	20	0.034	0.081	0.135	0.186	0.232	0.275	0.305	0.336	0.364
			25	0.042	0.101	0.172	0.242	0.294	0.357	0.412	0.468	0.518
			30	0.050	0.125	0.205	0.305	0.376	0.467	0.549	0.633	0.713
			35	0.059	0.151	0.251	0.387	0.481	0.588	0.675	0.773	0.866
80	10	13.6	5	0.009	0.017	0.023	0.028	0.032	0.035	0.038	0.041	0.044
			10	0.010	0.024	0.037	0.050	0.060	0.070	0.077	0.085	0.092
			15	0.014	0.032	0.054	0.073	0.091	0.108	0.121	0.137	0.151
		Pb 17/70	20	0.017	0.041	0.070	0.100	0.128	0.155	0.176	0.197	0.218
			25	0.021	0.052	0.092	0.134	0.168	0.212	0.254	0.299	0.343
			30	0.025	0.065	0.111	0.175	0.226	0.298	0.371	0.456	0.547
			35	0.030	0.080	0.140	0.234	0.310	0.409	0.502	0.623	0.758
80	10	7.8	5	0.015	0.029	0.040	0.047	0.054	0.059	0.064	0.069	0.074
			10	0.018	0.041	0.063	0.083	0.100	0.116	0.128	0.140	0.151
			15	0.023	0.055	0.091	0.120	0.148	0.174	0.193	0.217	0.237
		Pb 15/80	20	0.029	0.070	0.117	0.162	0.203	0.243	0.271	0.299	0.327
			25	0.036	0.087	0.150	0.213	0.260	0.319	0.372	0.426	0.477
			30	0.043	0.108	0.179	0.270	0.337	0.426	0.507	0.594	0.678
			35	0.051	0.131	0.221	0.348	0.439	0.547	0.637	0.742	0.845

IV.4.2   Mammography

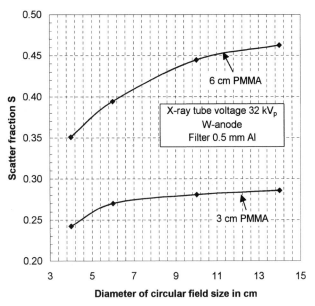

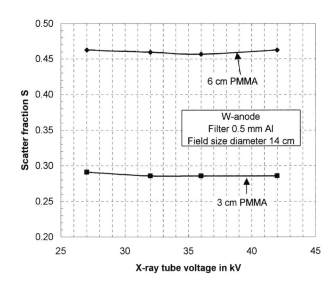

Fig. IV.4.13 Scatter fraction in dependence on the field size for a 3 cm and a 6 cm thick PMMA-phantom (Diagram adapted from Barnes 1978)

Fig. IV.4.14 Scatter fraction in dependence on the X-ray tube voltage for a 3 cm and a 6 cm thick PMMA phantom (Diagram adapted from Barnes 1978)

Table IV.4.2.1 (equivalent to table III.2.2): Typical values of the geometrical and physical characteristics of mammography anti-scatter grids

Grid	Grid ratio	Strip density	Primary radiation transmission	Scattered radiation transmission	Total radiation transmission	Selectivity	Contrast improvement factor	Bucky factor
	r	N	T_p	T_s	T_t	Σ	C_{if}	B
Pb 4/27	4	27 cm^{-1}	72%	23%	49%	3.10	1.48	2.05
Pb 5/31	5	31 cm^{-1}	72%	22%	47%	3.35	1.55	2.14
Pb 3/80	3	80 cm^{-1}	55%	18%	40%	3.06	1.37	2.5

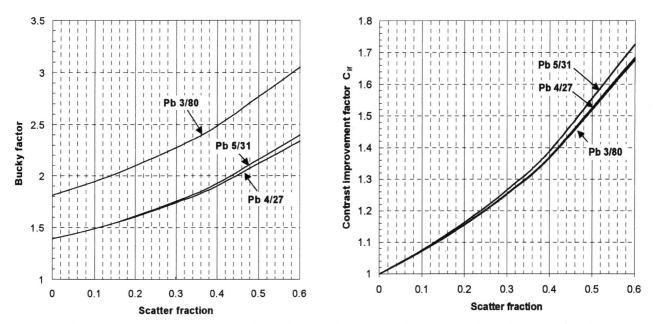

Fig. IV.4.15 Bucky-factor and contrast improvement factor: Geometrical and physical characteristics of the grids shown in the diagrams are listed in table IV.4.2.1

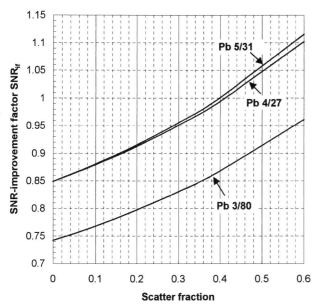

Fig. IV.4.16 Improvement of the signal-to-noise ratio: The geometrical and physical characteristics of the grids shown in the diagramm ar listed in Table IV.4.2.1

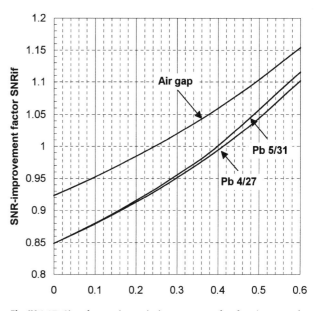

Fig. IV.4.17 Signal-to-noise ratio improvement by the air gap technique in comparison to that by the anti-scatter grids. Assumed geomtrical parameters: SID = 65 cm; x = 7 cm; a = 5 cm. (Compare Fig. II.5.8)

IV.5 Miscellaneous

IV.5.1 Penetration and Absorption of X-rays

Table IV.5.1 (equal to table II.4.1): Typical values for the attenuation ratio of material between the patient and the X-ray image receptor according to IEC 61223-3-1 (1999)

Material (components of x-ray equipment)	Attenuation ratio *
Patient support	1.25
Front panel of film changer	1.25
Anti-scatter grid	1.43
Automatic exposure control (AEC)	1.1

*Measuring parameters are 80 kV and an annuating layer of 25 mm Al

Table IV.5.2 (equal to table II.4.2): Gradation of the exposure parameters tube voltage (in kV) and tube current time product (in mAs) as exposure points (in EP)

EP	kV	mAs	EP	kV	mAs	EP	kV	mAs	EP	kV	mAs
-10		0.1	0	40	1	10	60	10	20	102	100
-9		0.13	1	41	1.25	11	63	12.5	21	109	125
-8		0.16	2	42	1.6	12	66	16	22	117	160
-7		0.2	3	44	2	13	70	20	23	125	200
-6		0.25	4	46	2.5	14	73	25	24	133	250
-5		0.32	5	48	3.2	15	77	32	25	141	320
-4		0.4	6	50	4	16	81	40	26	150	400
-3		0.5	7	52	5	17	85	50	27		500
-2		0.63	8	55	6.3	18	90	63	28		630
-1		0.8	9	57	8	19	96	80	29		800
									30		1000

Exposure points EP can be calculated by the relations: factor $= 10^{\frac{EP}{10}}$ and $RP = 10 \cdot \log(\text{factor})$ which is equivalent to a gradation according to $(^{10}\sqrt{10})^n$. "Factor" means the ratio between two image receptor dose levels.

IV.5.2   X-ray Detectors

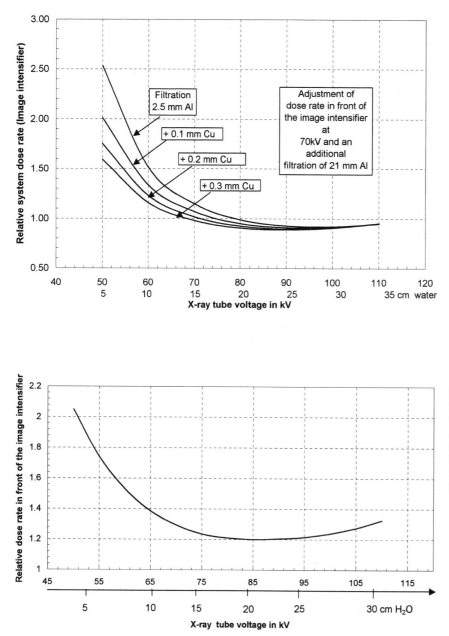

Fig. IV.5.1 Voltage response of an image intensifier calculated for increasing water phantom thickness and different additional Cu-filtration. As dose rate regulation curve the so called anti sowatt curve was assumed. (see Chaps. II.6.2 and II.6.5; Fig. II.6.7 and Fig. II.6.12 and Fig. IV.5.2)

Fig. IV.5.2 Voltage response of an image intensifier measured with increasing water phantom thickness by using the „anti sowatt" regulation curve. (see also Fig. IV.5.1 and Fig II.6.12). In contrast to Fig. IV.5.1, scattered radiation that transmits the antiscatter grid is included (identical to Fig. II.6.10)

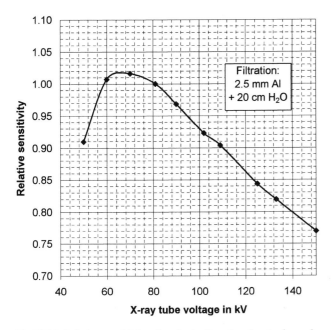

Fig. IV.5.3 Relative sensitivity of an ionisation chamber in dependence on X-ray tube voltage, measured in combination with a 20 cm thick water phantom arranged in front of the chamber (see Fig. II.4.10). Reference tube voltage is 81 kV

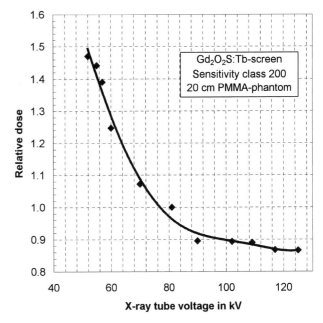

Fig. IV.5.5 Voltage response of a Gd_2O_2S: TB film-screen system: Measurement was carried out with the PMMA-phantom arranged near the focal spot

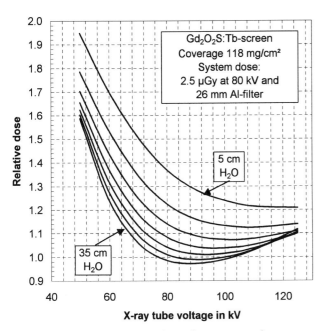

Fig. IV.5.4 Voltage response, i.e. relative dose necessary for constant optical density, of a Gd_2O_2S: Tb-screen (see Section II.6.1) calculated in dependence on X-ray tube voltage for a 5 cm, 10 cm, 15 cm, 20 cm, 25 cm, 30 cm and 35 cm water layer. Scattered radiation is not considered. (see data file SN_GOS.xls on CD-ROM)

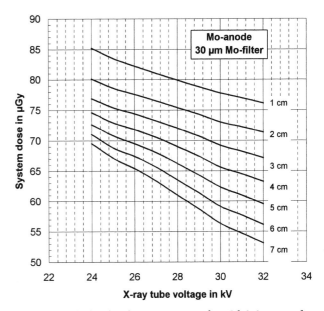

Fig. IV.5.6 Calculated voltage response of a Gd_2O_2S-screen for standard breast. Thickness from 1 cm to 7 cm; Mo-anode/Mo-filter system

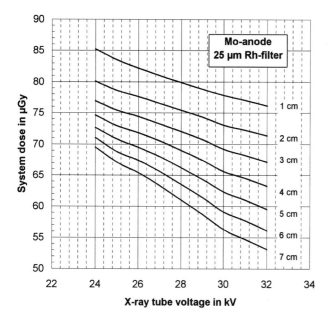

Fig. IV.5.7 Calculated voltage response of a Gd$_2$O$_2$S-screen for standard breast. Thickness from 1 cm to 7 cm; Mo-anode/Rh-filter system

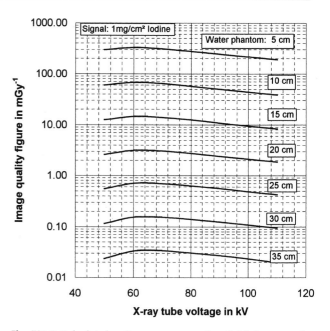

Fig. IV.5.9 Calculated voltage response of a Gd$_2$O$_2$S-screen for standard breast. Thickness from 1 cm to 7 cm; W-anode/Mo-filter system

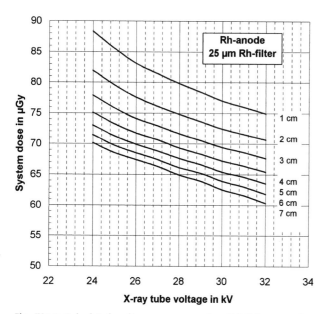

Fig. IV.5.8 Calculated voltage response of a Gd$_2$O$_2$S-screen for standard breast. Thickness from 1 cm to 7 cm; rh-anode/Rh-filter system

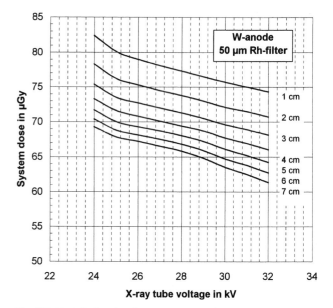

Fig. IV.5.10 Calculated voltage response of a Gd$_2$O$_2$S-screen for standard breast. Thickness from 1 cm to 7 cm; W-anode/Rh-filter system

IV.5.3 Image-Quality Figures

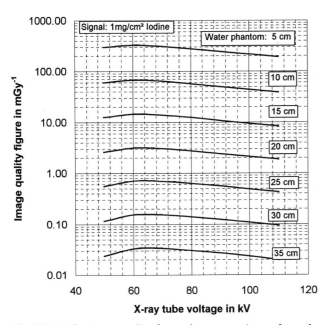

Fig. IV.5.11 The image quality figure shows a maximum for each phantom thickness which lies slightly above 60 kV independent of the phantom thickness. The maxima are caused by increased attenuation of the X-radiation above the K-edge of the iodine contrast medium. Coverage of the input screen of the image intensifier is assumed as 180 mg/cm² Csl. (s. data file SN_Csl.xls)

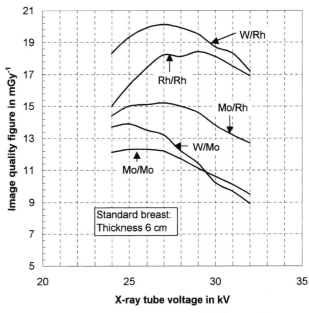

Mo/Mo:	Molybdenum anode;	0.03 mm Molybdenum-filter
W/Mo:	Tungsten anode;	0.06 mm Molybdenum-filter
Mo/Rh:	Molybdenum anode;	0.025 mm Rhodium-filter
Rh/Rh:	Rhodium anode;	0.025 mm Rhodium-filter
W/Rh:	Tungsten anode;	0.05 mm Rhodium-filter

Fig. IV.5.13 For each of the anode-filter combinations, the image quality figure (IQF) = (SNR)²/AGD shows a maximum for the imaging of a 6 cm thick standard breast phantom. The best compromise between contrast and dose is given with the W/Rh-system. (see data file SN_MDGOS on CD-Rom)

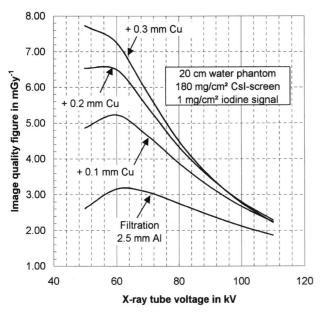

Fig. IV.5.12 IQF in dependence on X-ray tube voltage (data file SN_Csl.xls on CD-Rom). (see Section III.3.2 and Fig III.3.3

IV.6   Patient Dose Estimation

IV.6.1   General X-ray Diagnostics

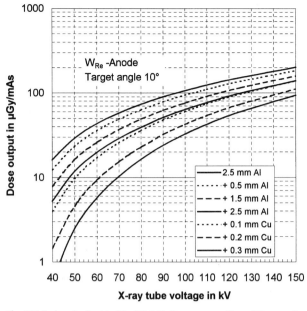

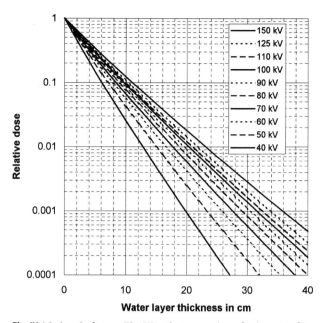

Fig. IV.6.1 (equivalent to Fig. III.1.3): Dose output Y_{100} of X-ray tube assemblies with W_{Re}-anode at a target angle of 10° and various additional filtration; focus distance of 100 cm

Fig. IV.6.3 (equivalent to Fig. III.1.2): Attenuation of primary radiation in water at X-ray tube voltages (ripple < 1%) from 40 kV to 150 kV; total filtration 2.5 mm Al, added filtration 0.1 mm Cu

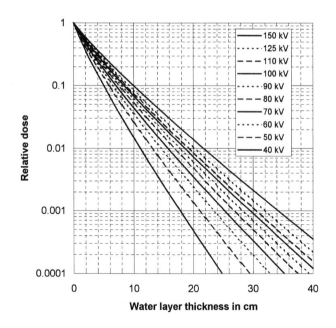

Fig. IV.6.2 (equivalent to Fig. III.1.1): Attenuation of primary radiation in water at X-ray tube voltages (ripple < 1%) from 40 kV to 150 kV; total filtration 2.5 mm Al

Table IV.6.1.1 (identical to Table III.1.1) Backscatter factors for water and PMMA*) in dependence on radiation quality and field size adapted from (Petoussi et al 1998)

| Radiation quality | | HVL | Mean Energy | Backscatter factors for various field sizes | | | | | |
| X-ray Tube voltage | Total filtration | | | 10 x 10 cm² | | 20 x 20 cm² | | 25 x 25 cm² | |
kV		mm Al	keV	Water	PMMA	Water	PMMA	Water	PMMA
50	2.5 mm Al	1.74	32.0	1.24	1.33	1.26	1.36	1.26	1.36
60	2.5 mm Al	2.08	35.8	1.28	1.36	1.31	1.41	1.31	1.42
70	2.5 mm Al	2.41	39.3	1.30	1.39	1.34	1.45	1.35	1.46
70	3.0 mm Al	2.64	40.0	1.32	1.40	1.36	1.47	1.36	1.48
70	3.0 mm Al + 0.1 mm Cu	3.96	44.0	1.38	1.48	1.45	1.58	1.46	1.59
80	2.5 mm Al	2.78	42.9	1.32	1.41	1.37	1.48	1.38	1.50
80	3.0 mm Al	3.04	43.7	1.34	1.42	1.39	1.51	1.40	1.52
80	3.0 mm Al + 0.1 mm Cu	4.55	48.2	1.40	1.49	1.48	1.61	1.49	1.63
90	2.5 mm Al	3.17	46.3	1.34	1.43	1.40	1.51	1.41	1.53
90	3.0 mm Al	3.45	47.0	1.35	1.44	1.42	1.53	1.42	1.55
90	3.0 mm Al + 0.1 mm Cu	5.12	51.7	1.41	1.50	1.50	1.62	1.51	1.65
100	2.5 mm Al	3.24	48.1	1.34	1.42	1.40	1.51	1.41	1.53
100	3.0 mm Al	3.89	50.0	1.36	1.45	1.44	1.55	1.45	1.57
100	3.0 mm Al + 0.1 mm Cu	5.65	54.8	1.41	1.50	1.51	1.64	1.53	1.66
110	2.5 mm Al	3.59	50.8	1.35	1.43	1.42	1.53	1.43	1.55
120	3.0 mm Al	4.73	55.4	1.37	1.46	1.46	1.58	1.48	1.60
120	3.0 mm Al + 0.1 mm Cu	6.62	60.1	1.41	1.50	1.53	1.64	1.54	1.67
130	2.5 mm Al	4.32	55.6	1.36	1.44	1.44	1.55	1.45	1.57
150	2.5 mm Al	4.79	59.1	1.36	1.44	1.45	1.55	1.46	1.58
150	3.0 mm Al	6.80	64.9	1.39	1.47	1.50	1.61	1.52	1.63
150	3.0 mm Al + 0.1 mm Cu	8.50	69.2	1.40	1.48	1.53	1.64	1.55	1.67

*) Polymethylmethacrylate (acrylic glass. i.e. Plexiglas)

Table IV.6.1.2 Typical values for the attenuation ratio of material between the patient and the X-ray image receptor

Material (components of X-ray equipment)	Attenuation ratio* (IEC 61223-3-1: 1999)	Attenuation ratio* (DIN 6809-7: Draft 2002-06)
Patient support	1.25	1.25
Front panel of film changer	1.25	1.25
Anti-scatter grid	1.43	1.4 – 1.6
Automatic exposure control	1.11	1.10

* Measuring parameters are 80 kV and an attenuating layer of 25 mm Al
* Fore more detailed information see IEC 1999 and DIN 6809-7 draft 2002

Table IV.6.1.3 Typical values of the selectivity of anti-scatter grids in dependence on the X-ray tube voltage and for the reciprocal primary radiation transmission (1/Tp) at 100 kV according to IEC 60627 (1978)

| Type of Anti-scatter grid | Selectivity Σ* | | | * |
	60 kV	75 kV	100 kV	100 kV
Pb 8/40	8.0	7.1	5.2	1.56
Pb 12/40	13.4	12.3	9.5	1.56
Pb 15/80	11.0	8.7	5.4	1.35
Pb 17/70	19.4	15.4	9.7	1.58

* Characteristics are valid for anti-scatter grids with Al-cover and paper interspace medium. The reciprocal primary radiation transmission $(1/T_p)$ agree with the attenuation factor m_{grid}, which is affected not much on X-ray tube voltage (< 10%) in the tube voltage range from 60 kV to 125 kV.

Table IV.6.1.4 (identical to table III.1.10 – III.1.15): Tissue-air ratios T_a (from Säbel et al 1980)

Table 1: Total filtration 2.6 mm Al; Tube voltage: 60 kV; Half value layer: 2.2 mm Al

Depth in cm	Field size in cm²			
	10 x 10	15 x 15	20 x 20	30 x 30
0	1.269	1.280	1.280	1.280
1	1.120	1.166	1.166	1.200
2	0.917	0.960	0.965	0.982
3	0.723	0.763	0.770	0.797
4	0.563	0.611	0.623	0.642
5	0.442	0.490	0.502	0.525
6	0.349	0.393	0.405	0.429
7	0.273	0.315	0.326	0.349
8	0.215	0.253	0.263	0.285
9	0.169	0.203	0.213	0.233
10	0.133	0.162	0.170	0.190
12	0.082	0.105	0.110	0.126
14	0.051	0.067	0.072	0.085
16	0.031	0.043	0.046	0.056
18	0.019	0.029	0.030	0.037
20	0.013	0.018	0.021	0.025

Table 2: Total filtration 2.6 mm Al; Tube voltage: 70 kV; Half value layer: 2.6 mm Al

Depth in cm	Field size in cm²			
	10 x 10	15 x 15	20 x 20	30 x 30
0	1.257	1.303	1.314	1.314
1	1.200	1.246	1.246	1.269
2	1.030	1.061	1.061	1.090
3	0.831	0.875	0.878	0.906
4	0.672	0.711	0.723	0.755
5	0.541	0.584	0.600	0.632
6	0.434	0.480	0.498	0.529
7	0.349	0.394	0.413	0.443
8	0.280	0.325	0.342	0.371
9	0.225	0.266	0.285	0.311
10	0.181	0.218	0.237	0.261
12	0.117	0.147	0.162	0.182
14	0.075	0.099	0.112	0.128
16	0.049	0.067	0.077	0.090
18	0.031	0.045	0.053	0.063
20	0.021	0.031	0.037	0.045

Table 3: Total filtration 2.6 mm Al; Tube voltage: 80 kV; Half value layer: 3.0 mm Al

Depth in cm	Field size in cm²			
	10 x 10	15 x 15	20 x 20	30 x 30
0	1.303	1.360	1.360	1.360
1	1.246	1.314	1.314	1.314
2	1.070	1.177	1.166	1.166
3	0.885	0.994	0.993	0.989
4	0.722	0.825	0.833	0.840
5	0.591	0.682	0.697	0.715
6	0.483	0.565	0.584	0.609
7	0.395	0.466	0.489	0.518
8	0.323	0.386	0.409	0.441
9	0.265	0.319	0.342	0.375
10	0.216	0.264	0.287	0.319
12	0.145	0.181	0.201	0.231
14	0.097	0.123	0.141	0.167
16	0.065	0.085	0.099	0.121
18	0.043	0.058	0.069	0.088
20	0.029	0.039	0.048	0.064

Table 4: Total filtration 2.6 mm Al; Tube voltage: 90 kV; Half value layer: 3.5 mm Al

Depth in cm	Field size in cm²			
	10 x 10	15 x 15	20 x 20	30 x 30
0	1.291	1.337	1.371	1.371
1	1.269	1.314	1.349	1.349
2	1.110	1.166	1.211	1.189
3	0.925	0.989	1.040	1.030
4	0.763	0.837	0.887	0.880
5	0.629	0.703	0.753	0.757
6	0.517	0.591	0.640	0.651
7	0.425	0.497	0.544	0.560
8	0.350	0.418	0.462	0.481
9	0.288	0.352	0.392	0.414
10	0.237	0.296	0.333	0.357
12	0.161	0.209	0.241	0.264
14	0.109	0.149	0.174	0.195
16	0.074	0.105	0.125	0.144
18	0.050	0.074	0.090	0.107
20	0.034	0.053	0.065	0.079

Table 5: Total filtration 2.6 mm Al; Tube voltage: 100 kV; Half value layer: 3.9 mm Al

Depth in cm	Field size in cm²			
	10 x 10	15 x 15	20 x 20	30 x 30
0	1.314	1.371	1.383	1.383
1	1.269	1.349	1.360	1.360
2	1.141	1.246	1.246	1.246
3	0.962	1.061	1.080	1.080
4	0.802	0.907	0.928	0.949
5	0.667	0.770	0.795	0.823
6	0.555	0.654	0.681	0.715
7	0.462	0.555	0.584	0.621
8	0.384	0.471	0.501	0.539
9	0.319	0.400	0.429	0.469
10	0.266	0.341	0.368	0.406
12	0.184	0.245	0.270	0.306
14	0.127	0.176	0.198	0.231
16	0.088	0.127	0.146	0.174
18	0.061	0.093	0.107	0.131
20	0.042	0.066	0.079	0.099

Table 6: Total filtration 2.6 mm Al; Tube voltage: 120 kV; Half value layer: 4.7 mm Al

Depth in cm	Field size in cm²			
	10 x 10	15 x 15	20 x 20	30 x 30
0	1.326	1.406	1.406	1.429
1	1.326	1.406	1.406	1.474
2	1.166	1.280	1.280	1.349
3	1.021	1.166	1.166	1.246
4	0.869	1.010	1.021	1.090
5	0.729	0.864	0.893	0.955
6	0.614	0.741	0.774	0.835
7	0.517	0.634	0.672	0.730
8	0.434	0.543	0.583	0.638
9	0.365	0.465	0.506	0.558
10	0.307	0.398	0.439	0.488
12	0.218	0.293	0.330	0.373
14	0.154	0.214	0.249	0.285
16	0.109	0.157	0.187	0.218
18	0.077	0.115	0.141	0.167
20	0.055	0.085	0.106	0.128

IV.6.2 Mammography

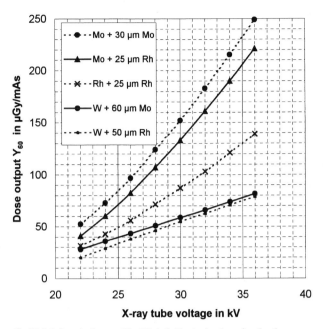

Fig. IV.6.5 (equivalent to Fig. III.1.4): Typical values for the dose output Y_{60} of mammography X-ray tube assemblies at an emission angle of 16° and various anode-filter combinations; focus distance 60 cm; X-ray tube voltages from 22 kV to 36 kV; 2 mm polycarbonate compression plate within the X-ray beam

Table IV.6.2.1 (identical to Table III.1.6) Conversion factor g in mGy/mGy, for conversion of entrance surface air kerma K_E of the breast into average glandular dose AGD, for various breast compositions and thicknesses, anode-filter combinations and tube voltages. For the mix the breast composition is 50 % adipose and 50 % glandular tissue by mass (adapted from Klein et al 1997)

Composition: 100 % adipose

Anode material	Filter thickness in μm	Filter material	Tube voltage in kV	g in mGy/mGy for breast thickness in mm							
				20	30	40	50	60	70	80	90
Mo	30	Mo	25	0.458	0.340	0.264	0.213				
Mo	30	Mo	28		0.371	0.292	0.236	0.197	0.168		
Mo	30	Mo	30				0.248	0.207	0.177	0.153	
Mo	30	Mo	32					0.215	0.184	0.160	0.140
W	60	Mo	25	0.497	0.373	0.292	0.236				
W	60	Mo	28		0.384	0.303	0.245	0.205	0.174		
W	60	Mo	30				0.258	0.216	0.184	0.160	
Mo	25	Rh	28				0.269	0.226	0.193	0.168	
Mo	25	Rh	30					0.234	0.201	0.175	0.154
Mo	25	Rh	32					0.242	0.207	0.181	0.159
Rh	25	Rh	28				0.282	0.238	0.204	0.178	
Rh	25	Rh	30					0.255	0.220	0.192	0.170
Rh	25	Rh	32					0.270	0.233	0.204	0.181
W	50	Rh	28				0.328	0.277	0.239	0.209	
W	50	Rh	30				0.339	0.288	0.248	0.218	
W	50	Rh	32					0.293	0.254	0.223	0.197
W	50	Rh	34					0.302	0.262	0.230	0.205

Composition: 100 % gland

Anode material	Filter thickness in μm	Filter material	Tube voltage in kV	20	30	40	50	60	70	80	90
Mo	30	Mo	25	0.365	0.241	0.176	0.136				
Mo	30	Mo	28		0.267	0.197	0.153	0.124	0.104		
Mo	30	Mo	30				0.162	0.132	0.111	0.095	
Mo	30	Mo	32					0.138	0.116	0.100	0.087
W	60	Mo	25	0.399	0.267	0.192	0.151				
W	60	Mo	28		0.278	0.205	0.159	0.130	0.109		
W	60	Mo	30				0.170	0.139	0.117	0.100	
Mo	25	Rh	28				0.178	0.145	0.122	0.105	
Mo	25	Rh	30					0.152	0.128	0.110	0.096
Mo	25	Rh	32					0.157	0.133	0.114	0.100
Rh	25	Rh	28				0.190	0.156	0.132	0.114	
Rh	25	Rh	30					0.170	0.143	0.124	0.109
Rh	25	Rh	32					0.181	0.153	0.132	0.116
W	50	Rh	28				0.223	0.183	0.155	0.134	
W	50	Rh	30				0.232	0.192	0.162	0.140	
W	50	Rh	32					0.196	0.166	0.144	0.126
W	50	Rh	34					0.204	0.174	0.150	0.132

Composition: mix with 50 % adipose and 50 % gland

Anode material	Filter thickness in μm	Filter material	Tube voltage in kV	20	30	40	50	60	70	80	90
Mo	30	Mo	25	0.407	0.284	0.213	0.168				
Mo	30	Mo	28		0.313	0.237	0.187	0.154	0.130		
Mo	30	Mo	30				0.198	0.163	0.137	0.118	
Mo	30	Mo	32					0.170	0.143	0.124	0.109
W	60	Mo	25	0.444	0.314	0.236	0.186				
W	60	Mo	28		0.325	0.246	0.195	0.160	0.135		
W	60	Mo	30				0.206	0.170	0.144	0.124	
Mo	25	Rh	28				0.216	0.178	0.151	0.130	
Mo	25	Rh	30					0.186	0.158	0.136	0.120
Mo	25	Rh	32					0.192	0.163	0.141	0.124
Rh	25	Rh	28				0.229	0.190	0.161	0.140	
Rh	25	Rh	30					0.205	0.175	0.152	0.134
Rh	25	Rh	32					0.218	0.186	0.162	0.142
W	50	Rh	28				0.267	0.222	0.189	0.164	
W	50	Rh	30				0.278	0.232	0.198	0.172	
W	50	Rh	32					0.237	0.202	0.176	0.155
W	50	Rh	34					0.246	0.210	0.183	0.162

Table IV.6.2.2 (equal to table III.1.7): Conversion factors g to calculate the average glandular dose AGD for different breast thicknesses from the entrance surface air kerma K_E (from CEC 1996)

Compressed breast thickness in mm						
HVL in mm Al	30	40	50	60	70	80
0.25	0.234	0.174	0.137	0.112	0.094	0.081
0.30	0.274	0.207	0.164	0.135	0.114	0.098
0.35	0.309	0.235	0.187	0.154	0.130	0.112
0.40	0.342	0.261	0.209	0.172	0.145	0.126
0.45	0.374	0.289	0.232	0.192	0.163	0.140
0.50	0.406	0.318	0.258	0.214	0.177	0.154
0.55	0.437	0.346	0.287	0.236	0.202	0.175
0.60	0.466	0.374	0.310	0.261	0.224	0.195
0.65	0.491	0.399	0.332	0.282	0.244	0.212

Table IV.6.2.3 (equal to table III.1.8): Conversion factors gPB for calculating the AGD for a 50 mm thick standard breast from the entrance surface air kerma KE measured at the tube loading recorded for exposure of the 45 mm thick PMMA (polymethylmethacrylate) standard phantom (from CEC 1996)

HVL in mm Al	g_{PB} in mGy/mGy
0.25	0.149
0.30	0.177
0.35	0.202
0.40	0.223
0.45	0.248
0.50	0.276
0.55	0.304
0.60	0.326
0.65	0.349

Table IV.6.2.4 (equal to table III.1.9): Typical half value layers HVL in mm Al for mammography units with different anode-filter combinations operated at various tube voltages (from CEC 1996).

Anode- and filtermaterial	Tube voltage in kV	HVL in mm Al	
		without compression-plate	with compression-plate (3 mm PMMA)
Mo + 30 μm Mo	25	0.28	0.34
"	28	0.32	0.37
"	30	0.34	0.38
"	31	0.35	0.39
"	34	0.36	0.40
Mo + 25 μm Rh	22	0.30	0.34
"	25	0.36	0.40
"	28	0.40	0.44
"	34	0.41	0.46
W + 60 μm Mo	22	0.33	0.37
"	25	0.35	0.39
"	28	0.37	0.41
"	30	0.38	0.42
W + 50 μm Rh	22	0.41	0.43
"	25	0.48	0.51
"	28	0.51	0.54
"	30	0.53	0.56
W + 40 μm Pd	22	0.36	0.40
"	25	0.44	0.48
"	28	0.48	0.53
"	30	0.50	0.55
Rh + 25 μm Rh	23	0.31	0.36
"	25	0.34	0.40
"	28	0.39	0.45
"	30	0.42	0.48

Subject Index

A

absorbed dose   30, 69, 85, 87, 90, 94
absorbed dose rate   31
additional filtration   3, 9, 27, 32, 33, 35, 52, 60, 62, 65
adipose breast tissue   36, 70, 91, 92, 93
air   80
air kerma   13, 28, 85, 86
air kerma area product   29, 88
air kerma rate   29, 31, 83
air-gap technique   41, 49, 99
Al attenuation equivalent   41
Al cover   41, 90
Al quality equivalent filtration   35, 96
ALARA principle V,   100
ambient light level   58
amorphous selenium   57, 60
angiography   3, 33
anode   7, 8, 10
anode angle   8, 9, 13
anode disc   7, 9, 49
anode/filter combination   27, 92, 93, 94
anthropomorphic phantom   36
anti-isowatt curve   111
antiscatter grid   3, 4, 40, 41, 46, 50, 51, 52, 54, 57, 63, 65,
    77, 87, 90, 99, 100
application limits   54, 55
artefact   40, 41, 50, 54
a-Si:H layer   60
atom   7, 10, 58
atomic cross section   15
atomic form factor   20
atomic number   7, 9, 11, 57, 62
attenuation   3, 37, 70
attenuation factor   35, 39, 41, 86, 87, 89
attenuation ratio   30, 39, 96
Auger electron   10, 19, 28
automatic exposure control system (AEC)   3, 39, 40, 42,
    54, 60, 61, 62, 87
average glandular dose (AGD)   91, 92

B

back-scatter factor   86, 89, 94, 96
barium   33
beam hardening   31, 33, 51, 80
beam stop   52
beryllium   10
binding energy   10, 13, 19
biological effectiveness   31
bone   80
bound electron   10, 18, 20
breast model   91
bremsstrahlung   10, 20, 25, 27, 28, 71
brightness   36, 62, 63, 70, 74
broad beam condition   37
Bucky-factor   40, 52, 53,

C

calcium tungstate screen   58
cancer   31
cancer induction   31
carbon fibre   40
cathode   7, 8
CCD   60
CDRAD-phantom   36, 78
cellular damage   30
cellular grid   50
central line   54, 55
central ray   8, 29, 87
chamber electrodes   62
characteristic curve   59
characteristic radiation   9, 11, 58
charge sensitive preamplifier   11
classical electron radius   20
coherent scattering   45
collimator   8, 12
compound   20
Compton attenuation coefficient   15, 16, 20
Compton continuum   12, 13
Compton edge   12
Compton effect   15, 16, 18, 20, 45
Compton recoil electron   17
Compton scattering   12, 16, 35

computed tomography VI, 29, 90
continuous X-ray spectrum   9, 10
contrast   3, 33, 50, 54, 59, 69, 70, 71, 72, 76, 79, 99
contrast improvement factor   49, 52, 53
contrast resolution   70
contrast-detail curve   78, 79
conversion efficiency   58
cotton fibre interspace material   40
Coulomb-force interaction   15
cover   3, 40, 87, 100
cross grid   50
CT dose index   30, 85
Cu quality equivalent filtration   35, 96
cut-off dose   42, 60

D

DC generator   10
DC-voltage   20
decentring   54, 100
deflection angle   18, 20
defocusing   54
density   15, 20
depth dose   87, 90, 94
detectability   74, 79
detective quantum efficiency (DQE)   3, 57, 70, 76, 77
detector   3, 57, 61, 87
deterministic   31
diaphragm   43
differential cross section   17
digital image receptor   57, 60
digital imaging   3, 41, 42, 53, 61, 63, 65, 70, 71, 99, 100
digital indirect technique   65
digital radiography   3, 62
direct technique   61
dose area product   28, 96
dose output   90, 96
dose profile   30, 85
dose V, VI,   3, 75, 99
dose-length product   29, 85, 88
dosimetry   V
dual energy   9
dual-screen cassette   57

E

effective attenuation coefficient   79
effective mass attenuation coefficient   80
effective dose V,   30, 88, 96
effective scatter point source   49
electrical noise   74
electrometer   12
electron   7, 74

electron-hole pair   11, 57
electronic equilibrium   30
element   20
elementary charge   10
emission angle   8, 11, 18
energy absorption   12
energy absorption coefficient   57
energy dependence   57, 60, 63, 111
energy distribution   37, 59, 60, 64
energy fluence   27
energy imparted   30
energy level   10, 16
energy response   3
energy transfer   16, 17, 18, 19
energy transfer coefficient   18
energy-spectrum   12
entrance dose   29, 40
entrance surface air kerma   29, 54, 86, 88, 89, 90, 91, 93,
entrance surface air kerma rate   29
entrance surface dose   29, 86, 88, 89, 93
entrance surface dose rate   29
entrance window   11
equipment attenuation factor   86, 89
equivalent dose   28, 29, 30, 31, 32,
escape fraction   13
excitation   57
exit dose   32
exponential attenuation law   3, 15, 70, 71
exposure   28, 37, 58, 69
exposure parameter   V, 3, 27, 36, 38, 96
exposure point   39
exposure rate   29, 31
exposure table   39

F

fan beam   49
fibro-glandular tissue   36
filament   7, 8
filament coil
film   57, 58, 59
film cassette   59
film cassette cover   3
film contrast   58
film-screen combination   42, 42, 59
film-screen system   32, 36, 38, 54, 59, 61, 69, 72, 77
filter material   33
filter technique   65
filtration   31, 32, 38, 59, 62, 70
fluence rate   29
fluorescence yield   11
fluoroscopic mode   65
fluoroscopy   62, 65, 79, 90, 91, 97
focal spot   8, 9, 12, 35, 42, 43, 45, 50, 69, 74

focal spot size   8, 9, 12, 41, 49, 72
focal spot track   7, 10
focus image receptor distance   54
focus object distance   45
focused grid   50, 54, 55, 100
focusing distance   54, 55, 89
focus-patient distance   94
free-in air   87

G

Gaussian radiation intensity distribution   72
Ge-detector   11, 13
geometrical characteristics   3
geometrical magnification   35, 41, 42, 49
geometry   41, 69
glandular tissue   69, 70, 91, 92, 93
gonads   31
gradient   77
gradient curve
granularity   69, 77
grid   3, 52, 54, 99
grid cut-off   54
grid lines   54
grid ratio   40, 54, 55, 99

H

half-value layer   13, 31, 32, 86, 93, 94
heel effect   49
hereditary   31
high-ratio grid   54
high-strip density grid   100
high-voltage transformer   12

I

II-TV system   65
image   3, 8
image contrast   32, 40
image display   73
image intensifier   41, 61, 62, 63
image intensifier output screen   32
image plane   46
image processing   63, 65, 69, 99
image quality figure   3, 70, 77, 79, 109, 110, 111, 112, 113
image quality index   77
image quality V, VI,   3, 9, 27, 33, 45, 47, 49, 53, 54, 57, 62,
   65, 70, 75, 78, 99
image receptor   3, 31, 32, 35, 38, 60, 62, 64, 69, 74, 76, 88
image receptor dose   27, 29, 32, 41, 59, 63, 64, 88, 89

image receptor dose rate   29, 63
incoherent scattering   45
incoherent scattering function   16, 20
indirect technique   62
infiltrating ductal carcinoma   36
information carrier   57, 74
inherent filtration   11
initial dosimetric quantity   88, 96, 97
intensification   58
intensifying screen   57, 58
intermediate layer   3, 35, 54
interspace material   3, 50, 51, 53, 90, 100
iodine   33
ionisation   41, 57
ionisation chamber   3, 28, 29, 40, 54, 61, 62, 85
ionising radiation   30
iris   63, 65

K

K shell   13
K-edge   37, 60
K-edge filter   32, 33
Kerma   28
Kerma rate   29
kerma spectrum   28
K-escape   12, 13
kinematics   16, 18
kinetic energy   10, 19
kinetics   16
Klein-Nishina   16, 17, 18
K-radiation   13

L

L shell   19
large-area parallel-plate ionisation chamber   43, 88
last image hold (LIH)   109
latency period   31
law of reciprocity   58
lead bar pattern   72, 73
lead disc   52
Leeds-phantom   36
line pairs per mm   73
line power stabiliser   12
line spectrum   10
line spread function   73, 77
linear attenuation coefficient   15, 36, 57, 70, 71, 76
linear energy absorption coefficient   62
low-contrast   69, 70, 75, 79, 99
luminescence   67
lung tissue   80

M

main amplifier   11
mammography   3, 4, 9, 11, 13, 20, 31, 32, 36, 47, 48, 49,
     57, 61, 64, 69, 77, 86, 90, 91, 94, 99,
mammography screening   91
mass attenuation coefficient   13, 15, 19, 20, 36, 45, 58, 71
mass energy-absorption coefficient   59, 60, 94
mass energy-transfer coefficient   20
mean energy   28
measuring electrode   61, 62
mediastinum   100
medical indication   36, 50, 65, 69
medical requirements   69, 99, 100
melting point   9
metal-ceramics construction   7
metal layer   41, 62
micro-ammeter   12
micro-calcification   69
mid-line dose   32
Mo/Mo anode filter system   32
Mo-anode   9, 13, 33
modulation   71, 74
modulation transfer function   71, 72, 73, 74, 76
molybdenum   112, 114, 115
momentum   16, 18
monitor   36, 63, 69, 70
mono-energetic photons   15
mono-energetic radiation   27, 79, 71
Monte Carlo   12, 91, 93
motion blur   27
multi-channel analyser   11, 12
multiple scattering   45
multiple slice scanner   30
multiple slit system   49
multi-pulse   27

N

narrow beam condition   38, 87
noise   3, 11, 53, 58, 69, 70, 74, 100
noise autocorrelation function   74
noise equivalent quanta (NEQ)   57, 76, 77
noise power spectrum   74, 80
nominal air kerma rate   29
nuclear photoeffect   15

O

optical density   36, 54, 58, 59, 63, 69, 72, 74
optical feedback   11
orbital electron   45
organ dose V,   3, 85, 88, 90, 91, 94

organ-dose conversion factors   91, 92, 93
output screen   63
over-table unit   43

P

paediatrics   3, 33, 40, 64, 69, 99
pair production   15
paper   50, 51, 99, 101
parallel grid   50
parallel-plate ionisation chamber   88
patient attenuation factor   86
patient dose V,   3, 4, 27, 28, 35, 40, 41, 58, 85, 88, 100
patient equivalent scattering phantom   46
patient exposure   9, 46, 54, 61
patient support   40, 41
patient table   87
peak tube voltage   27, 28
pencil ionisation chamber   29
phantom V,   36, 37, 62, 70, 79, 80
phosphor screen   37
photo peak   12
phosphor material   57
photo peak efficiency   12
photoelectric effect   11, 12, 15, 18, 19, 35, 45
photoelectron   41, 62
photon energy fluence   59
photon fluence   13, 27, 28, 60
photon scattering angle   17, 20
photon spectrum   3, 7, 28
photon-counting detector   76
PMMA-phantom   46, 48, 78, 85, 93
Poisson statistics   74
polychromatic X-radiation   71
potential difference   7
preamplifier   11
primary beam   9
primary beam filtration   33
primary photon fluence   37
primary photons   35
primary radiation   4, 8, 99, 100
primary radiation attenuation   36, 54
primary radiation contrast   100
primary radiation fraction   54
primary radiation intensity   46
primary radiation transmission   40, 41, 50, 53, 54, 77,
     90, 100
probability density function   74, 75
2-pulse   10, 27
6-pulse   10, 27
12-pulse   10, 27
pulse height   11, 12
pulse pile up   11
pulsed fluoroscopy   65, 66

Q

quality equivalent filtration   32
quantisation noise
quantum detection efficiency (QDE)   57, 58, 60, 76
quantum limited   69, 76
quantum mottle   69, 75
quantum noise   11, 53, 63, 74, 76

R

radiation dose   V, 85
radiation exposure   V, VI, 3, 4, 27, 49, 54, 65
radiation image   35, 57, 69, 76, 80
radiation intensity   73
radiation output   8, 29, 89
radiation protection V,   30, 50
radiation risk V,   85, 91
radiation detector   40, 46
radiation quality   27, 52, 64, 80, 85, 86, 88, 89, 94
radiation weighting factor   31
radiographic exposure V
radionuclides   12
random fluctuation   69
rare earth material
rare earth screen   58
Rayleigh scattering   15, 20, 45
receiver operating characteristic analysis
reciprocity law failure   58
recoil electron   17
reference detctor   57
reference levels V
relative atomic weight   15
relative biological effectiveness   30
relative depth dose   87, 90, 94
relative sensitivity   59
resolution limit   49, 73
rest energy   16
resting mass energy   30
Rh/Rh anode filter system
rhodium   11
rhodium anode   13
ripple   13, 52

S

scatter   38
scatter degradation factor   47, 48, 52, 99
scatter fraction   37, 40, 46, 47, 48, 53, 54
scatter free condition   40, 87, 89
scattered radiation V, 3, 15, 36, 37, 45, 46, 47, 49, 50, 99, 100
scintillator   58, 60

screen blur   57
screening VI,   91
secondary electron   62
secondary photons   35
secondary radiation   36
selectivity   49, 52, 53, 54, 77, 89, 90, 100
Selwyn coefficient   77
semiconductor element   41
sensitivity   3, 57, 58, 59, 60, 62, 64, 89
sensitivity class   64
sharpness   38, 69, 70, 72
shell   10, 11, 19
signal   69, 72, 75
signal-to-noise ratio   3, 33, 63, 65, 70, 75, 76, 80, 100
signal-to-noise ratio improvement factor   53
simulation calculation   V
sine function   73
single scattering   45
single-emulsion film   57
skin dose   31, 33
skin exposure   33
skin injuries   31
slice thickness   30, 85
slice-averaged axial air kerma   29, 90, 94
slit width   49
slot technique   49, 50
solid angle   12, 20
somatic cell   31
source-to-image distance (SID)   9, 41
source-to-object distance (SOD)
spatial frequency   57, 72, 73, 74
spatial frequency spectrum   73, 76
spatial resolution   41, 42, 58, 60
spectrometer   11
speed   64
speed of light   7, 16
spot-film device   39
spot-film operation   62
standard AGD   93
standard breast   91
standard deviation   74, 76
stationary grid   100, 104
statistical fluctuations   74
stochastic effects   31
stochastic radiation risk   31
stomach   112
strip density   51, 99, 100
strip height   51
stroboscopic effect   37
subtraction   46
Supervision mode   109, 110
surface entrance dose   31, 32
system dose   31, 54, 60, 63, 72, 80, 96
system dose rate   29, 63, 65

T

table top   3, 40, 54
target angle   9, 10, 11, 89
thermoluminescent dosimeter (TLD)   85
threshold level   31
tissue   30, 31, 35, 70, 85, 99
tissue substitutes   36
tissue weighting factor   31
tissue-air ratio   87, 90, 94, 95
tissue-equivalent   36
total equipment attenuation factor   86
total filtration   9, 11, 13, 38, 52, 85, 96
total linear attenuation coefficient   18, 57, 62
total radiation   46, 52
transmission of primary radiation   51, 52
transmission of scattered radiation   51, 52
transmission of total radiation   51, 52, 53
tube current   12, 38, 39, 62, 65
tube current exposure time product   40, 88, 89
tube load   33, 49
tungsten   7, 9, 10, 86
translucence   53
TFT   60
total radiation intensity   46

U

under-table unit   43
unsharpness   60

V

vacancy   10, 11, 19
variance   74
video camera   63, 65, 69

video chain   63
viewing station   69, 74
visual resolution limit   49, 73
voltage response   57, 59, 60, 61, 62, 64
voltage waveform   9
visibility threshold   79, 99

W

white noise   74
width setting   70
Wiener spectrum   74, 77
window level   70
W/Mo anode filter system   112, 113
W/Rh anode filter system   112, 113
W anode   9, 13
windowing technique   53, 75

X

X-ray examination V, 40
X-ray fluence   69
X-ray generator   9
X-ray image   69
X-ray power   7
X-ray source   11
X-ray spectrometer   11
X-ray spectrum VI,   3, 4, 9, 33, 38, 57, 60, 70
X-ray tube   3, 7, 10, 27, 35, 43, 49
X-ray tube assembly V, 7, 8, 29
X-ray tube voltage   9, 13, 27, 31, 35, 37, 38, 39, 53, 57

Y

yield of photoelectrons   62

Druck: Strauss Offsetdruck, Mörlenbach
Verarbeitung: Schäffer, Grünstadt